AF698514

Mirror of the World

In the late fifteenth century, the production of print editions of Claudius Ptolemy's second-century *Geography* sparked one of the most significant intellectual developments of the era—the production of mathematically-based, north-oriented maps. The production of world maps in England, however, was notably absent during this "Ptolemaic revival." As a result, the impact of Ptolemy's text on English geographical thought has been obscured and minimalized, with scholars speculating a possible English indifference to or isolation from European geographic developments. Tracing English geographical thought through the material culture of literary and popular texts, this study provides evidence for the reception and transmission of Ptolemaic-based geography in England during a critical period of geographic innovation and synthesis, one that laid the foundation for modern geographical representation. With evidence from prose romance, book illustration, theatrical performance, cosmological ceilings, and almanacs, *Mirror of the World* proposes a new, interdisciplinary literary and cartographic history of the influence of Ptolemaic geography in England, one that reveals the lively integration of geographic concepts through narrative and non-cartographic visual forms.

Meg Roland is currently Dean of Arts, Social Science, and Humanities at Linn-Benton Community College in Oregon. She was Professor of English at Concordia University, Portland, Oregon, and previously taught medieval literature and material culture at Marylhurst University.

Mirror of the World

Literature, Maps, and Geographic Writing in Late Medieval and Early Modern England

Meg Roland

Routledge
Taylor & Francis Group
LONDON AND NEW YORK

First published 2022
by Routledge
2 Park Square, Milton Park, Abingdon, Oxon OX14 4RN

and by Routledge
605 Third Avenue, New York, NY 10158

Routledge is an imprint of the Taylor & Francis Group, an informa business

© 2022 Meg Roland

The right of Meg Roland to be identified as author of this work has been asserted by her in accordance with sections 77 and 78 of the Copyright, Designs and Patents Act 1988.

All rights reserved. No part of this book may be reprinted or reproduced or utilised in any form or by any electronic, mechanical, or other means, now known or hereafter invented, including photocopying and recording, or in any information storage or retrieval system, without permission in writing from the publishers.

Trademark notice: Product or corporate names may be trademarks or registered trademarks, and are used only for identification and explanation without intent to infringe.

British Library Cataloguing-in-Publication Data
A catalogue record for this book is available from the British Library

Library of Congress Cataloging-in-Publication Data
Names: Roland, Margaret Mary, author.
Title: Mirror of the world : literature, maps, and geographic writing in late medieval and early modern England / Meg Roland.
Description: Abingdon, Oxon ; New York, NY : Routledge, 2021. |
Includes bibliographical references and index.
Identifiers: LCCN 2021004578 (print) | LCCN 2021004579 (ebook) |
ISBN 9780367560560 (hardback) | ISBN 9780367560584 (paperback) |
ISBN 9781003096245 (ebook)
Subjects: LCSH: Cartography–England–History–15th century. |
Cartography–England–History–16th century. | Ptolemy, active 2nd century.
Geographia. | Cartography in literature.
Classification: LCC GA793.5.A1 R65 2021 (print) | LCC GA793.5.A1
(ebook) | DDC 526.0942/09024–dc23
LC record available at https://lccn.loc.gov/2021004578
LC ebook record available at https://lccn.loc.gov/2021004579

ISBN: 978-0-367-56056-0 (hbk)
ISBN: 978-0-367-56058-4 (pbk)
ISBN: 978-1-003-09624-5 (ebk)

DOI: 10.4324/9781003096245

Typeset in Times New Roman
by Taylor & Francis Books

To Thomas Malory, William Caxton, and the Mandeville author: you asked for your readers' prayers and to think kindly of you; I have done so and been rewarded a hundred-fold.

To Claudius Ptolemy: you asked your readers to add to your great scholarship, and I have added my feather.

To the many scholars of cartography, book history, and literature whose work informed this project: your scholarship invited a long conversation between author and reader.

To rare book librarians everywhere: thank you for sheltering the maps of knowledge.

And, especially, to Conor and Nico for your unwavering enthusiasm and for being proud of your mom. With gratitude and love.

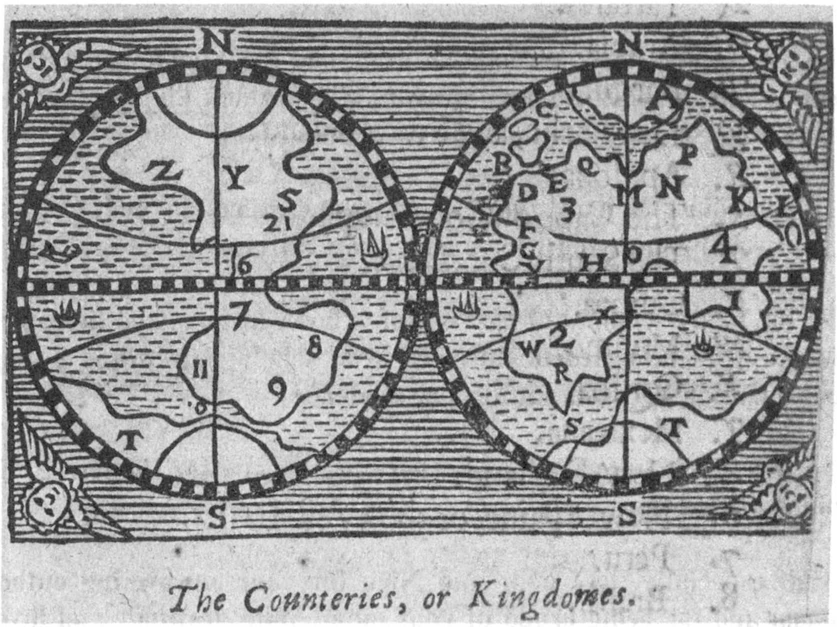

The Counteries, or Kingdomes.

The Compost of Ptolomeus, M. Parson for Henry Gosson 1600/38?
© British Library Board

Contents

Illustrations

Figures

Tables

Preface

This book, set in Times New Roman font, has been published by Routledge, a London-based publisher that is part of the Taylor & Francis Group, itself part of Informa. I finished this book on a MacBook Air in Portland, Oregon, but previously used a variety of laptops, largely made in China. When working with books and manuscripts in rare book libraries, I often took long-hand notes that I later transcribed, but I saved documents, images, and whole EEBO files on a 4 GB San Disk thumb drive. I wrote rough drafts of sections of the book with pen on a legal pad while some sections are "born-digital." I also kept a "book of the book," which grew into multiple journals that chronicle the discoveries, conundrums, and insights of the research, but they also chronicle the closure of two small liberal arts colleges where, collectively, I taught for 20 years.

Based in Northern Ireland, my kind editor, Michael Greenwood, supported the project and managed the peer review process; based on these reviews, I submitted final revisions from Napili Bay, Hawaii. Libraries and museums in the U.S., England, Germany, and Spain sent me digital images, usually in jpeg format, to illustrate the book, and Bridgerman Images secured two rare book images even while libraries were closed due to covid. Editorial assistant Stewart Beale, based in Milton Park, south of Oxford, evaluated all of the images to ensure the proper resolution for quality print reproduction. Edward Gibbons, Senior Production Editor and also based in Milton Park, saw the book through the production process, which includes copyediting, indexing, and proof corrections. The book will be printed largely on demand, with additional e-book and chapter downloads.

Various sections of the book were first written as conference papers given in Vancouver and Wolfville in Canada; in Los Angeles, Kalamazoo, New York, Boston, Washington, D.C., Seattle, and Portland in the U.S.; in Bangor, Bristol, London, and Oxford in the UK; in Rennes, France, and in Utrecht, The Netherlands. One summer, I followed the route of Arthur's Roman War campaign from London to Rome and back, recounted in my fragmentary and as yet unfinished blog, "Passionate Geography." Earlier versions of segments of two of the chapters appeared in the journal *Arthuriana* or as a chapter in collections on medieval geography or Malory.

I often printed out chapters or sections of chapters on my sturdy HP LaserJet printer. Twice, I went to Kinkos/Fed Ex and printed out "the whole book" to aid in my final editing process. The footnotes and bibliography follow the Chicago Manual of Style and Routledge's house style. Located in Glasgow, Scotland, Susan Dunsmore, copy-editor for Taylor & Francis, graciously regularized my occasional intermixing of British and American English, corrected errors, and made clarifying suggestions.

As Virginia Woolf noted, a room of one's own is a critical gift to a writer and for the last six months of the project I had a writing studio, the Pine Cone, in Corvallis, Oregon, which gave me time and space to lay chapters out on the floor and hours of solitude to think and write. I've also used innumerable web-based databases and archives as well as digital mapping projects. I looked up many queries on Google and, of course, in books held by university library collections or obtained through the Orbis Cascade Alliance, which enables library users in the Pacific Northwest to tap into the collective holdings of 37 academic libraries. Colleagues have generously read chapters and offered suggestions for revision, commenting on word docs or google docs via email.

The process of academic publishing is still a negotiation of authors, editors, printers, librarians, booksellers, and readers. You, reader, no doubt checked this book out of an academic library, purchased it online, or are reading a chapter online through a library's digital services, all part of the ongoing collective shaping of book culture in the twenty-first century.

Acknowledgments

Earlier versions of sections of this book have been published under the following titles: "After Poyetes and Astronomyers': English Geographical Thought and Early English Print," in *Mapping Medieval Geographies* (Cambridge: Cambridge University Press, 2013), "Malory and the Wider World," in *The New Companion to Malory* (Cambridge: D.S. Brewer, 2019), and "Malory's Sandwich: Marginalized Arthurian Geography and the Global Middle Ages," in *Rethinking Medieval Margins and Marginality* (London: Routledge, 2020).

As a faculty member at a small, liberal arts college on the West Coast of the U.S., fellowships, conferences, colleagues, libraries, and Early English Books Online created a lifeline to manuscripts and books from medieval and early modern England. I began this journey by registering for a Malory seminar with Dr. Christine Rose at Portland State University in which I first encountered the work of the editor, William Caxton. I then studied in the Textual Studies program at the University of Washington under the mentorship of Drs. Mona Modiano and Míċeál Vaughan. Dr. Modiano subsequently invited me to teach at UW's Rome Center, the invitation of a lifetime. My studies were further informed by a Cambridge University International Summer Programme, "Shakespeare and the Book," with Dr. John Lennard, in which we hand-set and printed a sonnet in the basement of the Cambridge University Library. I also studied at the London Rare Book School's seminar on "The History of Maps and Mapping," led by Dr. Catherine Delano-Smith and co-taught by Drs. Sarah Tyacke, Peter Barber, and Roger Kain. Subsequently, Dr. Delano-Smith generously invited me to present at the Maps and Society lecture series sponsored by the Warburg Institute.

I am grateful to the National Endowment for the Humanities for their support of a five-week seminar led by Dr. John King, "Reformations of the Book," which allowed me and a group of "seminarians" to study early printed books in Antwerp and at the Bodleian Library in Oxford (see printingbooksandmaps.blogspot.com). My research during the seminar structured this current book project, and the work of all the participants has rippled through Early Modern book studies. I am especially grateful to Dr. Anne Coldiron, a fellow seminarian, who has offered inspiration via her many books on early print as well as encouragement along the way. I am so thankful to the Huntington Library for the Gilbert and Ursula Farfel fellowship which allowed me to study the *Shepherd's Kalender* and Malory's *Myrrour* with time to wander the lovely botanical gardens. I also received support from UCLA's Center for Medieval and Renaissance Studies to participate in the "Mapping Medieval Geographies" conference, directed by Dr. Keith Lilley in honor of Denis Cosgrove.

The support I received from Marylhurst University led me to Rome, for which I will be eternally grateful. I received a Faculty Excellence Award which funded my participation in "The Language of Maps" colloquium organized by Keith Lilley, Queens University, and Nick Millea, Oxford University. I then followed the fictional journey of King Arthur from London to Rome and return, a journey of over 2,500 miles, making a large detour to see the Behaim globe in Nuremburg. Dr. Günther Görz, who led the Behaim Globe semantic web project, was a kind host, offering a personal tour of the globe, discussion, and dinner. Subsequently, Dr. David Plotkin, Provost at Marylhurst, supported my development of a study abroad program to London and Rome which has become woven into my work as a scholar and teacher. I hold a special place in my heart for students who joined me on those sojourns.

Librarians at myriad institutions have warmly supported my research. Nancy Hoover, Director of the Library at Marylhurst, and the late Kirk Howard, Reference Librarian at Marylhurst, gladly tracked down all manner of requests. I've been supported by the libraries and librarians of Concordia University, the University of Washington, Portland State University, Reed College, the Pierpont Morgan Library, the British Library, the Bodleian Library at Oxford University, and the Senate House Library. The Linn Benton Community College librarians miraculously secured an edition of Spenser's *The Shepheardes Calendar* for me during the early days of the coronavirus pandemic.

The following libraries and museums have generously provided the images that appear in this book: the University of Glasgow, Durham University Library, John Rylands Library (University of Manchester), Hereford Cathedral, the British Library, Freiburg University Library, the Germanisches Nationalmuseum, the Prado Museum, the New York Public Library, the Boston Public Library, the Bell Library, and the Huntington Library.

The counterpart to research in rare book libraries was the writing of the book. Mostly, I wrote at my home in Portland, Oregon, but I was fortunate to also spend time writing at my cabin in Welches, Oregon, and at the cabin of my friend, Perrin Kerns, at Lake of the Woods. These were special times of solitude and productivity. As a parent, writing sometimes meant being at the cabin with a pack of boys with ebullient spirits and soggy ski jackets. I learned to write in the spaces in between, and that is partly why this book took so long, but I wouldn't have it any other way. My parents were ever supportive of my research and would have celebrated its completion.

Thank you to many friends, including Karen, Matt, Don, Sharon, and Tom who encouraged me over the years. My sister Anne provided good company and insightful feedback on my Wynkyn de Worde paper as we drove through the woods of Maine one summer night on our way to the Maple Malory conference in Nova Scotia.

Good colleagues were generous, good readers. Though I worked as the only medievalist at Marylhurst University, I was inspired and connected to other medievalists through the International Arthurian Society and its triennial conference in Europe. Brilliant and warm-hearted Malory scholars have created a community of scholarly friendship that is beyond what any reader of Malory might imagine. Good fellowship, indeed. I especially want to thank Dr. Karen Cherewatuk for being a rigorous reader of the manuscript and for helping this to be a better book. I'm enormously grateful, too, to the anonymous peer reviewers of this book who made invaluable suggestions and caught ridiculous typos on my part. Any omissions or inaccuracies are my own, and I welcome readers, as William Caxton did, to "correcte and amende."[1]

My Marylhurst colleagues Perrin Kerns, Jay Ponteri, and Keri Behre read drafts, offered encouragement, and were the best colleagues one could wish for. Special thanks to my colleague John Caruso who was my academic writing partner for a number of years and who read many drafts at the Rain or Shine Cafe. Dana Hardy, my former student, became my editorial assistant and was an invaluable help in preparing the final bibliography.

I finished this book through the closure of two universities where I taught as an English professor, sequentially. Both institutions, over 100 years old, could not weather the financial crisis facing small liberal arts colleges in the U.S. This was followed by COVID 19 which, at my new institution, put additional pressure on humanities programs. It has been a tumultuous time in higher education, but I've been heartened by the capacious models of the late fifteenth century when writers and artists found inspiration in the intersection of science and the humanities.

Close friends and family journeyed with me, geographically and emotionally. My good friend Padeen Quinn joined me for the epic drive from London to Rome and, with her inimitable sense of humor, created adventures and hilarious stories. I'd like to thank my former husband, Jerome, for his support of my scholarly journeys and for generously holding down the home front at key times to allow me to pursue my studies. My sons Conor and Nico have been part of my geographic sojourn, joining me in London, Cambridge, Rome, or just at the doorway of my home office. They have offered heart-felt support and love over the many years it took to complete this book. *Grazie a tutti!*

Note

1 William Caxton, ed., *Myrrour of the World*, ed. and Introduction, Oliver H. Prior, *Caxton's Mrrour of the World* (Berlin: A. Asher & Co. for the Early English Text Society, Extra Series CX, 1913), 7.

Introduction

"Master Ptolemy": The Ptolemaic revival and the trace of Ptolemy's *Geography* in early English print culture

> But of one thing I am sure, that Ptolemy was first Master to them all, who hath set down so good and perfect rules of describing the Earth, be it whole or part, as in the opinion of most learned men, no better can be invented.
>
> – Thomas Blunderville, *A Briefe Description of Universal Mappes and Cardes* (1598)[1]

> No one surpassed him [Ptolemy] in genius or painstaking care.
>
> – Swiss humanist, Heinrich Loriti of Glaris, 1527[2]

Not until the 1968 photograph "Earth Rise," taken from the Apollo Space Mission, has the visual and spatial conception of the Earth been altered as profoundly as during the "Ptolemaic revival" in late medieval and early modern Europe. Fueled by the translation of Claudius Ptolemy's second-century *Geography* from Greek into Latin, continental production of new world maps rapidly co-developed with the medium of print. Ptolemy's *Geography* propelled a dynamic transformation of geographic thought, from the mid-fifteenth century, when the first printed editions of Ptolemy's *Geography* were published, to the late-sixteenth century as Gerardus Mercator's 1569 atlas and geographic model of the Earth gradually superseded that of Ptolemy.[3] In England, however, cartographic representation of the world lay fallow: England had been an active participant in medieval geographic culture, producing maps such as the Hereford, but the era of the *mappamundi* was passing. As Lawrence Worms succinctly sums up English cartographic practice in the early print era, "the history of map printing in the British Isles before [1555] is simply told: there was virtually none."[4]

Ptolemy's grid-based principles of latitude and longitude came to dominate European cartographic practices and conceptions of "the global," giving rise to one of the most significant intellectual developments of the fifteenth century, shaping early modern navigation, mapping practices, and imperialism. The period from the early 1400s to the mid-1500s was distinguished by the prestige and increasing influence of Ptolemaic geography in European humanist circles and, with the advent of print, the near-heroic printing of the folio-sized work, typically with regional maps and a map of the world. If a

DOI: 10.4324/9781003096245-1

concurrent cultural process took place in England, however, it was largely undocumented by English-produced cartographic artifacts; that is, by English world maps. Without English-made globes or world maps, the impact of Ptolemy's text on English geographical thought has been minimized by scholars who describe English geographical thought from the period as "illogical," "chaotic," "belated," or "out of date," or the English as "slow coaches in world adventure."[5] No study has previously utilized literary texts and early print culture to document the unique transmission of Ptolemaic-based, vernacular spatial thought in England. A disciplinary gap exists between the work of print culture scholars and historical cartographers in terms of tracing the representation of space and time at the threshold of modern mapping practices.[6] The late fifteenth and early sixteenth centuries in England are often cast as a time of naïve or jumbled geographic expression, with neither the symbolic grandeur of the medieval *mappamundi* nor the mathematical accuracy of Renaissance cartography. Long overshadowed by the ascendancy of the linear, visual cartography of the Renaissance or the certitude of the Reformation, the era of what can be termed the "Ptolemaic revival," roughly the early fifteenth to mid-sixteenth century, has an ethos and aesthetic of its own—simultaneously heterotopic and pragmatic, poetic and spatial. Michel Foucault suggests the mirror as an example of what he calls "heterotopic" spaces—at once material and yet "unreal, virtual space" that transacts, for example, when a person gazes into a mirror, then reconstitutes the self that is reflected back.[7] In this regard, a mirror and a map are not so very different. Each is a material object that invites the viewer's gaze and that, in turn, requires a reckoning of the self in the world. The contestations of the simultaneously mythic and real animated the work of fifteenth-century cartography and geographic writing, offering a new "mirror of the world," to draw upon the title of William Caxton's 1480 text of the same name. The medieval genre of the *speculum*, or mirror, was associated with political conduct guides, such as the 1331 *Speculum regis* or Machiavelli's 1513 *The Prince*, as well as encyclopedic works that inspired wonder at the world. With his two editions of *The Myrrour of the World*, Caxton adds another association with the genre of mirror literature: a realistic geographic description of the world.[8]

Mirror of the World: Literature, Maps, and Geographic Writing in Late Medieval and Early Modern England takes up the curious absence of world map production in England in the late medieval and early Tudor era. Interrogating the material culture of literary and popular texts, this study seeks evidence other than English-produced world maps for the reception and transmission of Ptolemaic-based geography among English readers during a critical and as yet little-understood period of English geographic history. The "poetics of global space"[9] during the Ptolemaic revival, marked by the "rediscovery" of Ptolemy's geographic work and concomitant new representations of a global Earth and mathematically-derived spatial relations, reveal a surprising wealth of geographic representation embedded in English literary texts, including Thomas Malory's late medieval Arthurian romance, *Le Morte Darthur;* William Caxton's editions of *The Myrrour of the World* and the *Description of Britain*; Wynkyn de Worde's world landscape illustrations; the inscription of passages from Mandeville's *Travels* on the Behaim globe; John Rastell's theatrical production (inclusive of the first English-made globe); Tudor cosmographical spectacles, and the multi-edition *Kalender* tradition, also titled *The Compost of Ptholomeus*, which eventually produced an early, and little-known, English printed world map. Looking beyond the limited definition of cartography as map *production*, English literary and popular culture reveals a lively participation in the dissemination of Ptolemaic-based geographies and suggests a mutually constitutive relationship between literature, early print culture, and geographical thought in fashioning fresh perspectives of a global world. Within the context of the extraordinary influence of Ptolemy's *Geography* in European humanist print culture, the under-explored transmission of Ptolemaic-based geography in England can be identified in a wide range of early print English texts.

As critical theory and maritime studies propose a "radical elsewhere, outside space and time,"[10] and as cultural geographers consider the possibility of polyvalent rather than fixed-scale analyses, the transitional poetic geographies of pre-Mercator England offer a historical model of "fluid cartographies,"[11] one that accommodated mixed-scale, hybrid models of a global Earth in imaginative writing and geographical thought. Through the lens of book history and historical cartography, *Mirror of the World: Literature, Maps, and Geographic Writing in Late Medieval and Early Modern England* uniquely integrates theoretical and disciplinary approaches to write an innovative literary and cartographic history of the influence of Ptolemaic geography in England, one that informs postmodern constructs of space and place and that offers alternative models of non-instrumental spatial analyses.

* * *

Although England had a vigorous history of map production during the medieval period, a cartographic quiet in terms of English representations of the world ensued.[12] Peter Barber notes that there is "little evidence" of a significant cartographic presence in late fifteenth-century England, and that practical, administrative "map consciousness" was not widely present in England, a premise Catherine Delano Smith and Roger Kain contest, arguing that by the early fifteenth century, the practical use of maps were "diffusing into society at large."[13] On the Continent, however, the production of editions of Claudius Ptolemy's *Geography*, inclusive of maps based on his mathematical projection, powerfully intersected with print technology. Ptolemy's geographic model posed a geo-centric Aristotelian construct long familiar to medieval readers, but its representation of the Earth was essentially mathematical: a graticule formed by latitude and longitude coordinates and with equatorial and ecliptic lines. Maps based on the Ptolemaic projection depicted land masses of the known world based on Ptolemy's use of spherical geometry and measured coordinates, outlines roughly familiar to today's maps (without, of course, North America). Maps drawn from the instructions in the *Geography*, "the quintessence of classical cartographical knowledge,"[14] presented a different visual representation of the world from the medieval *mappamundi* (Figure 0.1): Jerusalem was no longer placed as the sacred "navel of the world" and the map was no longer tied to a universal Christian history. Ptolemy's calculations—essentially lists of locational reckoning—and instructions on how to draw maps became the basis of early modern cartography. Visually distinct from its predecessors, Ptolemaic maps typically oriented the representation of the world to north, rather than the more familiar orientation to the east of most (though not all) *mappaemundi*. In addition, Ptolemy's *Geography* emphasized place names and regional detail that were actively studied by humanist scholars: translations from the Greek

Figure 0.1 Hereford Map, c. 1300
Source: By permission of Hereford Cathedral.

into Latin were undertaken in the early fifteenth century followed by cartographic printing, first, in Italy and then northern Europe. The period was distinguished by the prestige of Greek scholarship generally and the particular esteem for Ptolemaic astronomy, astrology, cosmology, and geography, securing Ptolemy's regard as "first Master of them all."

The *Geography* was primarily a list of place names, what we now call a gazetteer, as well as locational coordinates, theoretical commentary, and instructions on drawing maps based on the mathematical projections of the second-century mathematician and cosmographer. Ptolemy's calculations formed the basis of late medieval and early modern editions of the *Geography*, many of which included regional maps and a double folio-size world map drawn upon Ptolemy's "projection" or method of representing a spherical world in a two-dimensional depiction (Figure 0.2). Ptolemy's *Geography*, along with his other works, *Almagest* (a work of astronomy) and *Tetrabiblos* (a work of natural astrology or the "celestial influences on the terrestrial environment and human lives"),[15] exerted a powerful influence on European geographical thought in the mid-fifteenth and mid-sixteenth centuries when Ptolemy's text and maps were first produced as a proto-atlas in manuscript and print editions. Ptolemy's *Geography* found its way to Italy via the Byzantine scholar Manuel Chrysoloras who began a translation into Latin subsequently completed by his pupil, Jacopo d'Angelo de Scarperia by

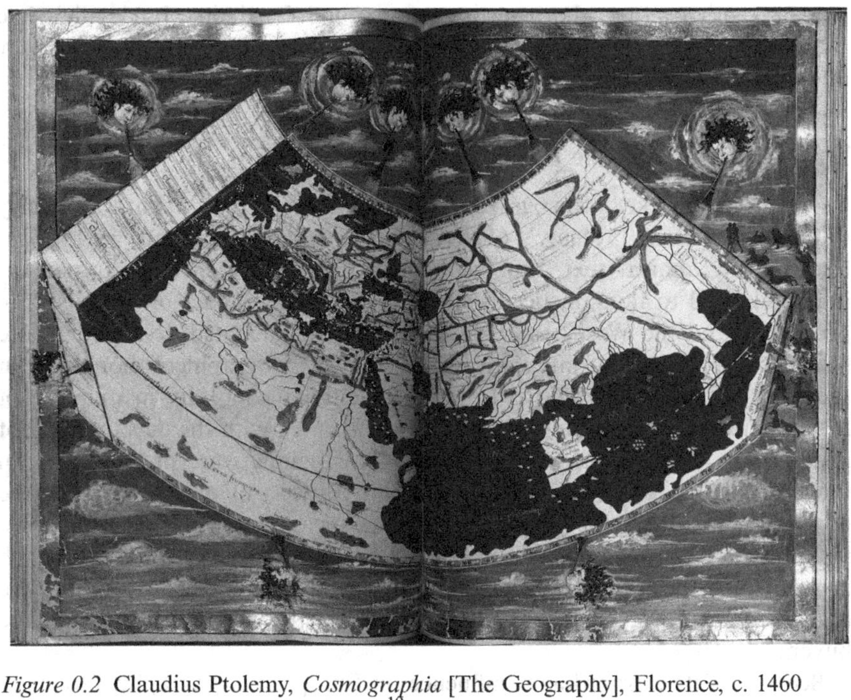

Figure 0.2 Claudius Ptolemy, *Cosmographia* [The Geography], Florence, c. 1460
Source: The New York Public Library.[19]

1410.[16] Manuscript editions in Rome and Florence were followed by print editions made in Vicenza (1475), Bologna (c.1477), Rome (1478, 1490), Florence (1482), Ulm (1482, 1486) and, by 1550, editions in Rome, Venice, Strasbourg, Nuremburg, Ingolstadt, Basel, Paris, and Lyons.[17] By the end of the seventeenth century, Ptolemy's *Geography*, more commonly entitled *The Cosmographia* by medieval printers, had been printed 41 times.[18]

From the commentary and locational coordinates that form the basis of Ptolemy's "tables" and projection of global space, typically 26 regional maps plus a world map were drawn and included as part of the work. If maps accompanied Ptolemy's original work, they have not survived. The world map depicted a spherical Earth in two-dimensional form, typically orientated to the north (as is now our convention). A Ptolemaic map or globe (minus the American continent) is known to have been part of Columbus' navigational tool kit late in the century. But well prior to that date, in tandem with the long-known discussion of the Earth's sphericity in texts such as Johannes de Sacrobosco's thirteenth-century *De sphaera mundi* (which drew upon Ptolemy's *Almagest*), Mandeville's *Travels*, portolan charts and rutters (coastal navigational maps and itineraries), and first-person pilgrimage and travel accounts, the accumulating impact of new geographic thinking as expressed in Ptolemy's *Geography* formed part of the restless geography of the fifteenth century. The common misconception that Europeans thought the world was flat until Columbus sailed west is disproved by the production of spheres in the thirteenth century, the assertion of a round Earth in Mandeville's late fourteenth-century *Travels*, [20] and by the vigorous interest in the *Geography* in the fifteenth century.

The *Geography* and the public sphere of print

As a primary marker of geographic sophistication in the fifteenth and early sixteenth centuries, Ptolemy's *Geography* was regarded not only as a scientific or geographical text, but, as Patrick Gautier-Dalché has argued, as intimately tied to humanistic discourse.[21] While its impact on Renaissance geography and the West has been recognized and debated, less evident are the uneven processes of integration into Western geographical thought.[22] In the early 1400s, Ptolemy's work began to circulate in Italy in manuscript form, then, with incredible synchronicity, coincided with the development of print. Maps of the world and of individual regions were produced as an integral part of both manuscript and print editions of the *Geography*. The text and maps of the *Geography* provided a means by which to visually re-imagine the physical space of the known world by virtue of the maps which accompanied the text. The *Geography* became a standard luxury book, one that attested to both the skill of the printer and the cosmopolitanism of the buyer.[23]

Because of its extraordinary influence on European cartography, Ptolemy's *Geography* has received greater critical attention than Ptolemy's other works, but these were equally regarded and, in the case of the *Almagest*, more widely circulated in England, as evidenced by literary references and university library holdings during the period. The *Almagest*, roughly translated as the "great treatise" or the "greatest matter," was a text concerned with mathematical astronomy, while the *Tetrabiblos* (*Quadripartitum* in Latin) took up the issue of astrology in relation to natural philosophy. The *Tetrabiblos* loosely informs the text of the *Kalender of Shepherds* and *The Compost of Ptholomeus*, widely disseminated, popular print almanacs. While Ptolemy's reputation now is solidly fixed as a central influence in the development of modern geography, Gautier-Dalché points out that Ptolemy was studied by humanists not only for his theoretical geography but in the broader context of classical authors (such as Solinus and Pliny), astronomy, and astrology.[24] Benjamin Weiss frames this interest as indicative that "interest in the *Geography* was not driven by mapmakers," but rather it was astronomers and mathematicians who sought out the text to address "longstanding astronomical and astrological problems."[25] It is this contextual reading of Ptolemy's *Geography* that suggests that English interest in popular texts such as *The Compost of Ptholomeus* was part of a geographic discourse and geo-poetic book culture in which Ptolemy's *Geography* was "read in an astronomical context instead of a cartographic one."[26] It is only retrospectively that the absence of English-produced world maps based on Ptolemy's *Geography* appears to be an absence of engagement with the milieu of fifteenth-century cartography.

English readers in the early sixteenth century, as evidenced by the popularity of texts such as the *Kalender of Shepherds* and *The Compost of Ptholomeus*, developed a conceptual geography within the context of astronomy, astrology, and prognostication—information fashioned from the authority of Ptolemy that also drew upon a geo-poetic aesthetic of "Poyetes and Astronomiers."[27] Henri Lefebvre discounted the mutual influence of humanism and science in the early modern period, but Gautier-Dalché makes a clear case that astrology "should never be overlooked in a study of the reception of Ptolemy's geography."[28] Such divisions, he argues, reflect a later disciplinary divide between poetics and science not pertinent to early humanist intellectual discourse. Literary and popular texts from this period exemplify the connections made by late medieval and early modern readers between geography, astronomy, astrology, and literature, a discourse not solely dependent on, but co-generative with, the emerging field of print cartography. For English readers and writers of this period, the "reading" of geography was not so much a break with the classical and medieval past as a confirmation and expansion. Geography did not represent a world "out there," but one intimately connected to the self through the practice of astrology, a practice framed by the interpretation of textual and celestial signs—the habitus of poets and astronomers.

Despite its impact among scholars and intellectuals, no edition of the *Geography* was printed in England during the fifteenth or sixteenth century.[29] England has a curious lacuna in its history of map production: while new maps based on Ptolemy's *Geography* were printed on the Continent beginning in the late 1470s, it was not until 1544 that Sebastian Cabot's world map was printed in London, followed by Richard Wright's (1599) *Chart of the World* and John Speed's (1627) *Prospect of the Most Famous Parts of the World.* But copies of Ptolemy's works certainly made their way to England via the book trade.

Due to the apparent absence of English engagement with the production of new cartographic representations, D.K. Smith has argued that in the fifteenth and sixteenth centuries, "England was a geographical backwater, separated from the technological advances of the continent."[30] E.G.R. Taylor made a similar observation, concluding that there was "a lag in English geography far behind the standard reached on the Continent."[31] Smith notes that by the end of the sixteenth century, England once again "had taken its place at the forefront of geography, creating the first national atlas in the west and producing maps as detailed and accurate as any in the world."[32] Were English readers, printers, and intellectuals temporarily uninterested in or technologically unable to reproduce geographic advances in the early print era? Kathy Lavezzo has suggested a geographic marginalism had long been cultivated in England as a form of cultural exceptionalism.[33] The limitations of English print technology and fierce international competition suggest another possibility.[34] Indeed, the absence of early English world maps may also reflect a certain business acumen on the part of English printers: virtually every continental printer who undertook the expensive production of an edition of the *Geography* went out of business shortly thereafter.[35] Whatever the reasons may have been, English printers did not take up the publishing of Ptolemy's *Geography* in the early print era.

With this "blank space" in the early print history of world map production in England, how can we trace the ways in which writers and readers in England synthesized and disseminated the emerging Ptolemaic paradigm and changing conceptions of the world? Were there ways in which English readers and printers participated in this changing geographic culture of the Continent?[36] English geographic thought during the 150-year period of the Ptolemaic revival cannot be traced along the surface of a printed English Ptolemaic map; however, it can be contextualized within the emerging English book trade and read in the pages of literary and astrological early printed books, embodied in the visual and material cultures of books. Using map *publication* as the sole basis for determining English geographic thought presents an inaccurate assessment of a reading market very much engaged in transnational exchange. Utilizing an interdisciplinary methodology that integrates cultural geography, book history, map history, and postmodern spatial theory, this study refutes the claim that England's engagement with Ptolemaic-based geography was a "belated transformation."[37]

Late fifteenth-century Europe was in the midst of a period of rapidly changing geographic knowledge and book technology, a co-development of print and cartographic technology analogous to the contemporary co-development of GIS technology and digital books. George T. Kimble argues that the merging of classical and medieval geography with new cartographic representations resulted in an "acute" geographic practice in the mid-fifteenth century such that "scarcely any two world-maps of this period give us the same world-view."[38] This perspective of "chaotic" geographic practices of the fifteenth century is formed by the powerful ascendancy of the Cartesian model and its emphasis on mathematical, coordinate-defined space. As speculative, hybrid geographical expressions, both Malory's *Le Morte Darthur* and the Behaim globe, in contrast, allow two time-space continua to co-exist within a single work. The poetical and the chronological are intertwined, creating a geographical otherness and simultaneous, heterotopic spaces that interpenetrate the physical, the remembered, and the imagined.[39] By 1475, print editions of the *Geography* included maps drawn on the new Ptolemaic principle. To grasp the enormity of such a paradigm shift in visual culture, Kimble suggests that the impact of the rediscovery of Ptolemy's *Geography* was "authentic, convincing, and vastly superior to the average textbook of the day."[40] But this estimation of superiority, of course, relies upon a dismissal of the poetic geographic practices of the Middle Ages in favor of the scientific and "realistic" maps based on measurable relative space. This premise has now come under pressure in the wake of postmodern cultural geography.

The common "boundary" of the year 1500 to distinguish the medieval and early modern literary periods must also be critically set aside.[41] This temporal turnstile failed to account for the continuities in literature and cartography during this transitional period and the way in which cartographic writing and images co-developed with early print in the incunabula period, from the mid-1470s to the early 1500s. Medieval texts continued to be actively translated, printed, and read during the sixteenth century. In this regard, the term "Ptolemaic revival" is a useful referent for the years that span the fifteenth century and early decades of the sixteenth, a period during which cartography, humanist scholarship, geographic writing, and print culture productively co-developed,[42] transforming medieval forms of literature and geographic writing. This influence only gradually waned with the publication of the atlas of the Flemish cartographer Gerardus Mercator in 1569, whose projection continues as the basis of modern world maps. Much of early Renaissance thought about exploration, travel, and new peoples continued to be influenced by the reading of romance and the medieval conception of the *mirabilia*—the grotesque and wondrous marvels of the *outremer*, the lands over the sea, and the *oukemene*, the unknown world.[43] Early print books were not, on a broad scale, newly written "Renaissance" works: the reading and printing of medieval works formed an active part of early modern print history, including early modern geographic and cartographic works.[44]

In the absence of English-made maps, the integration of continental Ptolemaic geography, broadly defined, can be traced via Arthurian romance, narrative geographies, book illustration, theatrical performance and court spectacles, as well as popular compendiums of astrological, spiritual, and health advice. In *The History of Cartography*, David Woodward acknowledges the overdetermination of the transition from the medieval to the modern as a moment of radical change or rupture.[45] In his study of the relationship between medieval conceptions of cosmos and the formation of urban space, Keith Lilley similarly observes that, "all too often human geographers have emphasized discontinuities between the medieval and the modern."[46] In literary studies, William Kuskin similarly traces the continuities between medieval text, early modern readers, and print production. Such observations all point to the consensus in geographic and literary studies: the "firewall of 1500"[47] is a myth. Thus, I employ the term "Ptolemaic revival" to refer to the transitional period when Ptolemaic-based geography and maps transformed medieval visual culture and the representation of "the world." Geographic knowledge in England in the later fifteenth and early sixteenth centuries was disseminated via multiple and protean cultural forms—a dispersed and many-layered process of social construction that integrated classical and medieval geographic traditions and emerging geographic discourse into a new medium—the public sphere of print.[48]

English early print culture

Medieval romances, with their common motif of expulsion-and-return, presented stories well-situated to redactions and revisions that resulted from the flux and accretion of geographic imagination and thought during the Ptolemaic revival.[49] Fifteenth-century romance such as Malory's Roman War account and travel narratives such as Mandeville's *Travels* continued to be vital parts of the reading experience in England during the sixteenth century and played a key role in shaping geographic writing and thought in England during this time. Notwithstanding the absence of English publication of world maps, English readers *were* interested in Ptolemy's work, at least in pseudo-Ptolemaic versions, as well as in geography's sister sciences: astronomy and astrology. Further, late fifteenth-century geographic thought accommodated the coexistence of what we now see as two distinct models: the history-centered *mappamundi* and the mathematically-based navigational map. Fifteenth-century "fluid cartographies" preceded instrumentalist geography and cartographically-inscribed colonialism,[50] allowing for a flexible reorganization of geographic representation in text and map as new information was integrated into a highly flexible and dynamic paradigm, what Ivo Kamps refers to as "the wonderful complex interplay between a growing desire for scientific precision and the utterly subjective nature of representation." As Violet Moller notes, classical scholarship such as Ptolemy's *Geography*, Euclid's *Geometry*, and Galen's medical texts were integrated into European culture in the late fifteenth century as part of a period of "assimilation and correction, rediscovery and rescue." Because this context of hybridity and assimilation has been poorly theorized, Zur Shalev has argued that a "historically nuanced reexamination" of the reception of Ptolemy's *Geography* is called for as part of the growing recognition of what Sean Roberts characterizes as "a vibrant and hybrid geographical culture that historians are only beginning to fully comprehend." English literature, printers, and readers are inextricably part of the geographic literary culture of the Ptolemaic revival.

Within this broader context of the circulation and synthesis of Ptolemy's work, we can begin to study English geographic thought not merely as an absence, but as a geographic discourse that emphasized astrology and astronomy, focused on narrative geography rather than world maps production, engaged in hybrid spatial practices, and participated in a transnational print culture. Evidence for such transnationalism includes book purchases, book lists from academic libraries, and the association of English humanists with their continental peers. For example, Duke Humphrey, a book collector in contact with Italian humanists and Milanese book suppliers and whose library became the foundation of the Bodleian Library, purchased *Pomponium Melam et Ptolomei cosmographium* in 1451, an early attestation to the presence of humanist geographic works in England. In addition, although the libraries of Oxford and Cambridge scholars and clerics focused primarily on theological works, they also held books concerned with matters of the quadrivium (music, arithmetic, geometry, and astronomy). Of these, books on astronomy (and in particular ones which take up the issue of the *compotes* or yearly calendar) figure more prominently, in part because of the connection between astronomy and the computation of movable feasts in the Church liturgical calendar. In addition, as Roger Lovatt has noted, works on astronomy generated interest because of the "links both with astrology and medicine." Evidence of this interest is readily apparent in the catalogues of medieval university libraries, and the accumulation of documentary evidence from the English book trade, including collectors, scholars, and humanists, suggests that a segment of English readers were actively engaged in a transnational book culture with their continental counterparts, a readership that would expand to "everyday" readers as part of the expansion of print culture in the sixteenth century.[51] Through English written and visual culture, the spatial and "cognitive mapping" processes of the new cartography were transacted.[52]

A primary focus of this study is to place geographical objects—maps, globes, and texts—within the broader context of print and reading culture in fifteenth- and sixteenth-century England. Early print scholar Anne E.B. Coldiron has emphasized the way in which the early print era, an "information revolution," created "a cross-cultural traffic in technology, ideas, and aesthetics."[53] Could English geographical culture have been truly "isolated" amidst a thriving—at least in London—cross-cultural literary traffic? Although the map trade was transacted through diffuse channels not perfectly synonymous with the book trade, there was, nonetheless, a significant overlap between the book and map trade, especially in terms of geographical books and books containing maps.[54] Coldiron points out that English printers during the period 1476–1557 had "powerful connections to France and the rest of the Francophone continent,"[55] as well as trading partners in Italy and the Low Countries, connections which obviate, at least in part, any intellectual isolationism posited for England during this time. At present, we lack a more detailed understanding of geographic thought in England in the early print era in large measure due to the paucity of documentation of the importation and ownership of continentally-printed maps and Latinate geographical texts; the trail is difficult to trace, as well, due to the fragile nature of maps themselves. World maps mentioned, for example, in an inventory of Cardinal Wolsey's goods do not specify if they are Ptolemaic-based maps and neither do the maps themselves survive.[56]

Making the case that the translation of French poetry was a vital part of early print production in England, Coldiron points out that "we are better equipped to understand early modern English literature if we read it in its full, polyglot, international context."[57] This same polyglot, international context offers the potential to reassess English geographical culture in the early print, or Ptolemaic revival, period. Translation, importation, and ownership practices challenge the tendency to equate *production* with *use*, and these networked practices posit cultural contact rather than isolation. A transnationalist book and cartographic milieu raises the possibility of a more active geographic culture within England than the map production record suggests. As Worms has pointed out, maps are "inherently international … requiring little in the way of translation" and thus are highly mobile objects of trade.[58] In a similar fashion, Ptolemy's *Geography* needed little translation: in addition to the accompanying maps, the *Geography* consists of a series of place names and short narratives and, as such, it circulated easily in international trade. To posit a uninterested readership for cartographic works in England minimizes the international nature of the English book trade during this time and, at least for English humanists, scholars, and printers, seems an improbable scenario.[59] The literary-historical record from the late fifteenth and early sixteenth centuries puts pressure on the geographical-historical record and suggests a new carto-bibliographic context for the otherwise puzzling absence of English geographic production during a period of dynamic cartographic representations in Europe.

Astronomy, astrology, cosmography, geography

A step toward gaining a better understanding of English geographic thought during the Ptolemaic revival is to recognize that the terrestrial strand of geography is one "to which geographical science is now perhaps too completely confined."[60] This perceptual "confinement" no doubt contributes to modern scholars' identification of England as largely deficient in terms of geographical development during the late medieval and early modern period. As Denis Cosgrove argues:

> So exclusively has the contemporary secular vision become limited to 'space-ship earth', that many students of geography scarcely appreciate the significance of planetary arrangements for patterns of climate on the earth's surface. Yet until very recently, geographical learning began with

> an aid of an orrery or celestial diagram … whose geometry binds terrestrial and celestial spheres.[61]

"Geography" (by definition, "earth-writing") was traditionally part of a threefold spatial hierarchy within locational studies, consisting of cosmography, geography, and chorography.[62] The great continental Renaissance cosmographers such as Sebastian Münster, Martin Waldseemüller, Peter Apian, and Oronce Fine all drew upon Ptolemy's *Almagest*, which described the constellations and planetary movements of the celestial sphere, as well as upon the *Geography*, which described the physical formations of the Earth. Reading the lack of English-produced maps as a lack of geographic sophistication is to conflate the three elements of geography as they were understood in the fifteenth and sixteenth centuries and to define the period through the lens of the post-Renaissance ascendancy of the visual. Narrative geography, rather than cartography, was far more influential throughout Europe than modern scholars have recognized and included an emphasis on Ptolemy not primarily as a geographer *but as an astronomer*.[63] In particular, there seems to have been a greater interest on the part of English printers and writers in the celestial and astrological components of the work of "Master Ptolemy."[64] For example, bemoaning the travails of marriage, Chaucer's Wife of Bath humorously (and spuriously) invokes Ptolemy's *Almagest* as supportive of her warning of marital angst:

> The same wordes writeth Ptholomee;
> Rede it in his *Almageste*, and take it there.[65]

Many English readers took her advice: they read Ptolemy's *Almagest*.

One of Caxton's earliest books printed in England, *The Dictes and Sayings of the Philosophers* (1477), was a compilation of maxims that had, through the translation and accretion process from Greek to Latin, become spuriously attributed to Ptolemy. "Tholome was a ryght wyseman," Caxton writes in his Preface, "and wele vnderstanden & in xpecyall in iiij scyencis that is to wytte geometrye/musike aritmetrik & astrologie/and he made many good bokis amonge the whiche one is called Almageste the whiche is of astrologye."[66] Caxton also mentions Ptolemy in his introduction to his edition of *The Myrrour of the World* as:

> a kynge moche subtil in Astronomye … a kynge of Egipte … [who] composed and made plente of right fair volumes and bookes, and many dyuerse in instrumentes by whiche was founde appertly all the gretnes of therthe, and the heyghte of the firmament, and how the sterres make their cours both by nyght & by daye.[67]

Ptolemy was actually a Roman citizen of Greek descent living in Egypt during the second century CE, but Caxton's reference to Ptolemy reinforces a common confusion between the cosmographer and one of the kings of the Greek royal family (which included Cleopatra) that ruled Egypt from 303 BCE to 30 BCE. Despite this confusion, Caxton makes it clear Ptolemy—be he king or cosmographer—was a geographical authority. Perhaps print technology in England lacked the technical capabilities to reprint the *Geography* with maps, perhaps importation met the relatively small demand for expensive editions of the *Geography*, or perhaps Ptolemy's *Almagest* and *Tetrabiblos* spoke more powerfully to the English geographic imagination. While England was not an innovator in cartographic production during this period, neither were English readers disengaged from the great integration of new geographic models of the world that marked the European Ptolemaic revival.

A geocritical methodology

In *Literature and Cartography: Theories, Histories, Genres*, Anders Engberg-Pedersen gathers together critical theorists whose work, in part, has "been instrumental in establishing literature as an integral part of the study of cartography."[68] In the same volume, Simone Pinet suggests that narrative and cartography "seem to mirror each other."[69] In particular, she co-locates medieval literature and visual culture as structuring thought in related ways, suggesting that the narrative aspect of maps and the diagrammatic aspects of narrative both imagine "a present and future" and that "cartographic operations [are] layered on medieval literary works."[70] Matthew Boyd Goldie has similarly suggested that medieval writing culture, in general, had far less disciplinary segregation. As Goldie notes:

> Medieval literature does not have the strong boundaries of genre that apply to later literature: authors in the period transition between what may be considered traditional poetic subjects and others, including science, with more ease than many authors in other periods of literature.[71]

The present study attempts to bring these geocritical perspectives to bear in a new reading of the transformation of late medieval geographic thought and to discover in these transitional English texts a productive integration of the cartographic and the literary, one that offers fresh insights into the subjectivity of cartographic practices and analyses. This study seeks to understand the role of narrative and early print in structuring and preparing English mapmakers and readers for a new conception of space in the early modern era. Through a wide variety of early print English texts, we can simultaneously rediscover a largely forgotten geo-poetical aesthetic and employ contemporary geo-critical methodology to reread the Global Middle Ages in English popular texts.

As Robert T. Tally Jr. has articulated, "geocriticism" as a critical approach brings "greater emphasis to space, place, and mapping in literary studies."[72] He proposes a useful distinction between the geo-poetics of the writer or work of literature and the geocritical methodology of the critic, a methodology that is interdisciplinary and attuned to postmodern spatial theory. As Bernard Westphal has argued, geocriticism "operates somewhere between the geography of the 'real' and the geography of the 'imaginary.'"[73] During the Ptolemaic revival, the quasi-fictional world of Mandeville and the fictional world of Malory suggested a world of porous boundaries and spatialized narrative that, in turn, harbored structured cartographical thinking, an interweaving of the emotional and the spatial, a passionate geography.[74] For this study, I examine literary and popular works within a geographical context and, conversely, geographical thought within the context of book culture; in other words, the mutual transaction of geography and literature, of maps and texts.

There is much in common between the polyvalent, heterotopic spaces of medieval literature and those theorized by postmodern geographic analyses. Henri Lefebvre announced an "epoch of simultaneity,"[75] a critical perspective that would hold in abeyance hierarchical, rationalized space.[76] Despite Foucault's characterization of a single medieval conception of space as one of hierarchical "emplacement," his proposal of heterotopias bespeaks fluidities of space that are entwined in both late medieval romance and cartography. Reading late medieval romance, geographical writing, playscripts, and astrological almanacs takes us into "ambiguous, heterotopic, uncontrollable"[77] spaces that postmodern readers will find uncannily familiar. The fifteenth century stands on the receding shore of the "sea voyage" of scientific rationalism, an intellectual point of embarkation, as Woodward has suggested, marked by both "continuity and change."[78] The shared aesthetic of poets and astronomers in fifteenth- and sixteenth-century geographic thought propels a reappraisal of the multiple geographic frameworks present in English print culture of the era and, further, offers innovative ways to think about the potential fluidity and "deep mapping" of the interdisciplinary field of the Spatial Humanities. In short, antecedents of the current interest in alternative geographic frameworks and humanistic cartographies can be found in fifteenth-century geographic thought.[79]

The plan of the book

Engberg-Pederson calls for a supplementation of *Literature and Cartography* with "additional histories and contextualizations."[80] The present study offers such an additional history and contextualization, offering a little-known "pre-history" to modern geographical practices, one that shares an ethos of simultaneity and polyvalence with postmodern geocriticism, at a moment of critical interest in querying the constraints and power structures of rationalized mapping.[81]Chapter 1, "Fluid Geographies: The Confluence of Medieval and Ptolemaic Space in Malory's *Le Morte Darthur*," considers Thomas Malory's fifteenth-century spatial and textual re-inscription of Geoffrey of Monmouth's twelfth-century *Historia Regum Britanniae*. Malory's Arthurian romance provides evidence of changing conceptions of spatial and mapping practices in the fifteenth century and narrates a cultural moment alive to the integration of alternative models of geographic thought. The time-based geographies of Malory's Roman War tale reveal a spatial reorganization that has not been contextualized within the emerging mathematically-based geography of the fifteenth century and the principles of Ptolemaic-based cartography. Written in the 1460s, Malory's work is renowned for its linear reordering of Arthurian events, "the birth, lyf, and acts" of King Arthur, a linearization that draws upon spatial and temporal realism. The Winchester manuscript version, thought to be closer to Malory's authorial text, both resists and claims England's Roman legacy as it details Arthur's fictional pan-European campaign to subdue Rome. Malory's densely layered narrative of socially produced political spaces is unfettered by "fixed scale," allowing complex scale modalities to border and intrude upon one another as Arthur and those in his network negotiate body, region, dream, Otherworld, and state. The Arthur that emerges in Malory's text is both powerful over and vulnerable to intersecting spatial imaginaries.

Despite critical misconceptions of Malory's "illogical" geography, Malory's narrative records the early stirrings of the adoption of Ptolemaic geography, including attentiveness to mathematically relational space, the pairing of ancient and modern place names, and the emphasis on urban sites. The comingling of romance, chronicle, and the rationalization of Arthurian geographic space suggests an in-process model of fluid geographies that shaped a hybrid, geo-poetic aesthetic for the late fifteenth century. Responding to the geographical sub-text, William Caxton's 1485 print version of the episode is the most edited section of the book. Caxton shapes his version of the Rome-bound Arthur as a figure operating in mathematical geographic space and, in so doing, prioritizes new linearized geographic knowledge and urban spaces, key signatures of Ptolemaic-based geography.

Chapter 2, "Cartographic Caxton: *The Myrrour of the World* and early English print," situates geographical texts printed by William Caxton—*The Myrrour of the World*, the *Description of Britain*, and the *Polychronicon*—as part of a geographic discourse unique to England's print culture of the later fifteenth century. In *The Description of Britain*, Caxton enacts a separation between the long-standing classical tradition of pairing history with geography and presents, instead, a new commodification and spatialization of English geographic thought. Through his two editions of *The Myrrour of the World*, Caxton is the first to popularize print illustrations of globes for English vernacular readers. Caxton is also renowned for his adoption of the practice of "rubrication," the provision of a table of contents in the front matter of a book. With signatures and indexing, this practice constitutes a "mapping" of the spaces of the book. Caxton's cartographies of the book serve as evidence for the way in which the underlying linear framework of Ptolemaic-based mapping found expression in late fifteenth-century English book culture. Caxton's successor, Wynkyn de Worde, subsequently infused the English Arthurian romance tradition with the cartographic aesthetic of early Netherlandish landscape painting in his illustrated edition of *Le Morte*

Darthur (1498). Closely associated with emerging cartographic practices, early Netherlandish landscape painting utilized linear perspective and a "bird's-eye view" to depict distant townscapes, atmospheric skies, dramatic outcroppings, and birds in flight to create a sense of a vast, expanding landscape. The iconography of de Worde's illustrated Malory reproduces the new relationship to space found in late fifteenth-century maps and early Netherlandish landscape painting, creating a cartographic transaction between Malory's romance and early modern cartography.

Chapter 3, "The equipollent Earth-apple: Mandeville's *Travels*, the Behaim globe, and globes in Tudor England," suggests a conceptual link between the Nuremberg-made Behaim globe (1492) and English geographic thought. The Behaim globe, the oldest extant terrestrial globe, is both geographically pragmatic and poetic. Though the globe theorizes an oceanic, trade-based Earth based on Ptolemy's geographic principles, passages from Marco Polo and John Mandeville are inscribed across the surface of the globe. Although Mandeville's English identity remains contested, his metaphor of the round Earth as an apple is reconstituted as Martin Behaim's "Erdapfel," providing a unique instantiation of both "English" and transnational literary geography. This chapter presents an original study of the Behaim globe, arguing that it should be read as a three-dimensional edition of Mandeville's *Travels* and indicative of the hybridization of medieval and Ptolemaic cartographic expression in the fifteenth century. The chapter concludes with a brief review of the way in which the importation of globes into England brought Mandeville's assertion of the world's sphericity back to England in the form of prestigious geographic objects that graced the libraries of elite families. With the production of the Anglo-Dutch Molyneux globes in 1592, Elizabethan England claimed the globe as a symbol fitting of its ascendancy as a world empire.

While the nexus of geographic production and Elizabethan drama has been well established, Chapter 4, "The painted world: John Rastell's stage globe and geographic pleasure in early Tudor England," reveals a geographical aesthetic, long in advance of Shakespeare and the Globe Theatre. John Rastell's geographic court spectacle, *A New Interlude and a Merry of the Four Elements* (c. 1518) introduces a "fygure" to demonstrate the Earth's sphericity, a stage prop that should rightly bear the designation as the first three-dimensional globe produced in England, albeit an ephemeral one, decades in advance of the English-made Molyneux globes. In revealing the stage globe, Nature declares "What I have shewid, man, prynt well in thyne heart" (319–321), forming a new association between memory, early print, and cartography.[82] In addition, the phrase suggests the shared transmission streams of Ptolemaic geography and the arts in early modern England. During the early Tudor period, cosmological ceiling paintings commissioned for festivities at the Field of the Cloth of Gold (1520), the "figura" or map in the continental edition of Thomas More's *Utopia* (1516), the globes in Hans Holbein's painting, "The Ambassadors" (1533), and Thomas Elyot's

The Boke Named The Governour (1531), each exemplify the association between geography, painting, literature, and moral education in Tudor England. Rooted in wonder and moral approbation, early Tudor geographic knowledge flexibly incorporated geographic information within the art forms of literature, theater, and painting.

An obscure headwater of innovation both for early modern English literature and populist cartography, the multi-text known as the *Kalender of Shepherds* is a "miscellany" and almanac inclusive of astrological prognostication, prayers, light verse, and popular imagery. Chapter 5, "'After poyetes and astronomiers:' the *Kalender of Shepherds* and Ptolemaic geography in popular print," traces the multiple print editions of the the *Kalender of Shepherds* and its derivative, *The Compost of Ptholomeus.* Together, these texts serve as a decades-long cultural transcript of pseudo-geographic reading in England. The increasing appropriation of Ptolemy in image and attribution within editions of the *Kalender of Shepherds* narrate the reception and transmission of Ptolemaic geography for non-elite readers, inclusive of an implied female readership. Images of Ptolemy and astronomical subject matter accompany the *Kalender of Shepherds* and serve as potent agents of an increasing interest in Ptolemaic geography in England. English editions of the *Kalender* and *Compost* formed a print genre of cosmological vernacular in which astrological and geographical space are linked to humankind through the practice of reading. Through the cultural work of these image-rich texts, the English reader is situated at the center of the universe. Into this rich textual and visual polystream, populist cartography is introduced in the form of a world map, one meant for the very Everyman and Everywoman who illustrate the text and to whom the narrative is directed.

At the close of the sixteenth century, the *métier* of the *Kalender of Shepherds* continued in two strands: the geographic and the literary. The hybrid cohabitation of "poyetes and astronomiers" gave way, on the one hand, to the adoption of the Mercator projection in cartography (still the basis of modern world maps) and, on the other, to a poetic form of English national literature. In Henry Gosson's edition of *The Compost of Ptolomeus* (1600/1638?), a simple wood block world map illustrates the text and is accompanied by instructions on how to read the symbolic codes of the map. In contrast, Edmund Spenser's *The Shepheardes Calender* transforms the proverbial and calendrical almanac tradition into a new literary form of English eclogue. From this juncture, the disciplines of geography and literature largely part company until renewed interest in cultural and humanistic geography emerged in the late twentieth century.

In "Epilogue and Analogue: Medieval Geography and the Spatial Humanities," I consider the way in which fifteenth-century and twenty-first-century geographic world-views only appear to stand on opposite shores of Cartesian space and time. The co-development of fifteenth-century book technology and geographic production offers an analogue to the contemporary co-development of digital book forms and mapping, including GIS and Google Earth, at a time of renewed urgency in framing and constituting global spaces. *Mirror of the World* concludes with the assertion that fifteenth-century geographic writing offers a productive, hybrid model of a geo-poetic aesthetic that can transform earlier critical views of late medieval geographic thought as well as inform twenty-first-century literary geography and the Spatial Humanities. New digital geographic platforms provide the potential for "deep contingency" in geo-spatial analysis, provocative new forms of literary analyses, critical problem-solving, geo-editions that foreground spatially attuned readings of the representation and poetics of space, and the intersection of social justice and geographic representation.

Mirror of the World traces the transmission of Ptolemaic-based geography in England on the eve of modern cartographic practices and lays the foundation for a study of its instantiations in English texts, artifacts, and performances. Through a material history of manuscript, early print, maps, and globes, this study proposes a counter-thesis to a purported lack of geographical interest in England via the exploration of literature and popular texts during a key period of transformation in the representation of space. Evidence other than maps suggests that Ptolemaic-based geography was of intense interest to English readers, as it was to their continental counterparts. Subtle channels of geographic thought streamed through popular print in England just as surely as geographical rivers were sketched across the surface of a Ptolemaic map. The carto-print history of England notwithstanding, Ptolemy's shadow is still writ large across English geographic thought.

We, too, stand on the precipice of a time-space reorganization. How is it possible that the "universe," the very concept steeped in a poetic term, is revealing itself to be a "multiverse?" Like fifteenth-century readers, we are stretching our intellects and imagination to apprehend a cosmos more complex and infinite than our wildest dreams. To cast fifteenth-century thinkers as geographically naïve is to miss the beauty of an extraordinary moment in human history when constructs of space and time flexed so as to accommodate new knowledge constructs and a moment of exchange between poets and astronomers.

Notes

1 Thomas Blunderville. *A Briefe Description of Universal Mappes and Cardes and of Their Use* (London: Printed by Roger Ward, for Thomas Cadman, 1598).
2 Henricus Glareanus, *D. Henrici Glareani poetæ lavreati De geographia liber vnus* (Basle, 1527), G3r.
3 Mercator's cylindrical projection was a solution to aid navigation. *Nova et Aucta Orbis Terrae description ad Usum Navigantium Emendata*, 1569 (A New and Augmented Description of Earth Corrected for the Use of Sailors).
4 Laurence Worms, "Maps and Atlases," in *The Cambridge History of the Book in Britain*, vol. IV: *1557–1695*, eds. John Barnard and D.F. McKenzie (Cambridge: Cambridge University Press, 2002), 229.
5 The last quote is from Edward Said, *Orientalism* (London: Penguin, 1991), 94.

6 See Sean Roberts, *Printing a Mediterranean World: Constantinople, Florence and the Renaissance of Geography* (Cambridge, MA: Harvard University Press, 2013). Roberts identifies Ptolemy's *Geography* as marking a critical transition from manuscript to print and from medieval to modern cartography. See also Adam Barrows, *Time, Literature, and Cartography After the Spatial Turn: The Chronometric Imaginary* (New York: Palgrave, 2016).

7 Michel Foucault, "Of Other Spaces," trans. Jay Miskowiec, *Diacritics* 16:1 (Spring, 1986): 22–27; 24. Foucault also postulates the medieval period as "the space of emplacement," associating a far greater fixity that evidenced in fifteenth-century literature and maps.

8 Pauline Reid, "Through a Looking-Glass: Invention and Imagination in the Visual Rhetoric of William Caxton's *Mirrour and Description of the World* and Stephen Hawes's *Pastime of Pleasure*," *Rhetorica* 34.3 (Summer 2016),#: 268–300.

9 Denis Cosgrove, *Apollo's Eye: A Cartographic Genealogy of the Earth in the Western Imagination* (Baltimore, MD: Johns Hopkins University Press, 2011), 3.

10 See Iain Chambers, "Maritime Criticism and Theoretical Shipwrecks," *PMLA*, 125:3 (May 2010): 678–684; 679.

11 Ibid., 681.

12 Sarah Tyack, ed. *English Map-making 1500–1650* (London: British Library, 1983); Catherine Delano Smith and Roger J. P. Kain, *English Maps: A History* (Toronto: University of Toronto Press, 2000). Peter Barber notes two later English *mappaemundi*, the Aslake (c. 1350) and the Evesham (c. 1390) in "Mapmaking in England, ca. 1470–1650," in *The History of Cartography: Cartography in the European Renaissance*, vol. 2, Part 2, ed. David Woodward (Chicago: University of Chicago Press, 2007), 1589–1669.

13 Barber, 1578; Delano Smith and Kain, 28–29.

14 See Jean Michel Massing, in *Circa 1492: Art in the Age of Exploration*, ed. Jay A. Levenson (New Haven, CT: Yale University Press, 1991), 228–229; 228.

15 Alexander Jones, "Ptolemy's *Geography*: A Reform that Failed," in *Ptolemy's Geography in the Renaissance*, eds. Zur Shalev and Charles Burnett (London and Turin: The Warburg Institute – Nino Aragno Editore, 2011), 15–30; 16.

16 For a discussion of the migration of the *Geography* from Constantinople to Florence and its importance in Florentine Renaissance book culture, see Roberts. It is thought that Jacopo d'Angelo worked on the translation from 1406 to 1409 with a possible publication date of 1410.

17 For a list of editions of the *Geography*, see Patrick Gautier-Dalché,"The Reception of Ptolemy's *Geography* (End of the Fourteenth to Beginning of the Sixteenth Century)," in *The History of Cartography*, vol. 3, *Cartography in the European Renaissance*, Part 1, ed. David Woodward (Chicago: University of Chicago Press, 2007), 285–364. See also Henry Stevens, *Ptolemy's Geography: A Brief Account of All the Printed Editions down to 1730* (London: Henry Stevens, Son, and Stiles, 1908; republished by Theatrum Orbis Terrarum, 1972). Print editions were later produced in Strasburg by Waldsemüller, 1513, and in Lyon, 1535. To see facsimiles of early editions of the *Geography*, see, for example, R.A. Skelton, "Introduction," in *Cosmographia: Bologna 1477* (Amsterdam: Theatrum Orbis Terrarum, 1963) or *Cosmographia: Rome 1478* (Amsterdam: Theatrum Orbis Terrarum, 1966).

18 Zur Shalev, "Main Themes in the Study of Ptolemy's *Geography* in the Renaissance," in *Ptolemy's Geography in the Renaissance*, eds. Zur Shalev and Charles Burnett (London and Turin: The Warburg Institute – Nino Aragno Editore, 2011), 1–14; 6.

19 Translated from the Greek by Jacopo d'Angelo da Scarperia, Nicolaus Germanus, cartographer, the New York Public Library, Manuscripts and Archives Division.

20 For the impact and dissemination of the fourteenth-century *Travels* by Sir John Mandeville, see Iain Macleod Higgins, *Writing East: The "Travels" of Sir John Mandeville* (Philadelphia, PA: University of Pennsylvania Press, 1997).

21 See Gautier-Dalché.
22 Zur Shalev and Charles Burnett, eds. *Ptolemy's Geography in the Renaissance* (London and Turin: The Warburg Institute – Nino Aragno Editore, 2011).
23 As one example of this prestige, Gautier-Dalché cites a 1493 letter from Johannes von Reitenau, a tutor, to his employer: "After I had mentioned to him the great universal map printed in Italy, which I had seen with you, he desired to have the same [map] and was insistent that I write to you for it." Rupprich, *Der Briefwechsel des Konrad Celtis*, 115–116. See Gautier-Dalché, 347, n. 437. See also Lisa Jardine, *Worldly Goods: A New History of the Renaissance* (New York: Nan A. Talese, 1996).
24 See Gautier-Dalché. See also Ptolemy, *Tetrabiblos*, ed. and trans. F. E. Robbins, The Loeb Classical Library (Cambridge, MA: Harvard University Press, 1940). Robbins notes that in the Middle Ages, "no one thought it inconsistent with Ptolemy's reputation as a scientific astronomer that he should also have written upon astrology," ix.
25 Benjamin Weiss, "The *Geography* in Print: 1475–1530," in *Ptolemy's Geography in the Renaissance*, eds. Zur Shalev and Charles Burnett (London and Turin: The Warburg Institute – Nino Aragno Editore, 2011), 108.
26 Ibid., 108.
27 This phrase is from *The Compost of Ptholomeus*, Robert Wyer, printer (London, 1540?), STC 20480a, sig. O(iii)v.
28 Gautier-Dalché, 352.
29 Gautier-Dalché provides a comprehensive list of editions of Ptolemy's *Geography*. For a less comprehensive list, see also Stevens, *Ptolemy's Geography*.
30 D.K. Smith, *The Cartographic Imagination in Early Modern England* (Aldershot: Ashgate, 2008), 41.
31 E.G.R. Taylor, *Tudor Geography: 1485–1583* (New York: Octagon Books, 1968), 13–14.
32 D.K. Smith, 41.
33 Kathy Lavezzo, *Angels on the Edge of the World: Geography, Literature, and English Community, 1000-1534* (Ithaca, NY: Cornell University Press, 2006). Lavezzo argues that Englishmen during Cardinal Wolsey's time maintained a "distinctly medieval perspective," 120.
34 Worms, 229.
35 Weiss, 91.
36 For a discussion of the complexities of fifteenth-century nationalism, see Kenneth Hodges, "Why Malory's Launcelot Is Not French: Region, Nation, and Political Identity," *PMLA* (May 2010): 556–571. Hodges points to the porous sense of territory and identity between England and areas of modern France that were contested territory, in particular, the English duchy of Guienne, which includes parts of Aquitaine and Gascony.
37 Robert J. Mayhew, "Geography, Print Culture and the Renaissance: 'The Road Less Travelled By,'" *The History of European Ideas* 27:4 (2001): 349–369; 355.
38 George H.T. Kimble, *Geography in the Middle Ages* (New York: Russell and Russell, 1938, reissued 1968), 197.
39 For a discussion of heterotopia in postmodern fiction and medieval literature, see Paul Smethurst, "The Journey from Modern to Postmodern in the *Travels of Sir John Mandeville* and Marco Polo's *Divisament dou Monde*," in *Postmodern Medievalisms, Studies in Medievalism XIII*, eds. Richard Utz and Jesse G. Swan (Cambridge: D.S. Brewer, 2004), 159–179.
40 Kimble, 215.
41 See, for example, William Kuskin, "'The Loadstarre of the English Language:' Spenser's *Shepheardes Calender* and the Construction of Modernity," *Textual Cultures: Texts, Contexts, Interpretation* 2:2 (Autumn 2007): 9–33.

42 Ptolemy's *Geography* disappeared from medieval Europe's scholarly tradition prior to the fifteenth century but had continuity in Greek scholarship. For a discussion of the continuous presence of Ptolemaic scholarship in Byzantine culture and its transference to Florentine humanist scholarship, see Roberts.

43 See Steven Greenblatt, *Marvelous Possessions: The Wonder of the New World* (Chicago: University of Chicago Press, 1992).

44 Carol J. Chase and Joan Tasker Grimbert edited a special volume of *Arthuriana* dedicated to Late Medieval French Arthurian Romance and its continued popularity during the Renaissance. See their Introduction, 3–6, and Jane H.M. Taylor "'Hungry Shadows:' Pierre Sala and His Yvain," *Arthuriana* 19:1 (Spring 2009): 7–19.

45 David Woodward, "Cartography and the Renaissance: Continuity and Change," in *The History of Cartography*, vol. 3, ed. David Woodward (Chicago: University of Chicago Press, 2007), 3–98.

46 Keith Lilley, "Mapping Cosmopolis: Moral Topographies of the Medieval City," *Environment and Planning D: Society and Space* 22:5 (2004), 681–698. See also Keith Lilley, *City and Cosmos: The Medieval World in Urban Form* (London: Reaktion Books, 2009).

47 Kuskin, 11. See also William Kuksin, ed. *Caxton's Trace: Studies in the History of English Printing* (Notre Dame, IN: University of Notre Dame Press, 2006), 1–31.

48 For the role of translation and the public sphere of print, see Anne E.B. Coldiron, *English Printing, Verse Translation, and the Battle of the Sexes, 1476–1557* (Aldershot: Ashgate, 2009).

49 Helaine Newstead notes this common formula in her overview of English romances in "Romances, General," in *A Manual of the Writings in Middle English: 1050–1500*, fascicle 1 (New Haven, CT: Connecticut Academy of Arts and Sciences, 1967), 24.

50 While Chambers takes up the migratory realities and imaginative trope of oceans as a means to propose a "potential exit from Humanism," he does not look to the fifteenth century but exclusively to twentieth-century theorists such as Walter Benjamin and Gilles Deleuze. The sea journeys in medieval romance are suggestive of the heterotopic spaces Chambers explores as a possible antidote, as discussed in Chapter 1.

51 For the interconnection of medieval European literary history, see David Wallace, ed. "Preface," in *Europe: A Literary History, 1348–1418*, vol. 1 (Oxford: Oxford University Press, 2016), v–vi.

52 Edward W. Soja, *Postmodern Geographies: The Reassertion of Space in Critical Social Theory* (New York: Verso, 1989), 62.

53 Coldiron, 2.

54 Worms, 228–229.

55 Coldiron, 2.

56 See David Starkey, ed. *The Inventory of King Henry VIII: Society of Antiquaries MS 129 and British Library MS Harley 1419* (London: Harvey Miller, 1998), 287–288. Kathy Lavezzo also discusses the two maps listed in the inventory and observes that their location at Hampton Court, Wolsey's residence, makes it likely that there were part of Wolsey's map collection, 116.

57 Coldiron, 20.

58 Worms, 228.

59 As just one example, Desiderius Erasmus published an edition of the *Geography* in Basel in 1533; although this was only two years before the death of Thomas More, it seems likely that More's personal and intellectual exchange with Erasmus in the early part of the century would have included geographic topics and texts.

60 Denis Cosgrove, *Geography and Vision: Seeing, Imagining and Representing the World* (London: I. B. Taurus, 2008), 18–19.

61 Ibid., 19.

62 Ibid., 17.

63 Gautier Dalché. Kim H. Veltman similarly argues that astronomy and geography were interconnected and that astronomy has been overlooked for its contribution to the development of linear perspective in "Ptolemy and the Origins of Linear Perspective," in *Atti del convegno internazionale di studi: la prospettiva rinascimentale, Milan 1977*, ed. Marisa Dalai-Emiliani (Florence: Centro Di, 1980), 403–407.

64 For a discussion of the various astronomical and astrological works printed in the early print era in England, see Henry Stanley Bennett, *English Books and Readers* (Cambridge: Cambridge University Press, 1965), 116–118. They include Caxton's *The Myrrour of the World*, Trevisa's translation of Bartholomew's *De proprietatibus rerum*, the *Kalender of Shepherds, The Compost of Ptholomeus, The Boke of Demaundes of the Science of Phylosophye and Astonomye Betwene kynge Boctus and the Phylosopher Sydrake, De Astronomia*, Andrew Borde's *Pryncyples of Astonomye*, William Salisbury's *The Description of the Sphere of Frame of the Worlde*, and, in 1552, Anthony Ascam's scholarly work *A Lytel Treatyse of Astronomy.*

65 Geoffrey Chaucer, "The Wife of Bath's Prologue," in *The Riverside Chaucer*, 3rd edition, ed. Larry Benson (Boston: Houghton Mifflin, 1987), 107.

66 William Caxton, *The Dictes and Sayings of the Philosophers: A Facsimile Reproduction of the First Book Printed in England by William Caxton in 1477* (London: Elliot Stock/New York: J.W. Boulton, 1877), 102–103.

67 Oliver H. Prior, ed. *Caxton's Mirrour of the World* (London, Kegan Paul, Trench, Trubner & Co., Ltd., Early English Text Society, Extra Series, cx, 1913), 150.

68 Anders Engberg-Pedersen, "Introduction—Estranging the Map: On Literature and Cartography," in *Literature and Cartography: Theories, Histories, Genres*, ed. Anders Engberg-Pedersen (Cambridge, MA: MIT Press, 2017), 11.

69 Simone Pinet, "Diagrammatic Thought in Medieval Literature," in *Literature and Cartography: Theories, Histories, Genres*, ed. Anders Engberg-Pedersen (Cambridge, MA: MIT Press, 2017), 173–197; 173.

70 Ibid., 189; 191.

71 Matthew Boyd Goldie, *Scribes of Space: Place in Middle English Literature and Late Medieval Science* (Ithaca, NY: Cornell University Press, 2019).

72 Robert T. Tally, Jr. "Introduction: On Geocriticism," in *Geocritical Explorations: Space, Place, and Mapping in Literary and Cultural Studies*, ed. Robert T. Tally, Jr. (New York: Palgrave Macmillan, 2011), 1–9; 1.

73 Bernard Westphal, *Geocriticism: Real and Fictional Spaces*, trans. Robert T. Tally Jr. (New York: Palgrave Macmillan, 2007), 107.

74 Giuliana Bruno, *Atlas of Emotion: Journeys in Art, Architecture, and Film* (New York: Verso, 2002). Bruno bridges the academic and the personal, connecting visual culture with motion and emotion. See also my blog, available at: www.passionategeography.com.

75 Henri Lefebvre, *The Production of Space* (New York: Wiley, 1992), xix.

76 For the process that undergirds the development of modern cartography, see David Woodward, "Maps and the Rationalization of Geographic Space," in *Circa 1492: Art in the Age of Exploration*, ed. Jay A. Levenson (New Haven, CT: Yale University Press, 1991), 83–88.

77 Chambers, 679.

78 See David Woodward, "Cartography and the Renaissance: Continuity and Change," in *The History of Cartography*, ed. David Woodward, vol. 3: *Cartography in the European Renaissance*, Part 1 (Chicago: University of Chicago Press, 2007), 3–98.

79 David J. Bodenhamer, John Corrigan, and Trevor M. Harris, eds. *The Spatial Humanities: GIS and the Future of Humanities Scholarship* (Indianapolis, IN: Indiana University Press, 2010).

80 Engberg-Pedersen, 12.

81 See Denis Wood, "Cartography Is Dead (Thank God!)," *Cartographic Perspectives* 45 (2003): 4–7.

82 Prior to the introduction of print technology, the maxim to print something in your heart was used as a set phrase to evoke an important remembrance. John Lydgate, for example, in his *Fall of Princes* (1431–1438) wrote "Of whose five wounds print in your heart a rose." For Rastell, in light of his engagement in the early print business, the aphorism takes on a new meaning.

1 Fluid geographies

The confluence of medieval and Ptolemaic space in Malory's *Le Morte Darthur*

Malory's great legend of King Arthur and the Round Table appears to be the most English of narratives, with its tragic plot deeply embedded in the English landscape: Tintagel, Camelot, Glastonbury. And yet, Malory tells readers that he drew upon many French as well as English sources, occasionally legitimizing his account—if not always accurately—with the phrase "as it tellyth in the Frenshe booke" (221.3–4; C VI.17), immediately identifying a potential gap in a purely English geographic perspective.[1] The work soon reveals a far wider geographical scope than just the action circling around the court at Camelot[2] to include liminal spaces, time-space disruptions, Otherworldly intrusions, and sea crossings that insert "ambiguous, heterotopic, uncontrollable space"[3] into Malory's narrative of the arc of Arthur's birth, life, and death.[4] The overlapping geographic models of the late fifteenth century have often been cast as idiosyncratic or naïve, an amalgam of the medieval *mappamundi* and emerging cartographic practices. Postmodern geo-critical theory, however, suggests a useful lens through which to reassess late fifteenth-century geographic thought, one that disrupts the fixity of the map with "fluid cartographies."[5] Re-inscribing modernity as a rupture from the medieval tradition, Zygmunt Bauman, in his work, *Liquid Modernity*, suggested the disruption of the "stable and apparently invulnerable one-to-one correspondence" of time and space as the marker of modernity. This unstable correspondence, however, is fully operative in Malory's *Le Morte Darthur*.

The tale with the most wide-ranging geography in Malory's *Morte* is The Tale of Arthur and the Emperor Lucius (also known as the Roman War account) in which Arthur crosses the sea to campaign across Europe, successfully killing the Roman emperor Lucius. In Malory's version, Arthur also receives the crown of emperor at the hands of the Pope, in Rome. But references to a wider geography—both foreign and otherworldly—are frequent throughout the long prose romance. Journeys to or intrusions from these foreign or otherworldly places underlie the superstructure of the work and the restless quest to define individual and English national identity. These multiple geographic sites—symbolic, political, emotional, and personal—construct an Arthurian geographic aesthetic that is fluid and heterogenous.

The "geography of *Le Morte Darthur*" is attributable not only to Malory but also to the geography of his sources as well as to the nature of

DOI: 10.4324/9781003096245-2

geographical thought during the 1460s when Malory was writing his great synthesis of myriad Arthurian tales. At the juncture between medieval and modern geographical thought, representations of global space were undergoing visual and cultural redefinition: from the era of the *mappamundi*, with its Jerusalem-centered locus of Christian universal history, to the longitude and latitude coordinates of early modern maps based on the projections of Claudius Ptolemy and Gerardus Mercator. The temporal and spatial developments of the mid-fifteenth century are as much a part of Malory's *Morte Darthur* as are Camelot, London, and York, such that, in addition to English spaces and politics, the work reveals a surprisingly global focus as well as an increasing rationalization of space resulting from the cultural integration of Ptolemaic concepts and the production of Ptolemaic-based maps.[6] Felicity Riddy points out that Malory's emphasis on issues of English nationhood give the *Morte Darthur* "a specifically English orientation,"[7] but Arthur and his knights not only "take the adventure"[8] in the forests surrounding Camelot but venture as well to Rome, Sarras (Babylon), and multiple other locations overseas, including France and Ireland. Foreigners also intervene into English territory, including French allies, Saracen invaders,[9] and Roman ambassadors, as well as more enigmatic or Otherworldly figures such as the Lady of the Lake and Merlin. Malory had, we might say, a dual focus on England and on a wider world over which Arthur might plausibly be in command. Malory frequently implies that Arthur is destined to be the ruler of "all Crystendom" (100.6; C IV.1) and, over the course of the work, seeks to articulate the outer boundaries of what that geo-political space might encompass.[10] In addition to its Englishness, then, *Le Morte Darthur* sets the concern for English nationhood within a sense of an expanding, cartographic world as well as a militant Christendom, over which Arthur is to preside.

Fluid geographies, simultaneity, and geo-poetics

"Malory's geography" is not, strictly speaking, "Malory's." It belongs to a set of spatial practices under development in the fifteenth century that looked back to geographic traditions and shaped developing ones. However, Malory's conception of territory and space has tended to be treated as his own individual or idiosyncratic take on the world. For example, in his 1919 edition, Edward Strachey declares that, aside from the Roman War tale, Malory's "geography is fanciful enough."[11] More recently, Dorsey Armstrong and Kenneth Hodges describe Malory's geography as "tricky to work with, which is why we say he uses an (il)logic of space."[12] Such readings of Malory assess his geographical world-view from the perspective of post-Enlightenment geography rather than the hybrid or fluid geographies of the fifteenth century. As a result, the "logic" of the geo-critical context of the emerging spatial paradigm of the fifteenth century, clearly evident in Malory's work, fails to be identified or understood. Though a writer of legendary Arthurian romance, Malory's geographic framework is part of the spatial developments of his day and, in turn, his text interpreted this geographic framework for his readers. In *The New Companion to Malory*, editors Megan G. Leitch and Cory James Rushton value the "polyvalent vitality" of new critical approaches to Malory's great text, and Catherine Nall similarly argues that Malory's work has been most productively historicized when situated in its wider discursive context.[13] However, the politics of England's civil unrest in the War of the Roses, a primary focus of Malory scholarship, has tended to shoulder out a Global Middle Ages approach. Even though Armstrong and Hodges grapple with Malory's extra-national geography, they do not contextualize *Le Morte Darthur* within the discursive context of geographic thought in narrative and visual representation in the fifteenth century, in particular, the spatialized narratives and new maps based on the principles of Claudius Ptolemy.

Sheri Chriqui has noted that Armstrong and Hodges' investigation into Malory's geography pays critical attention to regional borders "as hyphenated, tenuous spaces in Arthurian literature, galvanizing interest in localized identities,"[14] but here again, Malory's world is treated as a world unto itself rather than one in dynamic conversation with the processes of the reimagination of a global world. Robert Rouse and Cory Rushton make clear that medieval Arthurian geography is "a complicated mixture of real places and mythic sites" and, that this mixture is "shaped and reshaped in often idiosyncratic ways for successive authors and competing narrative traditions."[15] They, too, suggest that the geography of many medieval writers was "idiosyncratic at best."[16] However, Malory was writing within the structure of a complex body of thought regarding geography and space during a dynamic period of change in which Ptolemaic maps developed and our "modern" world-view was set in motion. As an alternative methodology, Malory's geographic paradigm must be read as early evidence of a productive process of experimental geography and geo-creativity, terms that have arisen over the past decade in reference to transdisciplinary exchanges between geography and the humanities.[17] Indeed, Anders Engberg-Pedersen observes the recent "cartographic turn" and renewed scholarly interest in the "entwinement" between literature and cartography, a cohabitation at ease in late medieval geographical writing.[18] Aside from Malory's possible boots-on-the-ground participation in a military campaign to France, it is essential to consider the cartographic and spatial milieu that constituted the spatial imagination of this English "knyght-presoner" (144.3; C IV 29) writing in the 1460s.[19]

Rouse and Rushton point out Malory's interest in English place names and the way in which he "adds a significant layer of realism to his text when he mentions places he knows."[20] P. J. C. Field identifies Malory's "liking for real places," and his oft-noted penchant for inserting English place names into his text when he can.[21] According to Field, Malory added geographical references because "the authors of Malory's French sources set their stories for the most part in a vaguely realized Britain that he knew better than they did."[22] The exception to Malory's additive geographical realism is his version of Arthur's Roman War campaign across the Continent, about which Field suggests that "when Malory does not understand [the geography], his story may not fit the landscape it claims to be set in."[23] Certainly, it must be acknowledged that Malory's knowledge of European geography was fragmentary and incomplete, informed by military experience, word-of-mouth accounts, cultural situatedness and, to a large degree, reading. As scholars grapple with the complexity of "Malory's geography" in a work reliant on centuries of cultural geographies, multiple authors, and multinational sources, it is no wonder that "idiosyncratic" seems the only plausible adjective in relation to his world-view, geographic imagination, and use of place. However, Malory was writing as part of a cultural moment in which space was being linearized and rationalized, but it was a hybrid paradigm that allowed for simultaneity and accretion, a late medieval form of what is now defined as literary cartography or geo-poetics.[24]

In all of the analyses of Malory's geography, his geographic coherence—or lack thereof—is measured against our modern conception of geographical "realism," as we understand it in the twenty-first century. But in the fifteenth century, geographic thought—when not in service to functional navigation or wayfinding—could be simultaneously operative in literary, spiritual, and cultural registers and function as a more fluid, symbolic, and creatively-engaged spatial construct. Malory's geographical revisions and William Caxton's editing of the Roman War campaign are revelatory of the great changes in geo-spatial representation that mark the transition between the geographic frameworks of the Middle Ages and the early modern period. In Malory's and Caxton's versions of the Roman War account, we can observe the "chaotic empiricism" that Robert J. Mayhew associates with early Renaissance maps and geographical narratives which "fused new claims with old learning and lore."[25] In addition, the accumulative editorial processes of the Roman War versions produce overlapping paradigms of geographic knowledge that seem contradictory or "chaotic" in the aftermath of a progressivist history of cartography, but that attest, instead, to a geo-poetic aesthetic unique to the fifteenth century.

The written geographical tradition of the fifteenth century is a textual weaving of "stratification and juxtaposition,"[26] a fusion of literary traditions and personal experiences, visual and mental maps. Reflecting on the geographical accretion evident in the Post-Vulgate Cycle, one of Malory's sources,[27] Field notes that a later adaptor "unfortunately failed to eradicate some of the discrepancies between the new material and the old."[28] From an editor's perspective, this may make sense, but from the perspective of cultural geography, such "discrepancies" reveal a cultural layering of geographic thought embedded in the movement to and from Arthur's court. After a long disciplinary estrangement that began with the gradual adoption of Ptolemaic and Mercator-based cartographies, the intersection of literature and geography is once again being explored as a "zone of creative interaction,"[29] such that the cultural geographies written into *Le Morte Darthur* can be read with a fresh perspective. The departures, adventures, and returns, crucial to the genre of romance, provide both literary spaces and poetic cartographies at a moment when humanist or "modern" concepts of space and mapping were in formation. The topographical capaciousness of Malory's tales reveals deeply embedded and yet shifting cultural routes, a rumbling of tectonic plates from earlier eras and texts alongside new spatial practices such as the rise of locational geography and a fascination with place names, both ancient and contemporary. Placing Malory as part of this "in-flux" geographical context, rather than merely as a geographically idiosyncratic, illogical, or source-driven writer, reveals the ways in which his romance anticipated and even shaped geographic thought for his readers. Malory operates within a cluster of aesthetic and geographical conventions which exemplify fifteenth-century geographical thought and include the classical geographical writing tradition, medieval *mappaemundi*, travel narratives,[30] pilgrimage narratives, chronicles, and emerging mathematically-based cartographic practices. Despite this creative amalgam of written and visual culture, visual maps have since become the primary signifier for the purported "break" from medieval to modern geography, overshadowing the gradual intellectual assemblage that led to mathematically-ordered cultural geography and the re-imagination of the relationship between the world and the geographical self.[31] This process coincided with Malory's great Arthurian project—that is, the linearization of space and time.

Ptolemy and new representations of space

To understand Malory's geography, it is essential to place his writing within the intellectual and geographical developments at work in the fifteenth century. European humanists, scholars, and bibliophiles were keenly interested in Ptolemy's *Geography*, with print editions produced in Vicenza (1475), Bologna (c.1477), Rome (1478, 1490), Florence (1482), Ulm (1482, 1486) with additional editions printed in Rome, Venice, Strasbourg, Nuremburg, Ingolstadt, Basel, Paris, and Lyons by 1550.[32] By the time that Caxton printed his edition of Malory's *Le Morte Darthur* in 1485,[33] there can be little doubt that Caxton was well aware of the new intersection of print and geography—by then luxury editions of the *Geography* held "pride of place"[34] in the libraries of northern European nobility, including the Burgundian court with which Caxton had been affiliated.[35] But "Malory's geography" is inflected by the stirrings of Ptolemaic geography, even while it is unlikely that Malory ever saw a Ptolemaic-based map.

Malory assembled an array of sources and episodes regarding Arthur and his knights, then famously unraveled the medieval interlacement of the structure of his source romances, rearranging them in a chronology of the "birth, life, and acts of the said King Arthur."[36] In his final *explicit* (a brief note to the reader at the conclusion of a tale), Malory asks those who "redeth this book of Arthur and his knyghtes from the begynnyng to the endynge" to pray for him (940.21–22; C XXI.13).[37] Malory thus creates a new English

Arthuriad that progresses "from begynnyng to the endynge"—that is, in linear, chronological time.[38] Malory's literary process is a recognizable parallel to the cartographic process of the linearization of space and time as manifested in the new maps based on Ptolemy's *Geography*, and Malory's book serves as a co-agent in the transmission of the rethinking of space that marks the fifteenth century.

Malory has long been associated with the taint of medieval geographical naïveté and often portrayed as a nostalgic, backward-looking author. Colin Richmond, for example, has argued that Malory was "resistant to the vast changes in the politics and culture of his time"[39] and that Malory's text was marked by "paradigm-shift-resistant qualities."[40] Nothing could be further from the truth; indeed, one of the unrecognized reasons for Malory's continued popularity, I would argue, is not only the way his book looks back to a chivalric code of the past, but for the way it *looks forward* to an emerging conception of space and time, expressed in a geopoetic aesthetic that is at once cartographic and literary, spatially rational and emotionally complex. Malory's lengthy romance not only epitomizes the in-process rationalization of space that defines Ptolemaic geography, it also reveals the in-process development of a new literary form. Kevin Whetter refers to *Le Morte Darthur* as a "long-form tragic-romance"[41] and Cory James Rushton agrees, arguing that the capaciousness of the prose romance aligns it with attributes that will come to be associated with the novel.[42] Malory "did not set out to write a novel," Rushton suggests, "but for all intents and purposes he ended up doing so nonetheless … The novel, eventually, followed him down the path he laid out by chance."[43] It is fitting that a spatial metaphor describes the proto-novelistic form of *Le Morte Darthur*, for Malory anticipates both the coming of the novel and the coming of the modern map.

A methodology informed by the spatial humanities suggests an approach to the geography and spaces of the *Morte Darthur* that sets "Malory's geography" within the geographic aesthetics and practices of the fifteenth century, a cluster of geographic practices associated with the development of early modern cartography. Within its emotional, sweeping geography, *Le Morte Darthur* unravels romance time, provides "double names" for places, and specifies distances in English "myles." Further, Malory's use of polyvalent scale as a tool for analysis is critical to the way in which he evokes a vast sense of space both within and beyond the boundaries of England.

Unraveling romance time

Malory's romance contributes to the changing cartographic grammar of the fifteenth century by engaging in "the structuring process of visualizing time."[44] Malory re-structured his sources to unravel synchronous romance time, an attribute of Malory's work long-remarked upon but never placed in the context of cartographic development. The Vulgate Cycle and Post-Vulgate Cycle, romances upon which Malory drew, operate in a kind of equipollent geometrics of time, similar to the simultaneity of the representation of time on medieval *mappaemundi*. Malory's Arthurian world, however, partially steps out of this synchronous dream-time and into a narrative where characters also move across space in accordance with chronological time. For example, toward the conclusion of the Fair Maid of Astolat episode, Malory writes "So thys passed on all that wynter" (830.31; C XVIII.20), followed by a passage that continues this tracking of time: "Thus hit past on tylle Crystemassse" (831.1; C XVIII.21). The next episode, The Tournament at Winchester, begins with the phrase "So at aftire Crystemas" (832.1; CXVIII.21) and the following episode, The Knight of the Cart, continues this sequential chronological thread: "And thus hit passed on frome Candylmas [February 2nd] untyll Ester" (841.1–2; C XVIII.25). These time-based passages convey a sense of the inexorable, doom-laden, movement toward the book's conclusion—a narrative enactment of the astronomical reckoning that forms the basis of cartographic projection—that is, the measuring of time.

Malory further imposes a chronological sequence on his narrative through the use of his famed paratactic style—his continual use of phrases such as "And than," "And so," "So whan this was done," and "Than."[45] Malory's relentless use of these paratactic phrases provides a strong narrative ordering, placing locations and actions in sequential order across the expanse of the page. Malory also frequently notes how long a character journeyed from one place to another, for example, when he observes that Sir Melyas has ridden into an old forest where "therin he rode too dayes and more" (684.20; C XIII.13). When Gwenyvere realizes Launcelot is sleeping with Elayne, for example, "than she … myght nat slepe a foure or a fyve owres" (632. 31–32; C XI.8). Subsequently, Launcelot, mad with love grief, abides in Sir Blyaunte's castle "more than a yere and an halff" (645.29; C XII.2). There are many such examples of the specificity of time passing in the *Morte*. Of course, Malory also uses stock romance "time units" such as "a twelvemoneth" and Christian holy days such as Pentecost or the "utas" (a week, or octave, after a feast day) of Seynte Hyllary (151.34–152.1; C V.2).[46] His frequent temporal references structure a narrative that unfolds under a persistent attention to the arc of the hours of a day or the turning of a season. As Malory himself notes in the opening Tale of Uther and King Arthur, "And so the tyme drove on and all thynges redy ipurveyed …" (17.4–5; C I.10). Malory's romance incorporates equipollent, episodic narratives from earlier romances as well as chronicle-based episodes within a superstructure governed by the passage of linear time. Linearity in the *Morte*, however, was not complete nor homogeneous—while the overarching narrative takes place "from beginning to end," there is still ample space for romance time and an amalgamated aesthetic of fluid, passionate geographies, that is—the emotional register of literary and cartographic spaces.

Keith Lilley has described medieval urban forms and *mappaemundi* as replete with the cosmological symbolism of "a sacred geometry of circles and squares."[47] With the increasing production and use of portolan charts and Ptolemaic-based maps in the fifteenth century, the medieval center-enhancing space of Jerusalem no longer marks the map in quite the same way: portolan charts depict harbors and trade routes with a linearization of place names (toponyms) fanning out along the coast (Figures 1.1 and 1.2). Despite this linearization, portolan charts also represent differing scale across sectors of the map[48] in much the way that Malory's text does not maintain "consistent"

Figure 1.1 Roselli portolan chart, 1466
Source: Courtesy of the James Ford Bell Library, University of Minnesota.

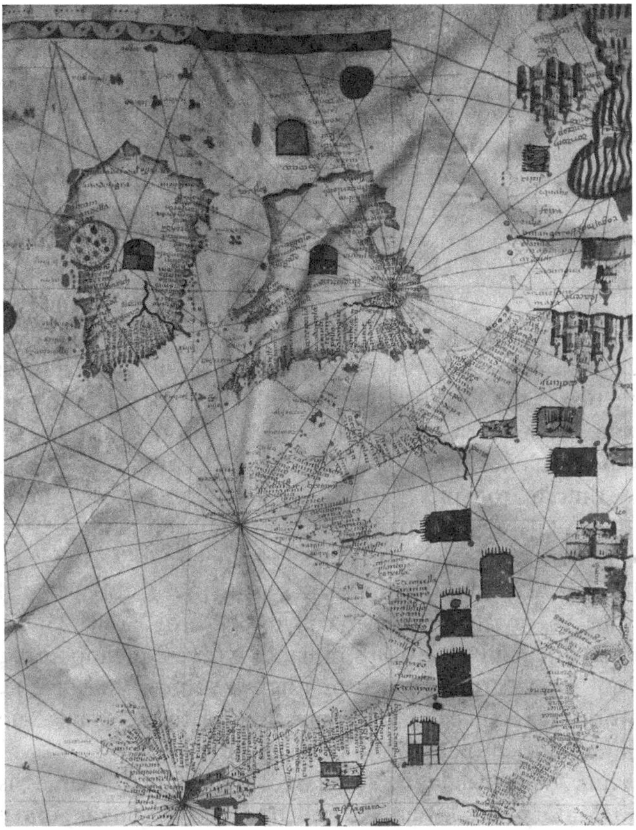

Figure 1.2 Roselli portolan chart, 1466, detail
Source: Courtesy of the James Ford Bell Library, University of Minnesota.

scale across the narrative. Portolan charts also use red ink to denote toponyms of greater importance to the mapmaker,[49] a rubrication scheme strikingly similar to the scribal practice in the Winchester manuscript of Malory's work in which red ink emphasizes names of characters and, occasionally, places.[50] As Armstrong and Hodges point out, many of the knights' names include a reference to their place of origin.[51] Thus, the red rubrication of names in Malory's text is also a rubrication of place, a form of gazetteer that invokes the disparate spaces from which the knights hail, including "Alardyne of the Oute Iles" (83.32; C.III.6), "Peryne de Mounte Belyarde" (64.33; C II.13) and "Sir Myles of the Laundis" (97.12; C III.15).

In Malory's sources, Mordred's usurpation during Arthur's Roman War campaign results in the dissolution of Arthur's realm, but Malory's famed deferment of the tragedy so as to stage Arthur's successful return opens up time and space for Arthur's kingdom to contain the various quests and adventures in a chronologic sequence while simultaneously decentering the cultural and political power of Rome. David Wallace has argued that Malory's "spaces of play"[52] break into a narrative of imperial conquest in the Roman War account, another instance in which combining romance time and chronicle time does not trouble the geographic aesthetics of the work. In Malory's version, Arthur's journey to the center of power does not destroy the periphery, as it did in Geoffrey of Monmouth's twelfth-century version, *Historia regum Britanniae*; instead, an island that had been on the far edge of a *mappamundi* now holds sway against the centripetal force of Rome.

The doubleness of place

Another element that Malory's narrative shares with fifteenth-century geographic practice is the identification of both the ancient, or legendary, name of a city and the modern name. As noted, Malory often adds specific contemporary place names for English locales, as if to ground the vague landscape of Arthurian romance in the soil and cityscape of fifteenth-century England. In the tale of the Fair Maid of Astolat, for example, Malory "georectifies" Camelot as at "that tyme called Wynchester" (807.12, C XVIII.10, see also 804.5; C XVIII.9), then tells of King Arthur's stopover at the town of Ascolat "that ys now in Englysh callyd Gylforde" (804.25; C XVIII.8). Malory infuses the landscape of his romance with a contemporary geographic specificity not found in his French or English sources and, in so doing, disrupts the world of romance with the present, creating a simulacrum of linear history from an ancient past to the political present.

This practice of coordinating local and ancient place names is not idiosyncratic to Malory—it was an integral part of the contemporary humanist fascination with Ptolemy's *Geography*. There is no evidence that Malory, writing in the 1460s, had access to a manuscript edition of the *Geography*, and the first print edition, in 1477, came well after Malory's date of composition and death. But the *zeitgeist* of the attention to place names, the Portuguese use of the astrolabe in determining latitude at sea, the development of portolan charts, and the impact of humanist interest in translating and producing maps based on Ptolemy's *Geography* all speak to an ethos of intellectual engagement with the paradigm of latitude and longitude and the inscription and overlay of a grid upon the world. Malory's impulse to layer new geographic knowledge over his source texts results in what we might call a consciousness of the doubleness of place also characteristic of the *Geography*: ancient and local place names provided a parallel reckoning of place as both timeless and bound by the measurements of time.[53]

Cartographic historian Patrick Gautier-Dalché has observed that during the first half of the fifteenth century, Ptolemy's *Geography* "was appreciated by Italian humanists for features other than those we see as constituting the originality of his work," that is, the resulting world and regional maps. "First and foremost," Gautier-Dalché argues, "the *Geography* was seen as a compendium of ancient place names."[54] In light of humanists' interest in linking ancient and contemporary place names, we can place what has been identified by literary scholars as an individual penchant of Malory's in the context of a developing geographic consciousness attendant to the double-life of place.

Myles

At times, Malory uses very specific distance to locate his characters. For example, Malory notes that Meleagant's castle was "within seven myle of Westemynster" (843.1; C XIX.1). In Hardyng's *Chronicles*, the *Prose Merlin*, and the Alliterative *Morte Arthure*, in contrast, distances are typically obliquely articulated, without the specificity we see in Malory. In Hardyng's *Chronicles*, for example, distance is generalized:

> With castels hyegh and toures at every myle
> Endlonge that walle with many other pyle.
> (Bk III, 994–995)[55]

Malory measures movement across specific spaces, describing Arthur or his knights traveling "ten myle" (17.9; C I.10 and 106.21; C IV.6) "within thre myle" (103.35; C IV.4), "half a myle" (134.20 and 26; C.IV.23), "six or seven myle" (814.33; C XVIII.13). Malory does not always provide specific distances in miles but when he does so, the effect is to ground the account in contemporary geographical detail as well as symbolic space. In the opening tale, for example, Arthur meets his allies Ban and Bors precisely "ten myle oute of London"(17.9; C I.10), the distance from the city suggesting both the specific distance and the incredible relief and gratitude Arthur feels toward these reinforcements in his bid for sovereignty.

Malory's French sources, if they gave specific distance, would have used the French measurement of leagues[56] and, in this regard, Malory anticipates by decades the 1528 English-language edition of *The Rutter of the Sea*, a navigational log that converted leagues from the earlier French source text (c. 1483) into English miles.[57] Malory's reckoning of miles serves as another layer of his translation process from French sources into English vernacular as well as serving as a late-medieval GPS system, providing detailed directional turns by the "half-myle." Additional studies of the distances articulated in Malory's sources may reveal the extent of his cartographic and symbolic spatial practice as he overlaid English topography onto his French sources.

Malory's Arthuriad, however, is not merely anticipatory of the coming scientific rationalism and the rationalization of space. The *Morte* is replete with emotion: fellowship, revenge, madness, grief. As Andrew Lynch poignantly argues, "the 'motion' in 'emotion' is crucial in Malory."[58] Lynch suggests that Malory treats "emotions *as* actions" and that "they are a mainspring of narrative direction."[59] This affective spatiality is what endears the *Morte* to readers and distinguishes it as both lyrical and cartographic, an intersectionality of narrative and heterogeneous spaces and an example of how the "transgressive movements of subjects," as Robert Tally, Jr. suggests, result in "the mobility of spaces themselves."[60] Examples in the *Morte* abound: when Uriens and Arthur fall asleep aboard a magical ship, for example, they are transported from a "grete foreste" to, respectively, "Camelot abbede in hys wyves armys" or a "durke preson" or a disruption of a "realistic" time-space continuum (106.18–107.32). Dorsey Armstrong and Kenneth Hodges argue that the oscillation between "real" and "natural" spaces in *Le Morte Darthur* creates "conflicted and conflicting geographic sensibilities," a conflict they identify as an "(il)logic of space."[61] While they are right to point out the oscillation, this geographic transgressiveness structures the very geopoetics of Malory's *Morte* and is entirely coherent with the fluid geographies of the fifteenth century.

Scale

Part of the aesthetic of fifteenth-century geographic practice included variable scale—that is, medieval maps did not necessarily adhere to a consistent spatial register. Typically seen as a form of geographic deficiency from the vantage of post-Enlightenment rationalism, multi-scalar perspective is capacious, open to interpretation, and allows for simultaneity. Malory's geography includes both "realistic" English landscape and ruptures of consistent scale.

Theories of non-scalar analysis have been a controversial thread in the "scale debate" within the contemporary field of human geography.[62] Such theories of space provide the shock of the familiar, however, to readers of Malory's Arthurian romance where different social processes can be "unevenly and complexly distributed across various scalar levels."[63] Cultural geographers such as Sallie A. Marston contest hierarchical scale in favor of "flat ontologies" as an antidote to the problems of globalism in the twenty-first century. Such non-linear scalar shifts are intimately familiar to readers of Malory's *Morte Darthur* as it moves from "mervylous dreme" to the locational specificity of Launcelot's river crossing "at Westmynster Brydge" in London (846.35; C XIX.4) or Arthur's men "redy on tho bankes" in the bustling port town of Sandwyche (152.30; C V.2). Indeed, having one's spatial imaginary shaken up is part of the enduring appeal of Malory's work. The text moves across global, local, national, personal, and imaginary boundaries in complex yet—if the longevity of the book is any testament—deeply moving ways that stir our emotions and offer multiple if sometimes contradictory perspectives on how to grapple with the complex human problems of government, divided loyalties, and familial networks. Malory, arguably, "tackles aggregations of power at multiple scales" and utilizes both "micro and macro-processes."[64] In so doing, he defies expectations of consistent scale and uses the "map" of his great narrative as an analytical tool. Cultural geographers such as Marston provocatively suggest that the rationalist-scientific construct of scale that arose out of late medieval and early modern Ptolemaic and Mercator-based cartography is in serious need of reconsideration. Manuel

Castells suggests, instead, the concept of a "space of flows,"[65] which resists categorizing spaces as discrete entities but as constituted by the relation of places, selves, and objects. Reading Malory's scalar "mash-up" invites modern readers to step outside our received paradigms of space and time and offers a literary model of spaces of flow.

Our own embeddedness in the dominant scalar framework, itself a fictive construct, is problematic, forestalling our ability to apprehend the spatial imaginary in which Malory wrote his geography—a world that was expansive and time-space flexible. Armstrong and Hodges argue that "Malory's use of what we like to call an (il)logics of space [is] a fundamental component of his text's structure."[66] Despite the transgressions of what we consider a "realistic" time-space continuum, the fluid geographies and polyvalent scale are uncanny and yet highly functional in Malory's Arthuriad. After the long dominance of the Mercator projection, the affective geographies of Malory's Arthurian landscape tolerate, perhaps even celebrate, spaces as porous rather than fixed as a means of exploring cultural and political power. Malory uses scale such that it can function not as a self-evident reality of geographic space but as a methodology for geo-political analysis. As a narrative instantiation of an alternative-scalar analysis, there is so much territory yet to be explored in Malory's *Morte Darthur*.

One of many examples of Malory's use of micro and macro spaces can be seen in the spatial modulations of the opening tale. Arthur must prove himself as legitimate patrilineal heir (when he steps into a public space before nobles, clergy, and commoners to pull the sword from the stone), as a medium between the terrestrial and magical geographies (when he tethers his horse to row across the lake with Merlin to receive the magical sword from the watery Otherworld), as a one-on-one combatant (when he almost slays Pellinore), as a passive dream-state body (when he "slepte merveylously sore" (107.24–25; C IV.6) in the ship of silk), and as global sovereign (when he rebuffs the Roman ambassadors). At either end of the opening episode, Malory toggles between issues of conception (Arthur's and Mordred's) and issues of international standing. It is no surprise that once Mordred survives Arthur's death-by-exposure edict, he is "cast up" upon the seashore (46.17; C I.27), the liminal space of the internal threat of conception and external threat from across the sea.

Malory's socially-produced political spaces reach from English provincial centers to the power center of Europe: Rome. He moves the reader from the public space of kingship to the most private spaces—the lovers' bedroom, the healing of a wound, the recounting of a dream. Shifting across macro- and micro-spaces, Malory eschews fixed scale, allowing spaces to border and intrude upon one another as Arthur and those in his network negotiate body, dream, region, nation, Christendom, and Otherworld—a pluri-verse of competing spaces. In so doing, Malory proposes a king who can repel threats imposed by both "false treson and by enchauntment" (117.5–6; C IV.12). Having contextualized Malory's Arthuriad as an active participant in fifteenth-century literary geography, three sites of geographic "entwinement" narrate critical junctures in *Le Morte Darthur*: (1) the establishment of the young Arthur as King; (2) the Roman War account; and (3) the concluding tale. A geocritical methodology situates the episodes and geography of the *Morte* as deeply engaged with and constructive of the geo-poetics of the fifteenth century. In defining geo-critical practices, Bertrand Westphal has articulated the productivity of attentiveness to "the relationship between place and space" as well as "the transgressiveness that is the guarantor of a fluid and non-permanent vision of territories."[67] Westphal particularly calls out that "fiction and its referent, which is imprinted somewhere in the real, are not inexorably separated."[68] This entwinement aptly constitutes the aesthetics of the fifteenth century in terms of geographic writing and cartography: fiction and geography were not separated into distinct fields.[69] Such transgressiveness forms the basis of Malory's fluid geographies and his mid-century transformation of Arthurian space and time.

The rise of Arthur

The opening tale of the *Morte Darthur* includes the story of Arthur's conception, the famed "Sword in the Stone" episode, the travails of the knight Balyn, the wedding of Arthur and Gwenyvere, and several additional adventures. The action quickly shifts from the unnamed power center of Uther Pendragon's court to the western, semi-autonomous region of Cornwall, a site of resistance to the sovereignty of Uther. At the siege of the castle at Tintagel, however, Malory's scale shifts to that of the body, at once the landscape of physical desire resulting in Arthur's conception and the immateriality of self as Uther shapeshifts to Gorlois, via Merlin's magic. Together, these micro-assemblages aggregate and renegotiate the self, the national, the regional, the international, and the Otherworldly.

Once Arthur establishes himself, at least provisionally, as heir and king, Camelot acts as a centripetal locus to which Arthur and his knights continually return for feasts, court life, and counsel. Sir Torre, for example, "departed and com to Camelot on the third day by noone" (91.22–23; C III.11); after wooing Gwenyvere, Arthur "brought her to Camelot" (95.7; C III.13); and after various adventures Arthur "rode to Camelot and helde there a grete feste with myrth and joy" (100.27; C IV.2). Widening waves of geographical contestation follow Arthur's coronation—Cornwall, Scotland, and Wales rebel against Arthur's claim to a unified kingdom. Arthur entreats his French allies, Ban and Bors, to assist him, and other references to French locales include Benwick, Nantes, and Calais.[70] Amid the focus on the internal ordering of the nation, "mo than fourty thousande" Saracens who made "grete destruccion" invade the north of England, laying siege to the Castell Wandesborow (30.9–11; C I.17) in Scotland.[71] Ban and Bors return to France to fight Claudas; a rebellious alliance regroups in the French cities of Surhaute and "Nauntres in Bretayne" (32.33; 33.14; C I.18); and an advance party "com into the courte," ambassadors that were "twelve knyghtes that were aged men, which com frome the Emperoure of Rome" (39.34–35; C I.23). It is a kingdom in continual motion.

Within this flux, five kings harry the northern boundary, entering "into the londis of Kyng Arthure" (100.33; C IV.2), causing Arthur to exclaim in sheer frustration, "Alas! … yet had I never reste one monethe syne I was kyng crowned of this londe. Now shall I never reste tylle I mete with tho kyngis in a fayre felde" (101.1–3; C IV.2). With his defeat of the five kings, Arthur presumably expels the Saracen invasion and repulses the initial diplomatic salvo of the Roman ambassadors (40.5–6; C I.23). The terrain of these early episodes stages an intense geographical restlessness and gauges Arthur's capacity to respond. Despite this bone-wearying unrest, the promise is held out that Arthur "hath the floure of chevalry of the wordle with hym" (101.26–7; C IV.2); and that it will be his destiny to quell the pagan forces of the East as well as supplant the imperial authority of Rome. The scale of the geographical framework is vast—nothing less than all of Christendom and extending into the Mediterranean and Middle East. Even Arthur's cruelty in condemning all male babies under four weeks old to exposure "in a shyppe to the se" (46.14; C I.27) contributes to the "transoceanic"[72] nature of Arthur's ascendancy to power. Within this global framework, there is the local scale of forest and hermitage as well as micro-spaces where individual knights hack and hew their unknown assailant at the level of the body. The struggle between Arthur and Morgan le Fay, and between Camelot and Avalon, provides symbolic alternative spaces of territory and flow. The "samyte" covered arm offers Arthur a potent sword from the watery Otherworld below, an object crossing from one spatial register to another. Thus, Arthur's drive to dominion is at once local, global, and Otherworldly.

For Malory, the geography of the Otherworld hermitages that bespeak Christian mysticism, and the terrain of dreams creates alternate geographical spaces from that of English political space. Arthur's negotiation of such multiple scalar frameworks is essential to the establishment of his fitness to rule, as well as to the fragility of that rule. Malory offers a nuanced and affective response to what is perhaps more incredible than any questing beast, white hart, or water-born sword—that an individual who stands, at one moment, precisely ten miles outside of London also possess the capacity to cross the scalar delimitations of spatial hierarchies. Thus, the king who "loked aboute the worlde" (106.33; C IV.5) in Malory's opening tale is multi-faceted, able to modulate between spatial registers of public and private selves.

Part of the work of the first tale is to establish Arthur, not only as the ruler of all of England, but as one whose success against masculine military incursions and feminine magic ensures that "all Crystendom shall speke of hit," a phrase reiterated in the first tale (100.6; C IV.1, 115.21–22; C IV.11, 121.29; C IV.15). Arthur's renown, in the future tense, will reach the edges of the known world. As the "whole book" unfolds, Arthur's conquests and that of his knights include Rome, Sarras and, ultimately, Jerusalem. While Rome functions symbolically as the seat of power for Christendom and imperial power, Sarras, which Malory locates as "in the partis of Babilonye" (788.21–22; C XVII.23), is the site where the Grail will be achieved in the tale of The Sankgreal, a place both of eastern mysticism and Christian spirituality. The international and Otherworldly allusions magnify the scale of Arthur's realm to that of world-atlas proportions and inoculate the drive for national sovereignty with imperial and spiritual aspirations.

At the end of this foundational tale of Arthur's beginnings, Malory's *explicit* declares his identity as a knight but also his location in prison: "Who that woll make ony more lette hym seke other bookis of Kynge Arthure or of Sir Launcelot or Sir Trystrams; for this was drawyn by a knyght presoner, Sir Thomas Malleorré, that God sende hym good recover" (144.1–4).[73] Confined to a room or set of rooms in Newgate Prison,[74] Malory effects a dramatic telescoping of the scale of the preceding tale: the solitude of a prison cell in contrast with Arthur's global incursions and forays magnifies the epic scale at which Arthur has operated. Intriguingly, Malory suggests that all of the action he has just relayed might not be all that transpired or all that is known: "lette hym seke other bookis," he suggests, at once declaring his limited textual resources but also that Arthur's geographical space might encompass even more territory than has just been related. The confined space of the author, speaking across unknown time and space, provides a momentary counterpoint to the restless geography of the text—a narrative pause, replete with the implicit hope that Arthur has, at least temporarily, stabilized the kingdom before taking up the campaign against Rome.

Periplus and pilgrimage: the rhizomatic itinerary of the Roman War

In the second tale of Thomas Malory's *Le Morte Darthur*, "The Noble Tale Betwyxt Kynge Arthure and Lucius the Emperour of Rome," the book's predominant genre of romance shifts to a pseudo-historical battle narrative and takes on the form of a *periplus*, a term that denotes "both a travel account and … a visual survey of the earth's imaginative or actual bounds."[75] Malory's Roman War account is drawn from a literary tradition that dates back to Geoffrey of Monmouth's twelfth-century *Historia regum Britanniae*, and possibly earlier.[76] Of all the tales that make up Malory's comprehensive "grete booke"[77] comprising the lineage, birth, life, and death of King Arthur, the Roman War account stands apart: it is the "most extended extraterritorial episode" of the work,[78] in which Arthur and his knights embark on a cross-European campaign from London to Rome. Heavily inflected with phrases from its alliterative source, it is also the most vigorously edited tale by William Caxton in his print edition. An intensely geo-spatial narrative, the tale provides a glimpse of a world-view undergoing active rewriting and mapping. As such, Malory's Roman War tale, and Caxton's editing of the text, provide a textual route by which to follow the dissemination and development of geographic thought in mid-to-late fifteenth-century English literary culture.

From Geoffrey of Monmouth to Malory, the Roman War narratives collectively produce a rhizomatic spatial representation of cultural memories: variations of a long Rome-bound journey that cut across the European continent narrate deeply-layered memories of contestation and connection, a semiotic chain of author, text, place, and mental map.[79] The various iterations of the geographical narrative provide a literary compendium of Arthurian space and how that space was symbolized, imagined, remembered, and re-written. The Latin form of an *itinerarium* was typically a narrative text rather than a cartographic image, listing place names along the route. Two visual examples of *itinerarium* in map form include Matthew Paris' *Chronica majora* and the Peutinger map, both of which are forms of "strip maps" in that they structure geographic representation as an itinerary. Though the Roman war narratives are verbal, not visual, they share a common language of *itineraria*, journeys represented textually or visually as a form of spatially-sequenced *place names.* As Gautier-Dalché, the preeminent scholar on the transmission of Ptolemy's *Geography* through European humanistic circles, has argued:

> The development of modern cartography has led us to overlook the importance of such lists [of place names] in the constitution of geographical knowledge, to forget that they are a fundamental means of ordering facts. The reading of Ptolemy's text and the study of his maps worked together in forming a determinate conception of the structure of terrestrial space.[80]

The conceptual structure of the Roman War narrative is one that is not substantively different from the lists of place names that formed the central core of late medieval editions of Ptolemy's *Geography.* Indeed, we see in the geography of the Roman War campaign a concomitant shift as that occurring in mapping practices: that is, a new emphasis on place-oriented relational time and space.

The round-trip journey between London and Rome is approximately 2,500 miles requiring a minimum of many months each way, and probably more if one considers the less-than-direct route of the town-to-town journey such itineraries necessitated. Strip maps, such as Matthew Paris's, constituted a form of "wayfinding" map[81] that may be the best cartographic analog for the literary itinerary. Oral, narrative mapping practices included highly localized knowledge that led the traveler to the next town rather than providing a big-picture trajectory. This form of narrative mapping practice is described by Rory Stewart in a contemporary memoir of his cross-Afghanistan trek in 2002. Describing his mapless walking journey, Steward observes the way villagers guided him from village to village:

> Much of the route they knew only from tales told in their father's guest rooms. But everyone had memorized a chant of names and villages along

> footpaths in every direction. This was a very useful map. It specified everything in terms of a man on foot: the best tracks, the distances that could be walked in a day, whom you should speak to in each village.[82]

This form of linear way-finding Stewart describes provides a glimpse into earlier cultural practices exemplified in the Roman Peutinger map, which survives only in a medieval copy. The Peutinger map visually represents wayfinding along Roman road routes rather than the universal histories that formed the visual basis of *mappaemundi*.[83]

Inscriptive foundation: "a very ancient book"

The Roman War account is one of the longest-enduring spatial narratives of English literature, older than Chaucer's renowned pilgrimage from the Tabard Inn to Canterbury and more culturally active across the centuries than the *Beowulf* epic, which disappeared from active readership for a millennium. In Geoffrey of Monmouth's *Historia*, the Roman War account was part of the larger foundational mythology of Britain which presented the establishment of the British kingdom as a result of the diaspora of the Trojan War. The tale's geographical mythos negotiated England's cultural relationship to Rome, the center of European power from antiquity through the early modern period.[84] This narrative of Arthur's "cartographic ambition"[85] exerted a special fascination for English readers from the time of Geoffrey's twelfth-century pseudo-history through Caxton's 1485 edition of Malory. In his prologue to *Le Morte Darthur*, Caxton claims that:

> many noble and dyvers gentylmen of thys royame of Englond camen and demaunded me many and oftymes wherfore that I have not do made and enprynte the noble hystorye of the Saynt Greal and of the moost renomed Crysten kyng, fyrst and chyef of the thre best Crysten, and worthy, Kyng Arthur, whyche ought moost to be remembred emonge us Englysshe men tofore al other Crysten kynges.[86]

Caxton goes on to refer to "the sayd noble kynge and conquerour Kyng Arthur" and notes that Arthur had been "kyng and emperor."[87] Caxton relates that he initially questioned the historicity of Arthur only to be rejoined by his noble patrons, providing Caxton the opportunity to enumerate the textual sources of his Arthurian publication, including Geoffrey's *Historia*.

Geoffrey, for his part, obfuscates his Roman source, referring only to "a very ancient book written in the English language" that recounted the deeds of British kings, purportedly given to Geoffrey by Archdeacon Walter Map.[88] Gordon Brooks has suggested, however, that the tale may have its origins in accounts of Roman military conquests and the Welsh tale of Mascen Weidig, both of which recount imperial ambitions and military campaigns against Rome.[89] Sent to defend and govern Britannia in 368 CE, the Roman general

Magnus Maximus reversed course and subsequently marched against his rivals in Rome. The tale may arise out of ancient power struggles played out across the geography between Rome and Londinium and woven into the English geographic imagination as a foundational narrative of heroic national resistance.

There is no single Roman War narrative, but many: a rhizomatic, centuries-long literary tradition akin to sandy paths intersecting and scattering across a coastal landscape. Whatever its origins, this potent tale of spatial and moral over-reaching forms an English literary tradition carried forth from Geoffrey in versions by Wace, a courtly Anglo-Norman verse account from the mid-twelfth century (*Roman de Brut*); by Laȝamon, a late twelfth-century English version (*Brut*); in the English alliterative poem *Morte Arthure* (Malory's most direct source) from the early fifteenth century, and in John Hardyng's Arthurian section of his *Chronicles of England*, written in the mid-fifteenth century. In Geoffrey's account, Arthur's ultimately devastating attempt to claim the crown of emperorship brings the Round Table to ruin. Arthur kills the emperor Lucius, who had demanded that Arthur pay "tribute" to Rome, but just as Arthur approaches the city of Rome, he receives news of treachery at home in England: Mordred has betrayed him politically and personally, staging a coup in Arthur's absence and engaging Gwenyvere in an adulterous affair. With this betrayal, Arthur's prophetic "wheel of fortune" dream is fulfilled: Arthur abandons his imperial aspirations and returns to England where, in battle with Mordred, he is fatally wounded. In turn, the ideals of the Round Table are shattered, victim of the inevitable turn of fortune's wheel. For Geoffrey, a Roman Catholic cleric, an attempted English conquest of Rome is never realized and, indeed, the folly of such geographical ambition is revealed.

Malory and Caxton, as his subsequent editor, draw upon this literary tradition. In the English literary imagination, the tale figures significantly as a form of an "inscriptive foundation,"[90] fusing the legend of King Arthur with English national identity and the claim of continental geographic space. As Tom Conley has argued, when "an emotive state is associated with the representation of a geographic space, there emerges an aesthetics and a politics of experience."[91] For an island nation at the edge of the former boundaries of the Roman Empire and of Latin Christendom, the affective impact of an English conquest of Roman and papal dominion suggests why this grafting of emotion and geography became such a deeply-embedded, centuries-long cultural narrative in England. The focus of this chapter is on two of those narrative paths: that of Sir Thomas Malory, surviving in a single manuscript copy from approximately 1470 and known as the Winchester manuscript,[92] and that of Caxton, England's first printer, who produced an estimated 500 copies of a print edition of Malory's work known as *Le Morte Darthur*. Of this print edition, a single copy is extant, held in the Pierpont Morgan Library in New York,[93] with an additional, imperfect copy in John Rylands Library in Manchester. To compare the manuscript and printed text, facsimile editions are available as well as the web-based Malory Project.[94]

In all of the pre-Malory Roman War accounts, Arthur's geographic expansionism results in the death of Arthur and the devastation of the English kingdom, a form of "eulogized space,"[95] grieving the unattained ideals of the Arthurian reign. In Malory's hands, however, this geo-imperial tale is told quite differently: it becomes the cornerstone of the successful establishment of Arthur's sovereign rule, his transformation into an ideal Christian ruler, and the beginning of a period of *Pax Arthuriana*. In Malory, Arthur fulfills the Roman conquest and the tragic end of the Round Table is forestalled, eschewing the previous tradition of the tale as evidence of Arthur as an "overweening hero."[96] Malory's *Le Morte Darthur* engages in a form of textual aphasia, as Thomas Crofts has astutely observed, in which the text seems to "forget" or obscure its textual predecessors and the inevitability of Arthur's death.[97] When Arthur's death comes in Malory's narrative, as it inevitably must, it centers instead on the liaison of Gwenyvere with Lancelot and the treachery of Agravain as the causative action. Adultery and treason, not aspirations of empire, are the seeds of ruin for Malory's Arthurian geographic conquest.

The multi-century re-telling of the Roman War account draws upon mappings, both material and those inscribed in memory, as well as the writers' and readers' conceptions of the English nation in relation to the Continent, to Rome, and to the East. As Ivo Kamps observes, "the metaphoric models of place … become inscribed in memory and shape how people conceptualize and experience their world."[98] Like a map, narrative geography functions as "a metaphorical paradigm"[99] and, as such, Malory's Roman War account functions as an enactment of geographic thought amid pre-modern nationalist politics in the fifteenth century, at the precise cultural moment when Ptolemaic maps enter into the scriptorium, print shops, private libraries, and universities of northern Europe. During this time, universal Christian history as spatially represented by *mappaemundi* [100] is gradually replaced by Ptolemy's *Geography* as the dominant geographic metaphor and visual representation of the Earth. Further, the subsequent transference of Malory's manuscript version of the Roman conquest into Caxton's printed edition suggests a "mythological moment transferred to geographical space,"[101] a transference that parallels and anticipates the developing geographic metaphor. Through Malory's revision and Caxton's editing, this pre-Ptolemaic tale crosses into early modern spatial practices. As Natalia Lozovsky has argued in relation to Carolingian manuscripts of Roman geographical texts, studies of the Roman tradition enabled "the new imperial people to assert their own place by taking over Roman knowledge about terrestrial space."[102] In a similar fashion, the foundational Roman source of Arthur's Roman War had been mystified by Geoffrey's oblique reference to Walter Map's "very ancient book," and Map's hypothetical book subsequently imprints a cultural map on the English geographical imagination.

English imperial ideology adopted and subsumed the very model of Roman geographical imperialism it sought to repudiate. It is, on the surface, an expulsion of Roman influence, but the "shadow" of the English Roman War narrative is a tale of Roman geo-spatial imperialism.

Each of the writers who have taken up the telling of the Roman War campaign have wrestled with its themes of ambition, violence, betrayal, and English geographical desire. Most critics, however, tend to focus on the *action* rather than the *space* of the tale. The symbolism of England's relation to Rome has certainly been acknowledged, but the intersection of the route and fifteenth-century geographical thought has been significantly under-theorized. The spatial humanities, a contemporary field of practice in which complex questions of geography and culture overlap,[103] provide a critical light by which to evaluate the itineraries of the Roman War narrative and how each instantiation negotiates the production of geographical and cultural space. The ground trampled and seas crossed in the various versions undergird and interpret the high dramatic action, forays, ambushes, knightly reconnoiters, slayings, and prisoner-takings. Countless journeys were made across this cultural route from late antiquity through the Middle Ages by Roman legionnaires, by diplomatic envoys, by merchants carrying cloth of gold from the Mediterranean, by pilgrims and legates journeying to and from Rome, by scholars with Latin texts bound for monastic libraries, and by couriers of books bringing romances and luxury editions of the *Geography* to English patrons and booksellers.

The Via Francigena: cross-European cultural routes

The early history of the route dates to Julius Caesar who attempted but failed to subdue Britain, followed by the establishment of forts and Romanized provinces in southern Britain after 43 CE. The Roman Empire reached its furthest extension into York and, by 128 CE, the northerly demarcation of Hadrian's Wall. In Geoffrey's account, once Arthur has rebuffed the Roman ambassadors, he gathers his allies, leaves the kingdom in the hands of Mordred, and departs for his formidable campaign toward Rome from the southerly port of Southampton. But in the Alliterative *Morte Arthure* and in Malory's subsequent version, Arthur first calls a parliament in York, the northernmost outpost of the Roman Empire and the place where allied nobles from throughout Britain would pledge their fealty to Arthur. In the *Morte Arthure*, Arthur sets this parliament just after the Utas (a term to denote the seventh day after a feast day) of St. Hilary, January 21st. The 200-mile upswing to York holds great symbolic power. Not only does it suggest the reach of England's sovereign power but it does so in the only place on English soil in which a historical Roman emperor was crowned, and the first Christian Emperor at that: Constantine. The historical Constantine was in York (Roman *Eburacum*), fighting the Picts alongside his father, Emperor Constantinus, in 306 CE when his father was slain.

In consequence, Constantine was acclaimed Emperor in York by his followers. As Arthur would need to do in the fictional tale, Constantine marched on Rome to consolidate his power. Thus, the otherwise inscrutable literary itinerary of holding a parliament at York before departing for his continental conquest owes much, I argue, to the force of Roman lore. By the time of the *Alliterative Morte Arthur* and Malory's version, Constantine's acclamation as Roman Emperor in York and Arthur's successful Roman campaign became conflated in a potent evocation of Romano-British Christian sovereignty and crusading. The initial staging of Arthur's campaign in York directly connects the military and Christian legacy of Rome to the claim of Arthur's English birth-right and his right to rule the Roman Empire. The legend of King Arthur ultimately synthesizes England's Romano-Celtic past: in the geography of the Roman War versions, Roman authority and Arthurian legend become inextricably linked.

In telling the tale of Arthur's campaign to Rome, The Roman War account largely coincides, geographically, with the well-traveled pilgrim route of the Via Francigena, linking Canterbury and Rome.[104] An ancient route, long established by the time Geoffrey of Monmouth first crafted his Roman War tale, the Via Francigena is documented in the tenth-century account of Sigeric,[105] Archbishop of Canterbury from 990–994, recounting his journey to Rome. The route began in Canterbury and moved across one of what is now referred to as one of the ancient "cultural routes" of Europe to the final destination of Rome.[106] From the countryside of Kent in England, travelers passed through Calais, Flanders, Burgundy, over the Alps, down the west coast of Italy, and into Rome. Some pilgrims continued on from Rome to Jerusalem.

Pilgrims sought out Rome in their thousands, especially in jubilee years: it is estimated that over 200,000 pilgrims gathered in Rome in the jubilee of 1300,[107] a spatialized embodiment of the centralized authority of Rome in relation to Christian Europe. In Malory's and Caxton's lifetimes, the Great Pentecostal Jubilee of 1450 was also attended by thousands of European pilgrims: pilgrims brought with them currency that supported the papal power structure and returned with indulgences and transformative cultural experiences.[108] Just past Malory's presumed compositional date but before Caxton's 1485 print edition, the jubilee of 1475[109] also brought thousands of European Christians to Rome. These jubilees reinforced and heightened the authority, and concomitant chafing under such authority, of a monarchical papacy and the unquestioned centrality of Rome as a religious and cultural center.

That pilgrimage or cultural routes formed part of the underlying geography of the Roman War route is attested early in the tale by the Lord of West Wales when he affirms his willingness to support and join Arthur on the campaign against the Emperor Lucius:

> For onys as I paste on pylgrymage all by the Poynte Tremble, than the vyscounte was in Tuskayne, and toke up my knyghtys and raunsomed

> them unresonablé. And than I complayned me to the potestate and the Pope himself, but I had nothynge ellys but plesaunte wordys … And therfore to be avenged I woll arere of my wyghteste Walshemen.
> (148.12–18; C V.2)

Malory's route for Arthur through France and Luxemburg (Barfflete, Paryse, Sessoyne, Lushburne, Flaundirs, Lorayne), through Almayne (154.6–7; C V.4, 164.18; C V.7, 175.23–24; C V.9), and south through Italy (Lumbardy, Myllayne, Porte Trembyll, Tuskanye, Vyterbe) (186.12 and 187.10–18; C V.12) generally follows that of the Via Francigena, with one notable exception: the Via Francigena passed by Lake Lausanne and crossed the Alps through the ancient Grand Saint Bernard Pass, once the Roman route that brought the forces of both Julius Caesar and Augustus Caesar north in pursuit of British subjugation. Following a route shift introduced in the *Morte Arthure*, Malory's Arthur routes through the mountain pass at "Godarte" (186.16; C V.12), the Saint Gotthard pass, reflective of route changes by Malory's lifetime. With its connective linking of "Almayne," Lake Como, and Milan, the mountain pass at Saint Gotthard came into regular use in the thirteenth century and so provides an "updated" routing of the Roman War campaign by its fifteenth-century redactors. Thus, Arthur is not merely traveling *to* Rome; he is traversing a deep cultural route across the Continent that reveals long-recounted textual and physical itineraries, spatial relations, cultural memory and symbolic geographies.

Over its long history, the geographic basis of the Roman War narrative engendered a strong revision of sources, an impulse to intervene, omit, and over-write the landscape of Arthur's journey, both physical and symbolic. In this regard, Malory is no exception. In each iteration, the Roman War route becomes a form of "textual archeology" in which the great spatial narrative of English military prowess on the Continent is rewritten across an actual landscape. Such topographies, while literary, are also historiographical. Historicized topographies, as Jennifer Summit has noted, "acknowledge the past is always with us, like the ground beneath our feet."[110] While issues of governance and kingship are certainly vital elements throughout *Le Morte Darthur*, the subject matter of English national identity set against continental geography distinguishes this tale from the rest of Malory's Arthuriad. Malory's version of the Roman War demonstrates the "reciprocal process of textualization and territorialization"[111] as England crafts its national identity vis-à-vis the Continent. Jennifer R. Goodman argues that while Sir John Mandeville and Marco Polo "presented themselves to their contemporaries as true travelers undertaking real journeys," in contrast, "the chivalric romance of the fifteenth and sixteenth centuries depict journeys of the imagination."[112] However, Malory's rewriting of Arthurian pseudo-historiography amidst the genre of romance, negotiates a narrowing gap between the actual journeys of the Via Francigena and the imaginative journeys of romance. In comparing Malory's Roman War narrative to his sources, Thomas

Crofts notes that Malory's version "seems to mark the entrance of *real-politik* into Arthur's court."[113]

Malory's account of the Roman War is derived primarily from the English Alliterative *Morte Arthure* and supplemented by Hardyng's *Chronicles*,[114] along with other minor sources.[115] In the Alliterative *Morte Arthure*, Arthur traverses the Continent from Normandy to Burgundy, where Arthur kills the Emperor Lucius. After the decisive victory, the Arthurian army moves north to Luxemburg, then begins its southward trajectory to Metz, Lucerne, the Gotthard Pass, Lake Como, and Milan. The route in Italy then passes through Spoleto, and on to Viterbo, never quite reaching Rome. In Malory's version, the site of the mortal battle between Arthur and Lucius is set in the "Vale of Sessoyne" (169.27; C V.8), long thought by scholars to suggest a location in northern France. This incongruous location, far north of the rest of the route, contributed to Malory's reputation as hazy on European geography and prompted Malory editor Eugène Vinaver to suggest that Arthur's doubling back across northern France and Flanders mirrored the route of Henry V's successful Agincourt campaign. Malory, Vinaver suggests, "could not help remembering the most remarkable English victory of the century."[116] Robert Kelly similarly suggests that Malory's route was indicative of a "pro-Lancastrian, pro-French" political statement.[117] In both cases, a political position is seen as overriding a "realistic" geography. However, William Matthew's linguistic and geographical identification of the town of Siesia, near Dijon, as Malory's intended battle site, clarified this geographic crux.[118] Matthew's identification of "Siesia-Sessoyne" as a more plausible site along an itinerary to Rome, strongly suggests that the misunderstood geography of Malory's text in this case may have less to do with political allusion than with scribal error or textual misreadings. Matthews' work in untangling Arthur's route through France makes clear that Malory's route was more geographically conscious than previously assumed.

Compression of the catalog: from kings to geographic regions

In addition to the overland continental geography provided in the Roman War tale, the catalog of allies of the Emperor reveals a multi-layered and inherited geography.[119] In the Alliterative *Morte Arthure* and in Malory's re-working of the tale, the ethnographic enumeration of Eastern kings begins to change in ways that can best be described as *cartographic*.[120] While *mappa-mundi*, such as the Hereford Map, were rich in ethnographic detail, populated with exotic animals, races, and monsters,[121] manuscript and print versions of Ptolemaic maps and portolan charts of the mid-to-late fifteenth century not only radically reshape the representations of landmass and coastlines, they also sweep clear the ethnographic quality of the *mappamundi*. There are no people on the maps of Ptolemy's *Geography*. The focus, instead, is on topographical features, such as mountains and rivers, and on city locations. It is not necessary that Malory actually saw an edition of Ptolemy; instead, it is clear that the conceptual linearization and cultural processes that underlay the development of Ptolemaic maps were similarly at work in literary culture.

The catalog of Eastern adversaries in the Roman War tale reveals a long shift away from an ethnographic *mappamundi* version of the tale and toward a version more akin to a geodetic, topographical world map. For example, in Geoffrey's early twelfth-century version, Lucius and the Senate of Rome gather "the Kings of the Orient" as their allies: "there duly assembled Epistrofus, the King of the Greeks; Mustensar, the King of the Africans; Ali Fatima, the King of Spain."[122] The catalog of eastern kings in the accounts by Geoffrey, Wace, and Laȝamon all include the individual names of the rulers, highlighting their noble and chivalric status. In Laȝamon's *Brut* (c. 1190–1215), a text that editor Mary Dictman believes Malory may have directly known,[123] the kings and dukes of the East are explicitly named as noble individuals as well as adversaries, for example, "Epistrod, King of Grece; Ethion, Duc of Boece … [and] Irtac, King of Turckie."[124] The result of such "imaginative geography and of the dramatic boundaries it draws," as Edward Said has argued, is a rhetoric both "anatomical and enumerative,"[125] a form of what Iain McLeod Higgins has described in relation to Mandeville's *Travels* as "an ethno-geographical survey."[126] In the fifteenth-century Alliterative *Morte Arthure*, however, the references to these Mediterranean allies of Lucius become less individual and more geographic, mentioning only that Emperor Lucius sent his summons "To Tartary and Turky."[127] Resisting his usual penchant for *adding* character names to unnamed figures in his source text,[128] Malory is similarly geographic, referencing only an anonymous "Kyng of Tars, and of Turkye" (151.5; C V.2). This key passage for establishing the context of the East-West anxiety of fifteenth-century England stands in marked contrast to a tendency John Withrington points out regarding Malory's treatment of minor characters elsewhere in the *Morte* in which he "frequently gave names to those unidentified individuals he found in his sources."[129] Here, the imaginative geography of England's military encounter with the known world moves away from amplification and augmentation of knightly characters and, instead, focuses on the geo-political confrontation of nations. Just as maps in the fifteenth century begin to transition from cultural ethnography to spatial projection, so too does the catalog of Eastern rulers in the fifteenth-century versions.

In the Alliterative *Morte Arthure*, the allied forces of Lucius are described as "a full gret nombyre," "legyons ynewe," and as "summes full huge,"[130] but in Malory, Lucius allies become partly monstrous: "fyffty gyauntys that were engendirde with fendis" (151.19; C V.2). Malory's version morphs into a loathsome, miscegenated horde, both a racial and territorial threat. The harrowing of the boundaries of Latin Christendom by Lucius's "horryble peple" (151.23; C V.2) highlights concerns that a Muslim invasion of Europe was imminent. The Turks had sacked Constantinople in 1453 and Mehmet II advanced into Turkey, just shortly before Malory wrote his romance. Malory's account of Lucius's defeat, therefore, provides a potent revenge narrative and a cathartic, if fictive, release from fears of Turkish incursion into Europe. Malory's grappling with the geo-spatial material of the Roman War account suggests a writer highly attuned to the geo-spatial potential of this ancient tale, an attentiveness that ultimately emphasizes the importance of European Christian unity in the face of national and religious fractures, positing Arthur as the symbolic antidote to the threat of a Turkish incursion.[131]

Malory's tale represents a geographically braided account of the cross-European movement of several sets of characters. It reads like a re-enactment of a military strategic plan complete with moves and counter-moves. For example, as the narrator moves back and forth between the coinciding actions of the two mustering armies, we read: "Now leve we Sir Lucius and speke we of Kyng Arthure (151.33; C V.2). The tale delineates the round-trip itinerary of the Roman ambassadors sent to "Carleyle" to deliver Lucius's demand for "truage," Arthur's pre-campaign preparations that move him from Carleyle to York to Sandwich, the march of Lucius and his allies from Rome through "Almayne" and on to the "Vale of Sessoyne" in France, the journey of Arthur and his troops across the English Channel, through Normandy, France, the Alps, and Italy. Arthur's individual journey also includes backtracking (from Mont St. Michel to Barfflete) and a northerly detour (to "Lushburne" and "Flaundirs") before his southern march toward Rome. Finally, the tale concludes with a non-specified route homeward, one that brings Arthur back to Sandwich Haven, his port of departure a year earlier and where, departing from his sources, Malory's Arthur is met not with the traditional attack by Mordred's troops but with a tender homecoming and celebration.[132]

Arthur victorious: Malory's temporal postponement

The most renowned change made to his source material is Malory's alteration of the traditional ending of the Roman War narrative in which Arthur came within reach of Rome but returned to England uncrowned in a desperate attempt to save his kingdom. In Geoffrey's twelfth-century account, in the *Morte Arthure*, and in Hardyng's *Chronicles of England*, Arthur's kingdom is undone by his campaign to Rome. In the *Morte Arthure*, Arthur plans to proceed to Rome and envisions his success:

> Than this roy royall rehersys theis wordes:
> "Now may we reuell and riste, fore Rome es oure awen![133]

Arthur includes specific plans for his coronation and occupation of Rome, plans that remain unrealized:

> We sall be ouerlynge of all þat on the erthe lengez!
> We will by þe Crosse Dayes encroche þeis londez,
> And at þe Crystynmesse Daye be crowned theraftyre;
> Ryngne in my ryalltés, and holde my Rownde Table,
> Withe the rentes of Rome, as me beste lykes.[134]

As in Geoffrey's account, the news of Mordred's usurpation abruptly ends the campaign without Arthur receiving the crown at the Pope's hand or the position of prime ruler of European Christendom. The implication in Geoffrey is that hubris precipitated the fateful rise and fall of the king's fortune. Malory, however, makes the campaign an inauguration of Arthur's reign. The only other source in which Arthur is not turned back on the eve of his coronation is that of Hardyng's *Chronicles*, a minor, but important, secondary source for Malory.[135] In Hardyng's *Chronicles*, the campaign brings only a temporarily triumphant conclusion: Arthur, flush from his victories across France and Italy, enters Rome and is crowned at St. Peter's Church:

> At [the] Capytole, in [the sea] imperiall,
> They crowned hym with crownes thre of golde
> As emperoure moste principall,
> And conquerour that daye most worthy holde;
> Wher then he fested the citee manyfolde,
> [Of Rome the byshop, and all his cardinals,]
> The senatours, with other estates als.[136]

But despite Arthur's coronation in Rome, Hardyng's account ultimately hews to the traditional narrative trajectory of Arthurian tragedy: Arthur immediately receives word of Mordred's treachery and returns to a fractured nation, following the moral arc of Geoffrey's tale.

Malory, however, eschews the Roman War narrative as marking the end of Arthur's reign and instead makes it the cornerstone of Arthur's ascendancy to power both domestically and abroad. Malory recasts the Roman War tale as an early episode that demonstrates the consolidation of Arthur's power. Although this was once seen as an individual innovation on Malory's part, several scholars have suggested the possibility that Malory drew on an episode from the French Vulgate in which Arthur's military support for his French allies segues into his defeat of the Roman emperor Lucius early in his reign.[137] Malory's major source, the *Alliterative Morte Arthure*, does not present such an early, career-defining episode nor do any previous chronicle versions. Malory's Arthur is consistently guided "by the helpe of Oure Lord" (176.5; C V.8), and Malory's frequent insistence on Arthur's devout Christianity suggests Malory could not tolerate a defeat for his Christian hero against a near-pagan adversary, nor could Malory bear to leave Christendom in such a state of disarray. As such, the Vulgate episode offered an alternative conclusion to the traditional disastrous military encounter between Britain and Rome. Whether the Roman War account was seen as a first and separate tale from his Arthurian project, as Vinaver suggested, or whether it was part of a planned and integrated tale within the larger book, Malory's Roman War account holds the promise that Arthur's "achievement" of Rome ensures unity within Christendom and a favored relationship between the Pope and Arthur, as ruler of England and all of Europe.

Several years before Malory refashioned his version of the Rome War account, a model for the fusing of Arthur's military and ecclesiastical confirmation of political power took place in Rome. In the spring of 1452, Frederick III of Germany was crowned Holy Roman Emperor by Pope Nicolas V within the city walls of Rome, a ceremony attended by thousands. One year later, however, the fall of Constantinople at the hands of the Ottoman Mehmet II shook European leaders and fueled heightened anxiety about further Turkish incursions into Europe. The fall of this imperial Christian city and the death of Constantine XI in the siege signaled the fragile state of Christian territorial rule. In 1468, just as Malory was presumably writing *Le Morte Darthur*, [138] Frederick journeyed to Rome as part of an "imperial pilgrimage" that brought him to the gates of Rome through the Porta del Popolo, the northern gate of Rome through which European pilgrims passed into Rome. On Christmas Eve of 1468, the same day as Charlemagne's actual coronation as Holy Roman Emperor and Arthur's fictional coronation, Frederick attended mass at St. Peter's Basilica and discussed the Turkish threat with Pope Paul II. After mass, the Pope blessed Frederick, praying that God grant him a victory over the Turks.[139] Thus, in this meeting of Emperor and Pope, we find a symbolic model of European political and religious unity and an imperial pilgrimage motif that possibly informs Malory's Roman War account. The symbolic nature of this highly ritualized meeting of Emperor and Pope achieved greater literary longevity than military success in resisting the Ottoman expansion and unifying Christian Europe: Frederick was the last Roman Emperor to be crowned in Rome and proved largely ineffective in stemming the advance of the Turks.

Malory infused his inherited material with a religious model of European Christian unity,[140] and his account resonates with a growing geo-political enmity with the East with Arthur as the symbolic hope of the West. A contemporary reader of Malory could not help but connect the geographic and political journey of Arthur to Rome with the widely recorded imperial pilgrimage of Frederick. By virtue of his imperial pilgrimage, Frederick crafted a symbolic conquest of European space that resonates in Malory's account of Arthur's crowning: "And at the day assigned, as the romaynes me tellys, he was crowned emperoure by the Poopys hondis, with all the royalté in the worlde to welde for ever" (188.2–5; C V.12). Malory cannot indefinitely forestall the death of Arthur, but, at least temporarily, he suggests a pan-European Christian realm undergirds the mythos of Arthur.

Just as pleasure and penitence intermingled in pilgrimage journeys, a "seson" in Italy offered Arthur and his men its own recompense: revelling in his victory over Lucius, Arthur rested in the "Vale of Vysecounte"[141] with "vynys full plenteuouse" (187.18–19; C V.12) prior to his coronation. After his acclamation as Emperor, Arthur spends the winter in Italy where he "suggeourned that seson tyll aftir that tyme" (188.5–6; C V.12). Pragmatic concerns, of course, would prevent an early winter return across the Alps, but a season in the wineries of southern Italy no doubt also offered Arthur, his men, and Malory's reader's a welcome respite from military campaigning.

Malory and maps

Did physical maps inform Malory's Arthurian itinerary? Kelly suggests that Malory likely never saw an actual map but, rather, read verbal geographic descriptions in his sources as well as in the "standard" reference works such as Ranulf Higden's *Polychronicon.*[142] But from the mid-fourteenth century on, the immensely popular *Polychronicon* often "included a world map to illustrate its first book on natural history and geography."[143] As P.D.A. Harvey suggests, many copies of Higden's world map circulated independently and as an accompanying illustration in codices.[144] By the mid-fifteenth century, members of the gentry class such as Malory would likely occasionally encounter large-scale wall maps or maps in books.[145] If Malory of Newbold Revel is the correctly identified author,[146] he would also have had on-the-ground geographic knowledge from his participation in military campaigns in France as part of the Hundred Years' War.

It is less likely, however, that Malory would have seen a manuscript edition of Ptolemy's *Geography* with its newly reconfigured map of the world. Manuscript editions of the *Geography* were produced in Italy as early as 1409 and a few made their way to English monastic and academic libraries prior to Malory's death in 1471, although Ptolemy's *Almagest* was the far more likely volume to be in such collections.[147] Print copies of the *Geography* with accompanying maps were not produced until the mid-1470s, certainly too late for Malory to have acquired or come across. Unlikely to have seen a visual map based on the calculations of Ptolemy, Malory's geographic imagination,[148] instead, was influenced by narrative geographies, chronicles, and a nascent shift in world-view, time, and space. In turn, Malory's narrative unraveling of the multiple and conflicting time sequences of his sources identifies him as an active agent in the developing cartographic sensibilities of the mid-fifteenth century. Indeed, Malory's cartographic imagination is highly original, and his narrative anticipates changes in spatial thinking that precipitated the reorganization of visual cartography. Malory's grasp of this reorganizing paradigm can be attributed, in large part, to the accelerating "inter-reading" of chronicle and romance in England.[149] As a result, the time- and place-based chronicles were increasingly putting pressure upon the quasi-historical structures of romance. The cultural reorganization of time and space that undergirds fifteenth-century geographic thought, then, is evident in Malory's integration of Hardyng's *Chronicles* into his romance.[150]

In contrast, as a bookseller and manuscript dealer, Caxton almost certainly came across the new maps being produced by European print shops in Italy and Germany, a material and visual cartographic world-view not available to Malory. While maps based on the projections of Ptolemaic coordinates were too late to directly influence Malory's spatial and geographic sensibilities as he wrote his book of King Arthur in the confines of Newgate Prison, Caxton, on the other hand, was living and trading on the Continent just as the earliest editions of Ptolemy's *Geography* were published, both in print and manuscript forms, and purchased as luxury items for wealthy patrons' libraries. With the fall of Constantinople to the Turks in 1453, Greek manuscript codices, including geographical texts, were part of the westward dispersion of classical scholars and works,[151] and translations of Greek scholarship into Latin, including Claudius Ptolemy's *Geography*, became more widely available.[152] Transnational trade brought the English-born Caxton to Bruges where, by 1453, he held the position of Governor of the Company of Merchant Adventurers and was a leading figure in the Anglo-Continental mercantile milieu in one of the busiest trade cities in northern Europe, an active node in the Hanseatic League network.[153] It seems reasonable to assume that Caxton not only had a working geographic knowledge of the Continent, but that he was familiar with the printing initiatives of deluxe editions of Ptolemy's *Geography*. Then, just as Caxton was setting up his press in England, one of the most spectacular endeavors of the nascent print industry was underway: a print edition of the *Geography* published in Bologna in 1477 and in Rome in 1478. Editions followed in Ulm and Florence in 1482, informing the print-world *zeitgeist* and transnational book trade in which Caxton was immersed. Notwithstanding the absence of absolute bibliographical or historical evidence, it seems likely that Caxton either held a copy of the *Geography* in his hands or knew of it from his role in the book trade.

Not surprisingly, Caxton's subsequent publishing *œuvre* reveals an attentiveness to changes in cartographic thought and visual representation.[154] He actively integrated developing geographic knowledge into his publishing and editing projects, reflecting the hybridity of fifteenth-century geography and its synthesis of ancient, medieval, and humanist geographic thought, an amalgam of practices that included astronomical treatises, place name lists, rutters (navigational handbooks),[155] portolan maps, narrative itineraries, eyewitness accounts, and the earliest Ptolemaic maps. In addition, Caxton's print edition of *Le Morte Darthur* reveals the importance of the literary processes by which dissemination of geographic knowledge occurred. Just as contemporary cultural geographers overlay modern geographical coordinates on historical maps to better understand geo-historical events, so too does Caxton geo-rectify the "coordinates" of Malory's Roman War. Via subtle but significant textual emendations, Caxton's editorial work leaves indirect evidence that Ptolemaic-based cartographic images had begun to accumulate and inform English readers' mental maps. In ways specific to Caxton and in ways generalized to English book and reading culture of the late fifteenth century, Caxton brought Ptolemaic cartographies to bear on his version of Malory's Roman War.

Boredom and the Roman War

Despite its fascination for medieval writers and readers, the Roman War account has long been considered one of the least inspired tales of Malory's work, a tedious account of battles and place names splayed across a sometimes-nonsensical geography of Western Europe. C.S. Lewis's sheer exasperation with Geoffrey of Monmouth's text is near comical: Lewis describes the battle scenes in Geoffrey's account as "blatantly, stupidly false" and "unendurable."[156] Lewis castigates the narrative as one of "military operations scrawled over the whole map of Europe and excluded by all the history we know."[157] This critical view of the tale was widely shared: the account of Arthur's Roman War campaign with its many battle scenes had long been considered a distraction, a bore, or, at the very least, a minor tangent to the primary romance-based tales of the Arthurian legend. Andrew Lynch documented a century-long trend during which time editors "lined up to abuse the fight scenes, and did so with increasing force."[158] Lynch provides what became a litany of criticism:

> 'A tedious battle … all the details of which are minutely related' (British, cited in Parin 89); 'the greatest exaggeration' (Southey xxxi); … 'the general reader would find it too long, too monotonous, and too obscure' (Knowles ii); 'a catalogue of encounters, with but little variation of the familiar incidents' (Minto, cited in Parins 199); … 'a monotonous lack of inflection; an endless succession of single combats … whose details are repeated with wearisome iteration' (Morros, Introduction) …[159]

Such a response serves as ample evidence of the general critical reception of fifteenth-century texts that gave rise to David Lawton's ironic essay, "Dullness and the Fifteenth Century."[160]

Upon the discovery of the Winchester manuscript version in 1934, the tradition of bashing Malory's tale of Arthur and Lucius took a critical about-face. The manuscript found in the Warden's Library of Winchester College by Walter Oakeshott provided a textual surprise: in comparison to the version of the Roman War account contained in Book 5 of Caxton's print edition, the manuscript version was nearly twice as long—the proverbial "uncut" edition. Vinaver, the major twentieth-century editor of Malory's text, sharply criticized Caxton for having printed "a drastic abridgement" of Malory's original version and abandoned his in-progress edition based on Caxton's text to produce an edition based on the newly-discovered manuscript.[161] Lewis, however, dryly lauded Caxton for having had the good sense to edit and shorten "the whole dreary business."[162] Vinaver accused Caxton of textual malfeasance, arguing that Caxton's version of the Roman War account "reduced it to less than half its size, and while doing so rewrote it from beginning to end."[163] Caxton had, indeed, significantly compressed the narrative by tightening dialogue and battle scenes and "modernized" the style of the tale for his new print-reading audience.[164] Vinaver's textual critique reflected his hierarchy of authorial intention over such editorial "corruption." Theories of constructivist editing or "fluid texts,"[165] in contrast, as Jerome McGann argues, suggest that texts reveal "a ceaseless process of textual development and mutation."[166] Such acknowledgment of the "textual condition," or Derridean "essential drift" of texts,[167] reframes editorial intervention as an inherent part of a socially-constructed, material artifact rather than the corruption of an ideal authorial form.[168]

Subsequent editors, notably Helen Cooper in her 1998 student edition of *Le Morte Darthur*, followed Caxton's lead in shortening the Roman War account. In the interest of her edition's audience of students and general readers, Cooper executed "some slimming of the opening sections of the Roman War account."[169] Cooper's svelte Roman War trims descriptions of Malory's catalog of foes as well as battle scenes likely to try the patience of late twentieth-century undergraduates. "Few readers," Cooper explains, "are now likely to share Malory's passionate interest in the details of battle tactics and tournaments, and I have cut these generously."[170] Cooper's generosity is not so much directed toward the writer but toward a projected twentieth-century reader, wearily trudging through a long textual journey to Rome and back. Caxton's edited version of the Roman War chapter suggests the ways in which he shaped the geography of the Roman War both by cartographic principle and geographic imagination. "Various readers and audiences," as McGann has noted, "are hidden in our texts, and the traces of their multiple presence are scripted at the most material levels."[171] From Geoffrey to Malory to Caxton to Cooper, the Roman War account has foregrounded a geography across which battles were fought and text "slain" by successive redactors and editors. Each compression of altered textual space offers a new reading of European geographical space on the part of the writer or editor. The terrain traversed in Malory's and Caxton's Roman War accounts, as represented by the route and place names of the tale, symbolically and spatially grapple with issues of empire, race, and militant Christianity, a "free-floating imperial episode."[172]

For fifteenth-century audiences, in contrast to later critical opinion, the text's concern with geography and empire was riveting, such that the tale underwent radical reformations in order to capture the spatial ideology of the writer, redactor, or editor. The repetitive battle scenes in the Roman War account not only signal Malory's intense interest in knightly battle conduct, but an intense rewriting of geography as well. Unlike most of the tales in Malory's capacious romance, the scenes of the Roman War campaign take place across specific topographies, locations that have not been contextualized within the greatest reconception of geographical space since antiquity. To date, Malory's and Caxton's geographies have not been considered as part of the attendant spatial practices and mappings of late medieval exploration, trade, travel, and literary expression.[173] Works such as *Mapping Malory* struggle to interpret Malory's geography, relegating it as an "(il)logic of space" rather than part of the particular cultural geography of the mid-fifteenth century.[174] Changes in the route and place names in Malory's and Caxton's version are more than individual geographies; they narrate the development between medieval and early modern spatial practices. If, as

Jennifer Summit has argued, space can be read "as a repository of history,"[175] then the geographies of the Roman War versions provide a literary navigation of certain spatial histories of the fifteenth century. Malory's version of the tale recuperates a spiritual and temporal unification of the Holy Roman Empire under the English Arthur whereas Caxton posits a greater geographical verisimilitude based on Ptolemaic geography, both of which are connected to the development of new spatial representations of the fifteenth century. Caxton reorders a more historicized fifth-century Arthur as a typological anticipation of a medieval Crusade. Much the same as fifteenth-century Ptolemaic maps, Caxton's Arthur is unraveled in time as well as space.

"The City Unseen": Caxton and the new cartography of the Roman War

In Malory's parlance, "now we turn" to Caxton's remapping of the Roman War, an intervention that makes the case for Caxton's familiarity with Ptolemaic mapping principles and that puts pressure on the assumption that English readers were unfamiliar with or disinterested in the emerging European cartographic representations of continental and global space.[176] Caxton and Malory were both part of the fifteenth-century process of localizing place names and bringing the realm of romance into "real-time." In this regard, their geographic register was not radically different. However, Caxton's awareness of new geographic material is irrefutable. Caxton's narrative geographical and cartographical projects, discussed more fully in Chapter 2, are marked by the influence of fifteenth-century cartography and exemplified in his editing of the geographical aspects of the Roman War account in *Le Morte Darthur*. Caxton's edition of Malory's work illustrates the accelerating geographical emphasis in the mid-to-late fifteenth century on place-based geography and coordinate-based cartography. William Kuskin argues that Caxton's intervention in *Le Morte Darthur* was not a marker of "intellectual simplicity, but of ideological complexity."[177] Similarly, John Withrington and Thomas Crofts point to Caxton's historical, political, and linguistic ideologies, arguing for the influential role of Caxton's printing projects—including romance, histories, and his multi-volume Nine Worthies project[178]—in shaping and creating early modern English reading culture.[179] The ideological complexity Kuskin cites, however, does not explicitly take note of Caxton's geographical complexity. In a period during which the introduction of technologies of print and cartographic production overlapped, it should come as no surprise that Caxton, transnational merchant and London's preeminent printer, bookseller and book importer, was highly attuned to both. Caxton's editing of the Roman War account suggests that he is particularly attentive to the principles of Ptolemaic-based mapping, an emphasis that includes a foregrounding of linear spatial relations and coordinate-based place names.

Three years prior to Caxton's 1485 edition of *Le Morte Darthur*, a print edition of Ptolemy's *Geography*, the most influential cartographic text of the fifteenth century, was produced in Ulm, Germany. Early editions of the second-century CE text by Claudius Ptolemy contained grid-based "tables" of locational coordinates—the latitude and longitude of cities and place names—as well as maps drawn on the basis of Ptolemy's coordinate-based calculations. Typically oriented to north, these new maps derived from a mathematical "projection," the process of representing a sphere on a flat surface, rather than from universal Christian history. Each city or place listed in the *Geography* is specifically identified by mathematical coordinates. In a parallel textual process, Caxton provides specific, locatable cities across which the action of the Roman War takes place in his edition. Having traveled to Cologne, Germany, to learn the new print trade, Caxton brought this innovative technology first back to Bruges, where he was engaged in trade, then to London, where he set up his print shop in 1476.

In editing Malory's work, Caxton remaps the Roman War in several ways: he completes the transformation of the eastern allies of Lucius from individuals into geographical spaces, disentangles Arthur from anachronistic Christian historiography, alters the route to a more plausible trajectory through northern Europe, shows a marked preference for the identification of urban place names, and promotes a unified England acclaimed by an urban populace. In so doing, Caxton shows a strong bias toward a geographical realism and urbanism that mark the great cartographic transition of the fifteenth and early sixteenth centuries. For Caxton, the unnamed or "unseen" locales of Malory's quasi-romance of Arthur's Roman War become specific, recognizable urban spaces of the emerging early modern era: cities locatable on a map. The imperial dreamscape of the Roman War, inflected with geographical verisimilitude from its inception, was textually fertile ground for the overlaying of Ptolemaic-influenced humanist cartography of the late fifteenth century.

Geo-locating the Roman War

An overlooked contribution of Caxton's textual intervention is the way in which his edition provides evidence of English geographic knowledge in a period of dynamic cartographic change and signals the incorporation of the spatial practices of the maps associated with the *Geography* into English geographies. Caxton's editing of *The Myrrour of the World* and *The Description of Britain*, texts discussed in Chapter 2, demonstrate that it matters a great deal to Caxton that he publish geographically "accurate" texts. Caxton imposes a linearity and geographic order based on knowledge available to him via his participation in the book and map trade. The evidence that Caxton may have seen a Ptolemaic map, either in manuscript or print form, is only indirect, but Caxton's geographic sensibility in his editing of Malory's Roman War tale is essentially Ptolemaic—that is, he emphasizes geographic "realism" and a preference for city names over the vague locations of romance.[180]

In editing out key references from Malory's account that resonate with the sensibilities of the *mappamundi*, Caxton shifts the focus from Arthur as primarily connected to symbolic Christian history to a figure who also operates in the geographic space of the fifteenth century. In the Caxton edition, consequently, the socially constructed imaginative space of the Roman War becomes mapped in a new way: Caxton's editorial work shapes the centuries-old narrative of English imperial aspirations to function in realistic space and time. In so doing, Caxton produces a new space for the Roman War narrative, one in which narrative spatial relationships cohere with emerging cartographic practices.[181]

If Caxton's editing of Malory's Roman War account is geographically driven, why does his publishing *œuvre* omit any publication of maps or indeed the *Geography* itself? An active book trade in London, the Latinate nature of maps, and an English narrative geographical tradition all contribute to the lack of production of world maps in England in the fifteenth century. Although an inventory of John Rastell's print shop in 1536 provides evidence of an active map trade as part of Rastell's book business, as we shall see in Chapter 4, we do not know the details of the contents of the barrels of books imported by Caxton nor whether part of his trade might have included individual maps or, more likely, maps as part of books.

To achieve a greater geo-historical recalibration of the cartography of the Roman War narrative as well as to alleviate the potential for the readerly fatigue that, despite Caxton's efforts, dogged the narrative into the twentieth century, the Caxton version shortens the tale to approximately half the length of Malory's original. Within the framework of this shortened version, Caxton's edition maintains the basic geographic action of the narrative: Roman ambassadors intrude on Arthur's Christmas-time celebration; Arthur goes to York to hold a parliament before departing from Sandwich for the Continent; he and his troops traverse France where Arthur slays Lucius before crossing the Alps into Italy; Arthur is ultimately crowned Emperor in Rome; and, finally, he and his forces return to a unified period of *Pax Arthuriana* in England. However, a comparison of the manuscript, presumed to be an earlier version, with the print edition provides ample evidence of interventions made by Caxton not only to shape the well-known medieval tale into modernized fifteenth-century prose, but also to bring the tale in line with Ptolemaic conceptions of geographic space and geo-historical plausibility.

Spatialization

In comparison to the earlier accounts of the catalog of kings and rulers aligned with the Emperor Lucius in Geoffrey, Laȝamon and Wace, Caxton's edition continues the process of compression that began in the *Alliterative Morte* and that subsequently increased in the Winchester manuscript version of Malory's tale. Geoffrey's "Epistrophus, King of the Greeks," for example, in Malory's text becomes "the Grekis" (151.10; C V.2). But in Caxton's text, individuals fully transform, instead, to locations: "alle these were subgette to Rome and many moo as Grece Cypres Macydone Calabre Cateland portyngale..."[182]

Caxton's compact version accelerates the process of dampening the individual identity of the nobles and kings of the opposing army. In Caxton's version of the catalog, the title of king is frequently deleted in comparison to Malory's Winchester version; thus, the "kynge of Capydos" becomes simply "Capadoce," and "the kynge of Macidony" is shortened to "Macydone" (151.5, 151.11; C V.2). In so doing, Caxton brings to near completion a process of increasing cartographic spatialization that had been occurring in the tale over the course of the fifteenth century. What was once a catalog of royal adversaries from the East transforms to a geographic gazetteer of place names. The allies of Lucius become less a catalog of noble combatants than a verbal reading of a map.

Geo-historical realism

When Caxton receives Malory's text, "a copye unto me delyverd, whyche copye Syr Thomas Malorye dyd take oute of certeyn bookes of Frensshe and reduced it into Englysshe,"[183] he finds a version in which the traditional ending of the Roman War tale has been changed to align with Hardyng's *Chronicles of England*, in which Arthur successfully, although fleetingly, achieves the emperorship of Rome. Working with Malory's new version of Continental success, Caxton sought to bolster the quasi-historical nature of the tale as part of his overall push to edit the account of Arthur into a plausible geo-historical context. One of the major motivations for Caxton's revision of the Roman War account, according to Edward Donald Kennedy, was his interest in bringing Malory's French-influenced version of the story closer in line with that of the English source: Hardyng's *Chronicles.*[184] As Kennedy has pointed out, Hardyng's *Chronicles* was "one of the most widely circulated medieval works in England in the late Middle Ages" and contained the well-established story of King Arthur's Roman conquest. It is the "part of Malory's book that would have been most familiar to Caxton's readers."[185] Though William Matthews argues that Malory attempted to "move the Roman War from the narrow world of Arthurian chronicle into the wider Arthurian world of French and English romance,"[186] Caxton seems to temper the elements of romance in the tale, at least geographically. In his Preface to *Le Morte Darthur*, Caxton suggests that while the historicity of Arthur may be contested by some, the cultural legacy of Arthur is irrefutable: "in dyvers places of Englond many remembraunces ben yet of hym and shall remayne perpetuelly … Thenne, al these thynges consydered, there can no man resonably gaynsaye but there was a kyng of thys lande named Arthur."[187] In seeking to secure a historical identity for Arthur, Caxton also "geo-rectifies," as it were, Malory's Roman War geography in the context of new representations of cartographic space and, in the process, creates a new spatial identity for Arthur.

One attribute of Caxton's geographical identity for Arthur is the de-emphasis of places associated with Christian historical narrative in the

Roman War and the disentanglement of Arthur from Christian historical narrative—instead, Caxton locates Arthur in "real" time and space. One means by which the Caxton edition achieves this effect is to excise references to mythical or biblical geography in the Roman War account. Gone, for example, is Malory's reference to "Preter Johanes londe" (151.6; C.V.2). Though the figure of mythical Christian ruler Prester John continued to appear on late fifteenth century maps such as the 1492 Behaim Globe and was included as a principality of the East in editions of Mandeville's *Travels,* Caxton removes mention of Prester John from the catalog of Arthur's adversaries. Caxton thus addresses the diminishing probability of the existence of such a ruler and nation in the face of fifteenth-century exploration and cartography. There is no land of Prester John on maps based on Ptolemy's mathematically based geography, and Caxton similarly removes the reference from his verbal map.

Caxton also excised Arthur's stated intention at the end of the Roman War narrative to continue on to the Holy Land. In the Winchester manuscript version, once Arthur has subdued Rome, he declares his plan "to rule as me lykys," and then "to gete me over the salte see with good men of armys to deme for His deth that for us all on the Roode dyed" (187.32–35; C V.12). Despite this bold claim in the Winchester manuscript, Arthur instead rests in Rome for a "seson" until his troops, growing restless of their Roman holiday, press Arthur to return to England and "reles us to sporte us with oure wyffis" (188.5; 34–35; C V.12). Battle-hardened as they are, in the Winchester version of Malory's text Arthur's men have no stomach for a further campaign to wrest the Holy Land from the Turks. While Malory's vision of Arthur-as-crusader aligns with Frederick III's pledge to Pope Paul during his imperial pilgrimage to defend Christendom against the Turks, Caxton's removal of this passage, as has been suggested by Kuskin, keeps the action of the quasi-historical Arthur distinct from that of Godfrey of Bouillon, a leader of the First Crusade and the subject of one of Caxton's other volumes in his Nine Worthies publishing project.[188] Caxton clarifies Arthur's historical moment by placing Arthur in a pre-Crusade historical context and reserving the confrontation with the Turks over the Holy Land for Arthur's dispersed knights at the end of the "whole book." Caxton's deletion of the Holy Land references within the Roman War tale presents a historically possible and, to the point of this study, a geographical probable Arthur—one whose success has greater certitude and clarity than the more tentative yet unfulfilled global aspirations of the conclusion to the Roman War campaign in the manuscript.

Although the final pages of the Winchester manuscript do not survive, in Caxton's print version the rumor that Launcelot's remaining four knights lived out their days as hermits is refuted and the claim put forward that they went to the Holy Land where they "dyd many bataylles upon the myscreantes or Turkes" (940.14–15; C XXI.13). The intention Arthur mused upon earlier in the Roman War tale—that of going over the "salte see" to revenge "His deth that for us all on the Roode dyed" (187.33–35; C V.12) is now partially fulfilled. The Caxton edition continues with the claim that the knights "dyed upon a Good Fryday for Goddes sake" (940.16; C XXI.13).[189] Cartographically, the culminating space of Jerusalem suggests a return to the center-enhancing space of the medieval *mappamundi* but it is only a fragile and temporary hold for, as Malory's and Caxton's readers well knew, Jerusalem was slipping from realizable Crusade aspirations and from the center of the map.[190]

Re-routing the Roman War

Caxton's version of the Roman War also suggests a different cartographic sensibility than Malory's by virtue of his alteration of the route of Arthur and his campaign across Europe. In the Roman War tradition, dating back to Geoffrey's Latin text, Arthur's landing site on the Continent was at *barbefluvius*, the coastal city of Barfleur in Normandy. In the *Morte Arthure*, however, the place name of "Barflete" is substituted,[191] possibly an Anglicized version of the French town name, and Malory follows this change in the Winchester version: after Arthur's night-time journey across the Channel "they had a syght of the bankys of Normandy, and at the same tyde the kynge aryved at Barfflete and founde there redy many of his grete lordis" (154.6–8; C V.5). The location of Barfleur on the Cotentin peninsula of Normandy makes geographic sense: it is part of the historic province of Gaul (which Geoffrey identified as the region in which Arthur landed), and it is close to the impending hand-to-hand combat between Arthur and the rapist-giant at Mont Saint-Michel, located along the northern coast of Normandy. But the change in place name to "Barfflete" creates a measure of geographic uncertainty, and Caxton apparently attempted to address this locational conundrum. In Caxton's edition, Arthur and his troops "had syghte of londe and saylled tyl they arryued atte Barflete in Flaundres and whanne they were there he fond many of his grete lordes ready …"[192] Why Caxton inserts the location as "in Flaundres" has been the subject of speculation,[193] but it is clear that Caxton, who lived in Bruges and who certainly knew Flemish coastal geography, was not ignorant of the geographical distinction between Flanders and Normandy. If the change is not the result of scribal or compositor error, Caxton may have chosen to foreground a more plausible commercial and political sea route: Sandwich to Flanders was indeed the more likely crossing for Rome-bound itineraries. In addition, before Arthur leaves Sandwich, the Winchester manuscript narrative recounts that the Roman ambassadors, fleeing Arthur's wrath at the content of their missive, depart Sandwich and "toke theire shyppyng and passed into Flaundres and Almayne" (149.34-150.1; C V.2). Caxton's edition maintains this route, stating that the ambassadors departed from "Sandwyche and passed forthe by flaundrys Almayn the montayns and all ytalye,"[194] a route that Caxton reinforces in his edition when Arthur himself crosses the Channel.[195] After suggesting the more pragmatic route-based itinerary, the Caxton edition eliminates the return of

Arthur to "Barfflete" once his battle at Mont Saint-Michel is done, a return that occurs in the Winchester version. Such a backtracking would have made no logistical sense given a landing site in Flanders, suggesting a geographical principle being exercised in Caxton's revised narrative. In Caxton's text, Arthur lands in Barfflete, purportedly on the coast of Flanders, marches westward to Normandy, then proceeds south into central France. Thus, the two terminal points of Arthur's continental confrontation in Caxton's text are consistent: Flanders and Rome, the beginning and end point for the common English route to Rome, and Caxton's editorial intervention suggests an attempt at geographic realism and coherence.

In addition to the change in Arthur's continental landing site, Caxton also modifies Arthur's route. Malory's route in the Winchester version takes Arthur and his noble knights "streyghte into Lushburne and so thorowe Flaundirs and than to Lorrayne … and sytthen he drew hym into Almayne and unto Lumbardy … [a]nd so into Tuskayne" (175.23–26; C V.5). The surprising north-western return to Flanders is an apparent addition by Malory, one not found in the *Morte Arthure* where the route is from Burgundy to Almayne (Germany), Luxembourg, Lorraine, Lucerne, and Lombardy, before reaching Tuscany en route to his destination of Rome. In his print edition, Caxton "simplifies Arthur's route," as Matthews points out, such that the route now takes Arthur from "the grete bataylle acheued ageynste the Romayns," which occurred on the plains of Burgundy, on a route via "Lorayne braban and Flaundres and sythen retorned in to hault Almayn and so ouer the montayns in to lombardye and after in to Tuskane."[196] While neither routes qualifies as a direct path to Rome, Caxton accommodates Malory's northwestern return to Flanders while providing a less circuitous route, one that at least attempts a geographical sequencing of the regions traversed. By eliminating Luxemburg from the route, adding Brabant, and shifting Lorraine to directly after the battle, Caxton proposes a sequential arc along the borders of the territories held by the Dukes of Burgundy up until the recent Burgundian Wars (ending in 1477 with the death of Charles the Bold). Caxton quickly moves the action from the successful slaying of the emperor Lucius in Burgundy and the return to Flanders, as presented in the manuscript version, to the advance of the campaign south. Arthur and his men march through Germany, Lombardy, and on to the final military victory of his long campaign—the crushing of the Tuscan city that leads to capitulation of Milan. While Caxton's specific reasons for route changes cannot be ascertained with any certainty, in combination, they suggest an awareness of routes and geographic relationships based on commerce routes familiar to transnational merchants such as himself.

Early modern urbanism

Despite realigning Arthur's continental campaign, Caxton uncharacteristically amplifies an apparent Malorian geographic entanglement: the extension of

Arthur's traditional route through Italy to the city of "Virvyn," thus creating the greatest Arthurian geographic *non sequitur* of the legendary campaign.[197] Generally, Caxton concretely identifies city names in an effort to ground the tale in geo-historicism rather than romance: in the Winchester manuscript, for example, once Arthur slays the giant at Mont Saint-Michel in northern France, he moves his battalions into a "fayre champayne," a non-specific location. In Caxton's edition, however, this stock location of romance is changed to an actual place name, the city of "Champayne."[198] This follows a tendency by Caxton, in mirroring the location-based attribute of Ptolemaic geography, to emphasize city names. In the *Morte Arthur*, once Arthur passes into Italy, he moves through the Lombardy region, passing near Lake Como, sending Sirs Florent and Floridas on an advance mission toward "þe cete vnsene,"[199] or unseen city. Their successful "enbushment" of that city leads to the subsequent capitulation of Milan. In the Winchester manuscript version of Malory's text, however, the city is specifically identified as "Virvyn." In *An Index of the Arthurian Names in Middle English*, Robert Ackerman identifies the Winchester's use of "Virvyn" as a corruption of "þe cete vnsene" from the *Morte Arthure*.[200] Vinaver similarly argues that there was "little doubt" that "Virvyn" is "either a misreading of *vnsene* or an attempt to coin a place name."[201] For his edition, Caxton is constrained by the narrative function of the capitulation of the city of Virvyn as the causative precursor to Milan's surrender to Arthur's advancing army. Faced with this narrative necessity, Caxton solves the textual dilemma by altering the city name to "vrybyn," or modern-day Urbino. Identifying the city as Urbino, however, causes Arthur and his troops to inexplicably lurch the campaign route south to a city on the central east coast of Italy before taking north-central Milan. Geographically, it makes no sense at all. In addition, the well-known traditional pilgrim and trade route through Italy, the Via Francigena, traverses south from Milan through west-central Italy en route to Rome with no such detour to Urbino. Caxton seems to inexplicably abandon his usual preference for georealism in Arthur's route. This choice can be seen, however, as part of Caxton's preference for the authority of an actual geographic location even if in this case it sacrifices route efficiency. As was evident in Caxton's identifying the city of Champagne, his logic seems to dictate that it is better for Arthur to pass through an identifiable city than traverse a hazy, unspecified landscape of romance. Caxton's choice of an actual Italian city, Urbino, may seem, at first glance, to suggest a cartographic carelessness on Caxton's part, but it is congruent with his remapping of the Roman War as a whole: with his strong preference for city-based geography, Caxton transforms as many "unseen" cities of Malory's romance as possible into actual cities and place names that feature prominently in the locational tables of Ptolemy's *Geography*. As Gautier-Dalché has made clear, the *Geography* was valued as much for the identification of classical place names as it was for the maps that were drawn based on his "tables" of mathematical coordinates.[202] Indeed, in the mid-fifteenth century, humanistic study of the *Geography* held this very promise: that through the pairing of obscure Greek or Latin place names with their contemporary places names, the identification of such "unseen" cities could be made. Thus, the cartographic impulse that underlies Caxton's carto-linguistic theory of the Roman War itinerary aligns with the Ptolemaic.

One speculative possibility to resolve the geographical conundrum of Malory's addition of "Virvyne" is the location of a small city 10 miles north of Milan, plausibly on the route from Lake Como. Dating back to Roman and medieval villages, Erba-Incino fits the narrative itinerary of the tale. Could this be the "cete vnsene" from the *Morte Arthure* that subsequently is transformed into Malory's "Virvyn" and Caxton's "vrbyne?" The possibility of Erba-Incino as the "unseen city," suggests, as readers of Matthews' article on Seisia-Sessoyne discover, that Malory's sometimes quizzical geography in both the Winchester manuscript and Caxton print may have a more grounded rationale than first assumed: the cruxes can arise from scribal errors, editorial changes, or geographical locations now obscure to modern scholars. Through the scribal and editing processes, Malory and Caxton each attempted to provide a version of the tale within the boundaries of their individual geographic knowledge of Europe. The geographic entanglement of the "unseen city" is not necessarily evidence of Malory's or Caxton's geographical naiveté, but an invitation to tease out the geographical trace within the multi-layered process of textual and geographical engagement of the fifteenth century. Further layers of topographical and mental itineraries are invoked by other routes referenced within this rhizomatic episode, including the remembered pilgrimage route of the Lord of the West Wales via Pontremoli in the Apennines ("all by the Poynte Tremble") (148.112–113; C V.1); the Roman ambassadors who must return to Sandwich ("by Watlynge Strete and no way ellys") (149.19–20; C V.2); and the northern march of the Emperor Lucius and his troops ("And so Lucius com unto Cullayne") (151.25; C V.2). These passages remind us that there is not one route in the Roman War episode but many, a situation complicated by statements of planned or recollected routes that traverse the Continent between London and Rome.

The return journey

Caxton's final geographical editing of Malory's text in contrast to the Winchester manuscript version comes as Arthur returns triumphant to England and provides a nuanced expression of his city-based geography. In the Winchester version, Arthur leaves Rome in "good governaunce" and "passyth thorow the contreyes of all partyes" before he "passed over the see unto Sandwyche haven" (189.7–13; C V.12). At news of his arrival, Gwenyvere "mette with hym at London … For there was never a solempner metyng in one cité togedyrs, for all maner of rychesse they brought with hem at the full" (189.14–17; C V.12). The large-scale continental space of the Roman War campaign has now inverted to an intimate domestic space—the space where romance can reassert itself. In the Caxton edition, however, the concluding centrality of London is exchanged for a dispersed and public Arthurian return. Caxton's Arthur "came ouer the see and londed at sandwyche ageynste whome Quene Gweneuer his wyf came and mette hym and he was nobly receyued of alle his comyns in euery cyte and burgh and grete yeftes presented to hym at his home comyng to welcome hym with."[203] Whereas the Winchester version suggests a private, domestic celebration in London, Caxton broadens his usual preference for a specific city and offers, instead, an Arthur for every city and burgh in Britain, a homecoming which specifically cites the "comyns" as ratifying Arthur's position as king and emperor with gifts, a welcome that underscores the increasing emphasis on urbanism. Similar to the earlier versions of Geoffrey, Wace, and Hardyng, Caxton's Arthur is reinscribed as a political and public subject. This stands in contrast to the Winchester version's more private sphere return of Arthur to Gwenyvere. In 1485, the year Caxton printed *Le Morte Darthur*, the strife between the factions of Lancaster and York was acute and the emphasis in Caxton's text on the acclamation by the commons for the newly exalted emperor of the Roman Empire suggests a hope for a unified Britain.

Shared geographic spaces in versioned literary texts

More work remains to better hone our understanding of Malory's and Caxton's symbolic "terrestrial textual strategies,"[204] but it is clear that Caxton seeks—at times strains—to wrangle the fluid and multi-layered geographies of the tale into a route that could be plotted on a fifteenth-century map. Just as Summit argues that Petrarch's poetry served to Christianize pagan Roman sites, Malory's narrative as instantiated in the manuscript and in Caxton's edition similarly "does not erase the former sites but retains them."[205] Thus, the geography of the Roman War narratives cannot be "mapped" as a single temporal moment but functions, even in an individual text, as a collective site of sedimentation and accretion of geographical knowledge, routes, and place names. As McGann has argued, "As the process of textual transmission expands … the signifying processes of the work become increasingly collaborative and socialized."[206] Vinaver's edition of the Winchester manuscript, while it sought to eliminate the trace of Caxton's editorial "intervention," in fact "enters the field, then, not by supplanting the Caxton text with one that is more 'authoritative' (least of all 'definitive'), but by supplementing it with a new version."[207] A thick reading of the texts proposes that the process of renaming and re-routing imbues new and accumulated significance to "a shared geographic space," one represented in the textual spaces of manuscript and print.[208] Matthew Sparke refers to such processes as "the ongoing process of spatial production, negotiation, and contestation."[209]

Periphery and center, redefined

First recorded approximately five hundred years after the retreat of the Roman legions, the Roman War narrative persisted for another five hundred years in English literature. The Roman War account, a fantasy of reverse conquest, provided a geographic narrative for English readers in which Rome was repudiated and absorbed into English imperial aspirations. The "implications of positionality" in terms of London and Rome and the cartographic "constructs of periphery and center" are fully reoriented in the Malory versions of Arthur's successful conquest of Rome.[210] As such, Malory's late-medieval version of the Roman War campaign and Caxton's editing of the text for print publication pave the way symbolically for the subsequent English break with Roman ecclesiastical authority, less than 50 years later. If, as McGann argues, "producing editions is one of the ways we produce literary meaning,"[211] Caxton proves to be a key reader of Malory's geography as well as of the accelerating changes in fifteenth-century English geographic thought at a key moment of geographical innovation and carto-literary dissemination of Ptolemaic constructs.

Caxton's subtle remapping of the Roman War is indicative of geographic practices broadly at work in late fifteenth-century European culture, practices which allowed—even encouraged—the incorporation of new information, including place names, itineraries, and geographic relationship. Geographic culture of the late fifteenth century showed a capaciousness for simultaneity and conflicting information. It was a process of accretion, doubleness, and paradox as new information was accepted alongside classical or traditional information. As Natalia Lovosky has argued, the cultural authority of classical geographic texts continued to exert their influence contemporaneously with the increasingly powerful authority of eye-witness accounts and discovery-based maps.[212] Caxton locates and reorganizes the Roman War within the new spatial order of the fifteenth century, a carto-literary text refashioned on the cusp of modern preferences for linearity and non-coevality.

Rome: **orbis** *and* **urbis** *of the western imagination*

After Caxton's edition of *Le Morte Darthur*, the cultural purchase of this particular tale of empire, religion, and geography begins to gradually wane, followed by the early modern absorption of Roman culture as a marker of English identity, a complex fashioning of national identity post-Reformation. For example, on the Agas map, first printed in 1555, the fulfillment of England's Roman lineage is expressed in the decorative cartouches in which London, with its ample natural resources and prime location, is claimed as the global successor to the legacy of Rome, as indicated by Geoffrey of Monmouth's *Historia*, because Britain was "founded by Brute the Trojan."[213] Here, Rome becomes culturally conquered, an instantiation of "colonial-turned-national cartography."[214] By Shakespeare's time, the tale of the Roman War campaign is folded into his script of *Cymbeline* in which England is invaded by Roman troops but successfully defeats them. King Cymbeline, unlike Arthur, magnanimously agrees to pay the demanded tribute to Rome anyway.

Roman rhetoric, literature, and architecture reinforced the conception of the Eternal City as the center of the earthly sphere,[215] and for the two millennia since, the "city of the Caesars and the popes, has figured in the Western imagination as the paradigmatic global city."[216] Cosgrove argues that the bronze globe held in the hand of Constantine's colossal statue from the later years of the Western Empire would become "one of the mirabilis of pilgrim Rome, a direct iconographic passage from antiquity to imperial papacy."[217] Rome—as both *urbis* and *orbis*—city and globe—was a deeply etched metaphoric and mythological association for English writers. Malory's Arthur returns from Rome an imperial ruler and, as we shall see in Chapter 3, John Mandeville returns from Rome with the imprimatur of the Pope, symbolically investing each of their returns to England, and by extension English geographical space, with the claim of a new geographical center and world order. Such readings took place in the context of growing English nationalism and England's complex relationship to Roman power as both a historical and ecclesiastical legacy.

Drawing on the work of C.G. Jung, Gaston Bachelard's meditation on the poetics of space includes the motif of a Roman foundation undergirding the layered history of the metaphysical space of a house.[218] In this same way, the Roman War campaign functions as an irruption of the Roman foundation into the "house" of England's national geographical mythology. Both in its textual inception with Geoffrey and in its spatial and textual reinscriptions over the next five centuries, the narrative works to resist Roman cultural colonialism and simultaneously to trace the physical evidence of England's Roman legacy. As such, the Roman War narratives reterritorialize "global space in the interests of repossession by the dispossessed,"[219] a process exemplified by the popularity of Roman themes in Shakespeare's plays and in the continued evocation of early modern London as inheritor of the legacy of Rome.[220] Over one thousand years after the actual occupation, Shakespeare's *Cymbeline* suggests that Britain becomes, not the conqueror of Rome, but the inheritor of its claims to empire and global grandeur.

Roman War redux

Until the 2013 printing of J.R.R. Tolkien's *Fall of Arthur*, his unpublished and unfinished metrical poem, Caxton's version of Arthur's Roman War campaign signaled the textual elegy for the narrative.[221] The tale became dormant, as in a dragon's lair, until Tolkien stirred the tale to a new geo-spatial purpose. Tolkien's Catholicism and English nationalism in the face of World War II invested the Roman War narrative with renewed potency to articulate Anglo-continental tensions across a literary-geographical space. *The Fall of Arthur* imaginatively refashions Arthur's eastward continental campaign toward Rome, a version based on Tolkien's deep familiarity with the accounts of Geoffrey, Wace, Laȝamon, *Morte Arthure*, Malory, and Caxton. Tolkien's son and editor points out that Tolkien's version departed "radically from the story of Arthur's last campaign overseas as told by Geoffrey of Monmouth and his successors."[222] This, however, *is* the history of the tale itself—a geographically-based narrative that gives rise to radical rewriting and remappings. In his version, Tolkien focuses on the threat of "the fall of Rome and Roman Christendom" and "the struggle of the Britons in the fifth century against the Germanic invaders,"[223] topoi with fresh political meaning in the context of the twentieth-century World Wars. In Tolkien's remapping of the Roman War, his Arthur "defends Rome, he does not assault it."[224] Like Malory and Caxton before him, Tolkien reinscribes the geo-political map of the Roman War with his own personal and cultural geographies and, in so doing, reactivates the near-millennial fluid geographies of the tale of Arthur's march on Rome. Tolkien's engagement with a remapping of the geography of the Roman War serves here as only as an amplifying testament to the power of the underlying geography, across generations, of this cross-continental cultural route. Tolkien's reading and imaginative re-instantiation of the longest-enduring spatial narrative of English literature—be it Arthur's conquest of Rome or "Saxon lands"—may well have informed his epic-scale *Lord of the Rings* as a narrative of resistance across an expansive geographic space.

Notes

1 All citations are from Thomas Malory, *Le Morte Darthur*, vol 1, ed. P.J.C. Field (Cambridge: D.S. Brewer, 2013), based on the Winchester manuscript. The second part of the reference provides Caxton's book number.

2 Malory, drawing on his French sources, primarily refers to Camelot as the seat of Arthur's court, whereas other English Arthurian texts favor Caerleon. See Robert Allen Rouse and Cory James Rushton, "Arthurian Geography," in *The Cambridge Companion to the Arthurian Legend*, eds. Elizabeth Archibald and Ad Putter (Cambridge: Cambridge University Press, 2009), 218–234; 219.

3 See Iain Chambers, "Maritime Criticism and Theoretical Shipwrecks," *PMLA*, 125:3 (May 2010): 678–684; 679.

4 For the role of the castle in structuring Arthurian space and community identity in Malory's *Le Morte Darthur*, see Molly A. Martin, *Castles and Space in Malory's* Le Morte Darthur (Cambridge: D. S. Brewer, 2019).

5 Chambers, 681.

6 David Woodward, "Maps and the Rationalization of Geographic Space," in *Circa 1492: Art in the Age of Exploration*, ed. Jay A. Levenson (New Haven, CT: Yale University Press, 1991), 83–88.

7 Felicity Riddy, "Contextualizing *Le Morte Darthur*: Empire and Civil War," in *A Companion to Malory*, eds. Elizabeth Archibald and A.S.G. Edwards (Cambridge: D.S. Brewer, 1996), 55–74; 64.

8 On the pattern of fragmentation and integration resulting from the movement of knightly adventure in *Le Morte Darthur*, see Jill Mann, "Taking the Adventure: Malory and the *Suite du Merlin*," in *Aspects of Malory*, eds. Toshiyuki Takamiya and D. Brewer (Cambridge: Boydell and Brewer, 1981), 71–96.

9 See *Arthuriana* 16/4 (2006), a special issue devoted to the subject of Saracens, ed. Jacquelyn de Weever.

10 For the seminal works on culture and space, see Henri Lefebvre, *The Production of Space* (New York: Wiley, 1992) and Gaston Bachelard, *The Poetics of Space* (Boston: Beacon Books, 1994, first published in English in 1964).

11 Sir Edward Strachey, ed. *Le Morte Darthur* by Sir Thomas Malory (London: Macmillan, 1919), xv.

12 Dorsey Armstrong and Kenneth Hodges, *Mapping Malory: Regional Identities and National Geographies* (New York: Palgrave Macmillan, 2014), 2.

13 Megan G. Leitch and Cory James Rushton, "Introduction," in *The New Companion to Malory*, eds. Megan G. Leitch and Cory James Rushton (Cambridge: Boydell and Brewer, 2019), 9; Catherine Nall, *Reading and War in Fifteenth-Century England: From Lydgate to Malory* (Cambridge: Boydell and Brewer, 2012).

14 Sheri Chriqui, "A 'Foreign' Queen in King Uther's Court: Fifteenth-Century Insular Xenophobia and Malory's Portrayal of Arthur's Mother," *South Atlantic Review* 81.2 (Summer 2016): 72–92.

15 Rouse and Rushton, "Arthurian Geography," 220.

16 Ibid., 219.

17 M.J. Dear, Jim Ketchum, Sarah Luria, and Doug Richardson, eds. *GeoHumanities: Art, History, and Text at the Edge of Place* (London: Routledge, 2011).

18 Anders Engberg-Pedersen, ed. *Literature and Cartography: Theories, Histories, Genres* (Cambridge, MA: MIT Press, 2017).

19 This self-reference only appears as part of an *explicit* in the Winchester manuscript; Caxton's text omits the *explicit*. For a discussion of Malory's incarceration, see Roberta Davidson, "Prison and Knightly Identity in Sir Thomas Malory's *Morte Darthur*," *Arthuriana* 14:2 (2004): 54–63.

20 Rouse and Rushton, "Arthurian Geography," 219.

21 P.J.C. Field, "Apparatus and Commentary," in *Le Morte Darthur* by Sir Thomas Malory, 2 vols., ed. P.J.C. Field (Cambridge: D.S. Brewer, 2013), 2:115. See also the "Index of Proper Names," in vol. 2 of this same edition. For a further discussion of Malory's geography, see also P.J.C. Field, "Caxton's Roman War," *Arthuriana* 5:2 (Summer 1995): 31–73, reprinted in *The Malory Debate: Essays on the Text of Le Morte Darthur*, eds. Bonnie Wheeler, Robert L. Kindrick, and Michael N. Salda. Arthurian Studies 47 (Cambridge: D.S. Brewer, 2000), 127–167.

22 Field, "Apparatus," 2:115–116.

23 Ibid., 2:116.

24 See Robert T. Tally, Jr., ed. *Geocritical Explorations: Space, Place, and Mapping in Literary and Cultural Studies* (New York: Palgrave Macmillan, 2011).

25 Robert J. Mayhew, "Cosmographers, Explorers, Cartographers, Chorographers: Defining, Inscribing and Practicing Early Modern Geography, c. 1450–1850," in *The Wiley-Blackwell Companion to Human Geography*, eds. John A. Agnew and James S. Duncan (London: Blackwell Publishing, 2011), 23–49; 28.

26 A phrase used by Dr. Jeffrey Blanchard, lecturer in Architecture, Cornell University, in lecture on ancient Roman architecture, Rome, Italy, 2013.

27 For Malory's use of the Post-Vulgate Cycle, see Ralph Norris, *Malory's Library: The Sources of the Morte Darthur*, Arthurian Studies 71 (Cambridge: D.S. Brewer, 2008).

28 Field, 2:2.

29 Doug Richardson, Sarah Luria, Jim Ketchum, and Michael Dear, "Introducing the GeoHumanities," in Michael J., Dear, Jim Ketchum, Sarah Luria, and Doug Richardson, eds. GeoHumanities: Art, History, and Text at the Edge of Place (London: Routledge, 2011), 3–4; 3.

30 The fourteenth-century *Travels* by John Mandeville, for example, continued to be widely read throughout the fifteenth century.

31 For a discussion of the "geographical self," see E.S. Casey, "Between Geography and Philosophy: What Does it Mean to Be in the Place-world?" *Annals of the Association of American Geographers* 91:4 (2001): 683–693. See also Steven C. Larsen and Jay T. Johnson, "The Agency of Place: Toward a More-Than-

Human Geographical Self," *GeoHumanities: Space, Place, and the Humanities* 2:1 (2016): 149–166.

32 For a list of editions of the *Geography*, see Patrick Gautier-Dalché, "The Reception of Ptolemy's Geography (End of the Fourteenth to Beginning of the Sixteenth Century)," in *The History of Cartography*, vol. 3, *Cartography in the European Renaissance*, Part 1, ed. David Woodward (Chicago: University of Chicago Press, 2007), 285–364.

33 Thomas Malory, *Le Morte Darthur* (Westminster: William Caxton, 1485), STC 801.

34 Lisa Jardine, *Worldly Goods: A New History of the Renaissance* (New York: Nan A. Talese, 1996).

35 George Doutrepont, *Inventaire de la "Librairie" de Philippe le Bon 1420* (Geneva: Slatkine Reprints, 1977). Note especially Item 199, *Cosmographia Tholomei*.

36 William Caxton, "Epilogue," in *Le Morte Darthur by Thomas Malory*, ed. Stephen H.A. Shepherd (New York: Norton and Company, 2004), 819.

37 The Winchester Manuscript is missing its final gathering; thus, this passage is attested only in the Caxton print edition.

38 For a discussion of literature and time in the context of contemporary geocriticism, see Adam Barrows, *Time, Literature, and Cartography After the Spatial Turn: The Chronometric Imaginary* (New York: Palgrave Macmillan, 2016).

39 Colin Richmond, "Malory and Modernity: A Qualm about Paradigm Shifts," *Common Knowledge* 14:1 (2008), 34–44.

40 Ibid., 35.

41 Kevin S. Whetter, *Understanding Genre and Medieval Romance* (Aldershot: Ashgate, 2008), 101.

42 Cory James Rushton, "Malory and Form," in *The New Companion to Malory*, eds. Megan G. Leitch and Cory James Rushton (Cambridge: Boydell and Brewer, 2019), 126.

43 Ibid., 143.

44 Homi Bhabha, *The Location of Culture* (New York: Routledge, 1994), 142.

45 Bonnie Wheeler, "Romance and Parataxis and Malory: The Case of Sir Gawain's Reputation," in *Arthurian Literature XII*, eds. James Carley and Felicity Riddy (Cambridge: D.S. Brewer, 1993), 109–132.

46 See Meg Roland, "The Utas of Seynt Hyllary: Time in the Roman War Campaign," available at: www.passionategeography.com.

47 Keith D. Lilley, *City and Cosmos: The Medieval World in Urban Form* (London: Reaktion Books, 2009), 40.

48 Roelof Nicolai, "The Premedieval Origin of Portolan Charts: New Geodetic Evidence," *Isis* 106:3 (2015), 517–543.

49 Tony Campbell, "Red Names on the Portolan Charts (1311–1677)," in *Map History/History of Cartography*, available at: www.maphistory.info/RedNamesCommentary.html#conclusions.

50 For a discussion of the Winchester manuscript rubrication process, see Thomas Crofts, *Malory's Contemporary Audience: The Social Reading of Romance in Late Medieval England* (Cambridge: D.S. Brewer, 2006) and Kevin S. Whetter, *The Manuscript and Meaning of Malory's* Morte Darthur: *Rubrication, Commemoration, Memorialization* (Cambridge: D.S. Brewer, 2017).

51 Armstrong and Hodges, 1.

52 David Wallace, "*Imperium*, Commerce, and National Crusade: The Romance of Malory's *Morte*," in *Medieval Literature: Criticism and Debate*, eds. Holly A. Crocker and D. Vance Smith (London: Routledge, 2014), 321–332; 324.

53 See J. Lennart Berggren and Alexander Jones, *Ptolemy's* Geography*: An Annotated Translation of the Theoretical Chapters* (Princeton, NJ: Princeton University Press, 2000).
54 Gautier-Dalché, "Reception," 359.
55 John Hardyng, *The Chronicles of John Hardyng*, ed. Henry Ellis (London: Rivington, 1812).
56 For example, in the Prose *Lancelot*, a maiden leads Lancelot a generalized distance, "a good league until they came to a shelter." See "Lancelot, Part V," in *Lancelot-Grail: The Old French Arthurian Vulgate and Post-Vulgate in Translation*, ed. William Kibler, general ed., Norris Lacy, 5 vols. (New York: Garland, 1993–1996), 3:216.
57 Robert Copland published *The Rutter of the Sea* in England in 1528, a translation of an earlier French rutter by Pierre Garcie. In Robert Wyer's editions of *The Compost of Ptolemy* in the 1530s and 1540s, a rutter was included and the French "leagues" were converted to English "myles."
58 Andrew Lynch, "Malory and Emotion," in *The New Companion to Malory* ,eds. Megan G. Leitch and Cory James Rushton (Cambridge: D.S. Brewer, 2019), 177–190; 178. See also Andrew Lynch, "What Cheer? Emotions and Actions in the Arthurian World," in *Emotions in Medieval Arthurian Literature*, eds. Frank Brandsma, Carolyne Larrington, and Corinne Saunders (Cambridge: D.S. Brewer, 2015).
59 Lynch, "Malory and Emotion," 178.
60 Tally, 6.
61 Armstrong and Hodges, 114.
62 Sallie A. Marston, John Paul Jones III, and Keith Woodward, "Human Geography without Scale," *Transactions of the Institute of British Geographers*, New Series, 30:4 (December 2005), 416–432; John Paul Jones III, Keith Woodward, and Sallie A. Marston, "Situating Flatness," *Transactions of the Institute of British Geographers*, New Series, 32:2 (April 2007), 264–276, and Simon Springer, "Human Geography Without Hierarchy," *Progress in Human Geography*, 38:3 (2014), 402–419.
63 Marston et al., "Situating," 265.
64 Ibid., 274.
65 Manuel Castells, *The Rise of the Network Society* (Oxford: Blackwell, 2000), 2nd edition, see 407–459.
66 Armstrong and Hodges, 16.
67 Bertrand Westphal, "A Geocritical Approach to Geocriticism," *American Book Review* 37:6 (2016): 4–5.
68 Ibid., 4.
69 For a discussion of the way in which medieval literature and science writing are far more porous than contemporary genres, see Matthew Boyd Goldie, *Scribes of Space: Place in Middle English Literature and Late Medieval Science* (Ithaca, NY: Cornell University Press, 2019), 4.
70 Peter Field identifies the author Thomas Malory as a knight from Newbold Revel, born in 1415. See Peter Field, *The Life and Times of Thomas Malory* (Cambridge: Cambridge University Press, 1993). If so, Malory was likely part of an English campaign in Bayonne, in the Gascony region of southwest France, which accounts for his familiarity with French geography.
71 See Barry Gaines, "Malory's Castles in Text and Illustration," in *The Medieval Castle: Romance and Reality*, eds. Kathryn Reyerson and Faye Powe (Dubuque, IA: Kendal/Hunt Publishing, 1984), 215–218.
72 Armstrong and Hodges, 9.
73 The *explicit* is omitted in Caxton's edition.

74 Anne F. Sutton, "Malory in Newgate Prison: A New Document," *The Library*, 7th series, 1:3 (2000), 243–262.

75 Denis Cosgrove, *Apollo's Eye: A Cartographic Genealogy of the Earth in the Western Imagination* (Baltimore, MD: Johns Hopkins University Press, 2011), xi.

76 Ibid., 50.

77 The reference to a "grete booke" in one of John Paston's letters is proposed as possibly Malory's *Le Morte Darthur*, see Karen Cherewatuk, "Sir Thomas Malory's 'Grete Booke,'" in *The Social and Literary Contexts of Malory's* Morte Darthur, ed. D. Thomas Hanks, Arthurian Studies 42 (Cambridge: D.S. Brewer, 2000), 42–67.

78 Wallace, 325.

79 For their influential theory of the non-hierarchical organizing structure of the rhizome, see Gilles Deleuze and Félix Guattari, *A Thousand Plateaus*, trans. Brian Massumi (Minneapolis, MN: University of Minnesota, 1987; London: Continuum, 2004).

80 Gautier-Dalché, 285.

81 Alan M. MacEachren, "A Linear View of the World; Strip Maps as a Unique Form of Cartographic Representation," *The American Cartographer* 13:1 (1986): 7–25. See also David Woodward, "English Cartography 1650–1750, A Summary," in *The Compleate Plattmaker: Essays on Chart, Maps, and Globe Making in England in the Seventeenth and Eighteenth Centuries*, by Norman J.W. Thrower (Los Angeles: University of California Press, 1978), 159–193.

82 Rory Stewart, *The Places in Between* (Orlando, FL: Harcourt, 2004), 190.

83 Richard A. Talbert, *Rome's World: The Peutinger Map Reconsidered* (Cambridge: Cambridge University Press, 2014).

84 The tale also has a more minor trajectory through the French romance tradition, though it is most often subordinate to Arthur's military campaign on behalf of, or in contestation with, Lancelot. For its appearance in three strands of the Vulgate Cycle (*Suite du Merlin, Lancelot*, and *Mort Artu*), see Norris, 53–54.

85 Sylvia Tomasch, "Introduction: Medieval Geographical Desire," in *Text and Territory: Geographical Imagination in the European Middle Ages*, eds. Sylvia Tomasch and Sealy Gilles (Philadelphia, PA: University of Pennsylvania Press 1998), 4.

86 William Caxton, "Prologue to *Le Morte Darthur*," in *Le Morte Darthur* by Sir Thomas Malory, 2 vols., ed. P.J.C. Field (Cambridge: D.S. Brewer, 2013), 2:854.

87 Caxton, 815.

88 Geoffrey of Monmouth, *The History of the Kings of Britain* (*Historia regum Britanniae*), ed. Michael A. Faletra (Plymouth: Broadview Editions, 2008), 41.

89 Gordon Brooks, paper delivered at the International Arthurian Congress, Bristol, 2013. See also J.B. Bury ed. *The Cambridge Medieval History* (Cambridge: Cambridge University Press, 1924), 238.

90 Tomasch, 5.

91 Tom Conley, "Early Modern Literature and Cartography: An Overview," in *The History of Cartography*, vol. 3, *Cartography in the European Renaissance*, Part 1, ed. David Woodward (Chicago: University of Chicago Press, 2007): 401–411; 407.

92 The manuscript is held by the British Library, BL Add. MS. 59678.

93 The book was controversially purchased by Ms. Abbey Hanscomb of New York in 1885 and later donated to the Pierpont Morgan Library. In consequence of her role in the out-migration of the text from Britain to the United States, Paul Needham refers to her as a "*femme bibliophile*" in his introduction to *Sir Thomas Malory, Le Morte Darthur, Printed by William Caxton*, by Sir Thomas Malory, facsimile edition (London: Scolar Press, 1976).

94 There are two facsimile editions of the text, one of the Winchester manuscript and one of the Caxton print. See *Sir Thomas Malory, Le Morte Darthur, Printed*

by William Caxton, by Sir Thomas Malory, Paul Needham, Introduction (London: Scolar Press, 1976), and *The Winchester Malory; A Facsimile Edition*, by Sir Thomas Malory (Oxford: Oxford University Press for the Early English Text Society, Supplementary Series 4, 1976). For digital images of the manuscript and printed book as well as transcriptions, see The Malory Project, directed by Takako Kato and designed by Nick Hayward, available at: www.maloryproject.com/. The Winchester ms is cataloged in the British Library as MS Add 59678. For a diplomatic, parallel-text edition of the Roman War tale, see Margaret M. Roland, "Material Malory: The Caxton and Winchester Documents and a Parallel-Text Edition," unpublished dissertation, University of Washington, 2002.

95 Bachelard, xxxiv.

96 Christopher Tolkien, "Introduction," in *The Fall of Arthur*, by J.R.R. Tolkien (Boston: Houghton Mifflin, 2013).

97 Crofts.

98 Ivo Kamps, "Foreword," in *Maps and Memory in Early Modern England*, by Rhonda Lemke Sanford (New York: Palgrave, 2002), xiii.

99 Naomi Kline, *Maps of Medieval Thought: The Hereford Paradigm* (Woodbridge: The Boydell Press, 2001), 87.

100 P.D.A. Harvey, *Mappa Mundi: The Hereford World Map* (London: British Library, 1996); Evelyn Edson, *Mapping Time and Space: How Medieval Mapmakers Viewed Their World* (London: British Library, 1998); Asa Mittman, *Maps and Monsters in Medieval England* (London: Taylor & Francis, 2006); John Short, *The World Through Maps: A History of Cartography* (Toronto: Firefly Books, 2003).

101 Cosgrove, 14.

102 Natalia Lozovsky, "Roman Geography and Ethnography in the Carolingian Empire," *Speculum* 81:2 (April 2006): 325–364; 362.

103 David J. Bodenhamer, John Corrigan, and Trevor M. Harris, "Introduction," in *The Spatial Humanities: GIS and the Future of Humanities Scholarship*, edited by David J. Bodenhamer, John Corrigan, and Trevor M. Harris (Indianapolis, IN: Indiana University Press, 2010).

104 For a contemporary walking journey along the Via Francigena, see Timothy Egan, *Pilgrimage to Eternity* (New York: Viking, 2019).

105 Veronica Ortenburg, "Archbishop Sigeric's Journey to Rome in 990," in *Anglo-Saxon England* 19 (1990), 197–246, and Alison Raju, *Via Francigena: Pilgrim Trail Canterbury to Rome* (Milnthorpe, Cumbria: Cicerone Press, 2013).

106 See the Council of Europe's web site on Cultural Routes initiative at: www.coe.int/t/dg4/cultureheritage/culture/routes/default_en.asp.

107 Robert Hughes, *Rome: A Cultural, Visual, and Personal History* (New York: Vintage Books, 2011), 58.

108 Ibid., 58.

109 Edmund O'Gorman, *Toward the Great Millennium Jubilee* (Leominster: Gracewing, 1998).

110 Jennifer Summit, "Topography as Historiography: Petrarch, Chaucer, and the Making of Medieval Rome," *Journal of Medieval and Early Modern Studies* 30 (2000): 211–246, reprinted in *Medieval Literature: Criticism and Debate*, eds. Holly A. Crocker and D. Vance Smith (London: Routledge, 2014), 304–320; 317.

111 Tomasch, 5.

112 Jennifer R. Goodman, *Chivalry and Exploration (1298–1630)* (Woodbridge: The Boydell Press, 1998), 45.

113 Crofts, 110.

114 For Malory's use of Hardyng as a source, see Edward Donald Kennedy, "Malory's Use of Hardyng's Chronicle: A Reconsideration," *West Virginia Philological Papers* 54 (2011), 8–15.

115 See Norris.
116 See Eugène Vinaver, ed., *The Works of Sir Thomas Malory* (3 vols), rev. by P.J.C. Field (Oxford: Clarendon Press, 1990), 1396.
117 Robert L. Kelly, "Malory's Argument Against War with France; The Political Geography of France and the Anglo-France Alliance in the *Morte Darthur*," in *The Social and Literary Contexts of* Le Morte Darthur, Arthurian Studies xlii, eds. D. Thomas Hanks Jr. and Jessica G. Brogdon (Cambridge: Boydell and Brewer, 2000), 111–133: 112.
118 William Matthews, "Where Was Siesia-Sessoyne?," *Speculum* 49:4 (October 1974): 680–686: 685.
119 Rouse and Rushton, "Arthurian Geography," 219. Regarding Caxton's editing of Malory's geography, see Meg Roland, "From 'Saracens' to 'Infydeles'": The Recontextualization of the East in Caxton's Edition of *Le Morte Darthur*," in *Re-viewing Le Morte Darthur: Text and Contexts: Characters and Themes*, eds. Kevin S. Whetter and Raluca Radulescu, Arthurian Studies 60 (Cambridge: D.S. Brewer, 2005), 65–77.
120 For a discussion of the "convergence of the three conceptual frameworks" of the *mappamundi*, portolan charts, and Ptolemaic maps, see David Woodward's extensive essay, "Medieval *Mappaemundi*," in *The History of Cartography*, vol. 1: *Cartography in Prehistoric, Ancient, and Medieval Europe and the Mediterranean*, ed. David Woodward (Chicago: University of Chicago Press, 2007), 286–320; see especially 314.
121 See Asa Mittman, *Maps and Monsters in Medieval England* (New York: Routledge, 2006) and Naomi Reed Kline, *Maps of Medieval Thought: The Hereford Paradigm* (Woodbridge: Boydell Press, 2001).
122 Geoffrey of Monmouth, 236.
123 Mary Dictman, "The Tale of King Arthur and the Emperor Lucius," in *Malory's Originality*, ed. R.M. Lumiansky (Baltimore, MD: Johns Hopkins University Press, 1964), 69.
124 *Laȝamon's Arthur: The Arthurian Section of La☒amon's Brut (Lines 9229–14297)*, eds. W.R.J. Barron and S.C. Weinburg (Austin, TX: University of Texas Press, 1989), lines 12658–12659.
125 Edward Said, *Orientalism* (New York: Vintage Books, 1979), 72–73.
126 Iain Mcleod Higgins, *Writing East: The "Travels" of Sir John Mandeville* (Philadelphia, PA: University of Pennsylvania Press, 1997), 203 and Iain Mcleod Higgins, *The Book of John Mandeville, with Related Texts* (Indianapolis, IN: Hackett Press, 2011).
127 *The Alliterative Morte Arthure: A Critical Edition*, ed. Valerie Krishna (New York: Burt Franklin, 1976), line 582.
128 Field, "Index of Proper Names," in *Le Morte Darthur*, 2:862–892; 861.
129 John Withrington, "'He Telleth the Number of the Stars; He Calleth Them All by Their Names:' The Lesser Knights of Sir Thomas Malory's *Morte Darthur*," *Quondam et Futurus: A Journal of Arthurian Interpretations* 3:4 (1993): 17–27; 17.
130 Alliterative *Morte Arthure*, lines 602, 605, 606.
131 See Meg Roland, "Arthur and the Turks," *Arthuriana* 16:4 (2006): 29–42.
132 See Meg Roland, "Malory's Sandwich: Alternate Arthurian Geographies and the Global Middle Ages," In *Re-thinking Medieval Margins and Marginality*, eds. Ann Zimo, Tiffany Sphrecher, Kathryn Reyerson, and Debra Blumenthal (New York: Routledge, 2019).
133 *The Alliterative Morte Arthure*, lines 3206–3207.
134 Ibid., lines 3211–3215.
135 For Malory's under-recognized narrative debt to Hardyng, see Edward Donald Kennedy, "Malory's Use of Hardyng's Chronicle," *Notes and Queries* 214 (1969): 167–170; Edward Donald Kennedy, "Malory's English Sources," in *Aspects of*

Malory, eds. Toshiyuki Takamiya and Derek Brewer (Cambridge: D.S. Brewer, 1981), 30–31; and Kennedy, "Malory's Use of Hardyng's Chronicle: A Reconsideration," 8–15. See also John Withrington, "Caxton, Malory, and the Roman War in the *Morte Darthur*," *Studies in Philology* 89:3 (Summer 1992): 350–366 and Masako Takagi and Toshiyuki Takamiya, "Caxton Edits the Roman War Episode: *The Chronicle of England* and Caxton's Book V," in *The Malory Debate: Essays on the Texts of* Le Morte Darthur, eds. Bonnie Wheeler, Robert L. Kindrick, and Michael N. Salda, Arthurian Studies 47 (Cambridge: D.S. Brewer, 2000).

136 Hardyng, Chapter lxxxii, 145.

137 For a summation of the scholars who have linked the "Claudas plot" of the Vulgate *Suite du Merlin* with Malory's departure from the Alliterative *Morte Arthure* and the chronicle tradition, see Norris, 54–55.

138 Though its exact date of composition is unknown, it has been established that Malory completed his book by 1471.

139 A commemorative medal by Bertoldo di Giovani was stamped for the occasion, depicting Frederick and the Pope in Rome. See the Frick Collection, available at: www.frick.org/exhibitions/bertoldo/12a.

140 D. Thomas Hanks, Jr. and Janet Jesmok, eds. *Malory and Christianity: Essays on Sir Thomas Malory's Morte Darthur* (Kalamazoo, MI: Medieval Institute Publications, 2013).

141 Field suggests this may be "an error for Vale of Viterbo." See "Apparatus," 2:892.

142 Kelly, 117.

143 Harvey, 35.

144 Ibid., 21.

145 Ibid., 35

146 See P.J.C. Field, *Malory; Texts and Sources*, Arthurian Studies XL (Cambridge: D.S. Brewer, 1998) and "Sir Thomas Malory, M. P.," *Bulletin of the Institute of Historical Records* XKVII (May 1974): 24–35. Also, Crofts, 14–23.

147 Elisabeth Leedham-Green, *Books in Cambridge Inventories: Book-Lists from Vice-Chancellor's Court Probate Inventories in the Tudor and Stuart Periods*, 2 vols. (Cambridge: Cambridge University Press, 1987).

148 For an analysis of Malory's geography in various points in the larger work, see Armstrong and Hodges.

149 For a discussion of the mutual influences of chronicles and romance, see Richard J. Moll, *Before Malory: Reading Arthur in Later Medieval England* (Toronto: University of Toronto Press, 2003). Moll notes that in the learned culture of late medieval Arthurian historiography, conscientious readers "critically compared the various narratives of the king to construct an historically authentic image of his reign," 80.

150 Meg Roland, "The Rudderless Boat: Fluid Time and Passionate Geography in (Hardyng's) Chronicle and (Malory's) Romance," *Arthuriana* 22:4 (Winter 2012): 77–93.

151 David Greetham, *Textual Scholarship: An Introduction* (New York: Garland, 1994), 311.

152 George H.T. Kimble, Geography in the Middle Ages (New York: Russell and Russell, 1938, reissued 1968), 200.

153 Wallace notes that "Round Table adventuring" remained popular among northern European merchants and that Arthur was "a revered figure along the trade route of the Hanseatic League," 322.

154 Mayhew, 28.

155 For discussion of one of the earliest English rutters, from the mid-fifteenth century, see Matthew Boyd Goldie, "An Early English Rutter: The Sea and the Spatial Hermeneutics in the Fourteenth and Fifteenth Centuries," *Speculum* 90:3

(July 2015): 701–727. See also Chapter 5 of this book, "After 'Poyetes and Astronomyers.'"

156 C.S. Lewis, "The Genesis of a Medieval Book," in C.S. Lewis, *Studies in Medieval and Renaissance Literature* (Cambridge: Cambridge University Press, 1998; originally published 1966), 19.

157 Ibid., 75.

158 Andrew Lynch, *Malory's Book of Arms* (Cambridge: D.S. Brewer, 1997), 82.

159 Quoted in Lynch, 82–83.

160 David Lawton, "Dullness and the Fifteenth Century," *English Literary History* 54:4 (1987): 761–799.

161 Vinaver, xxx.

162 C.S. Lewis, "The English Prose Morte," in *Essays on Malory*, ed. J.A.W. Bennet (Oxford: Clarendon Press, 1963) 7–28; 26.

163 Vinaver, xxx.

164 William Kuskin, "Caxton's Worthy Series: The Production of Literary Culture," *English Literary History* 66:3 (1999): 511–551; Jorge Fichte, "Caxton's Concept of 'Historical Romance' within the Context of the Crusades: Conviction, Rhetoric, and Sales Strategy," in *Tradition and Transformation in Medieval Romance*, ed. Rosalind Field (Woodbridge: D.S. Brewer, 1999), 101–113; and Fabienne L. Michelet, "East and West in Malory's Roman War: The Implications of Arthur's Travels on the Continent," *Multilingua* 18:2–3 (1999): 209–225.

165 See John Bryant, *The Fluid Text: A Theory of Editing for Book and Screen* (Ann Arbor, MI: University of Michigan Press, 2002).

166 Jerome McGann, *The Textual Condition* (Princeton, NJ: Princeton University Press, 1991), 9.

167 Jacques Derrida, "Sign, Signature, Context," *Glyph* 1 (1977): 172–197; 181.

168 Specific to Malory, constructivist editions include Stephen Shepherd's Norton Critical Edition, *Le Morte Darthur* (New York: Norton, 2003) and Kato's The Malory Project.

169 Helen Cooper, "Introduction," in *Le Morte Darthur*, ed. Helen Cooper (Oxford: Oxford University Press, 1998), xxv.

170 Ibid., xxv.

171 McGann, 10.

172 Wallace, 324.

173 See John Gillies, *Shakespeare and the Geography of Difference* (Cambridge: Cambridge University Press, 1994), Tom Conley, *The Self-Made Map: Cartographic Writing in Early Modern France* (Minneapolis, MN: University of Minneapolis Press, 1996); Sean Roberts, *Printing the World: Constantinople, Florence and the Renaissance of Geography* (Cambridge, MA: Harvard University Press, 2013). Roberts identifies Ptolemy's *Geography* as a critical transition point from the manuscript to print and from medieval to modern cartography.

174 Armstrong and Hodges, 2.

175 Summit, 307.

176 Kathy Lavezzo has argued that geographical otherness "promised both the exaltation and the marginalization of England during the Middle Ages." *Angels on the Edge of the World: Geography, Literature, and English Community, 1000–1534* (Ithaca, NY: Cornell University Press, 2006), 8.

177 Kuskin, "Caxton's Worthy Series," 511.

178 John Withrington, "Caxton, Malory, and the Roman War in the Morte Darthur," *Studies in Philology* 89:3 (Summer 1992): 350–366; see also Crofts.

179 See also William Kuskin ed. *Caxton's Trace: Studies in the History of English Printing* (Notre Dame, IN: University of Notre Dame Press, 2006).

180 The term "realism" is, of course, highly fraught. Here, it refers to a linear, measured distance approach rather than a primarily symbolic spatial relationship.

181 For a discussion of the development of cultural spaces, see Lefebvre.
182 Vinaver, 193.8–9 (Caxton version). Vinaver provides a quasi-parallel text in his edition, with Caxton's text provided at the page bottom. A digital facsimile is also available at The Malory Project.
183 "Caxton's Prologue to *Le Morte Darthur*," in *Le Morte Darthur* by Sir Thomas Malory, 2 vols., ed. P.J.C. Field (Cambridge: D.S. Brewer, 2013), 2:856.
184 See note 138.
185 Edward Donald Kennedy, "Caxton, Malory, and the 'Noble Tale of King Arthur and the Emperor Lucius,'" in *The Malory Debate*, 217–232; 223.
186 William Matthews, "A Question of Texts," in *The Malory Debate: Essays on the Text of Le Morte Darthur*, eds. Bonnie Wheeler, Robert L. Kindrick, and Michael N. Salda, Arthurian Studies 47 (Cambridge: D.S. Brewer, 2000), 65–107; 99.
187 "Caxton's Prologue," 2:855.
188 Kuskin, "Caxton's Worthy Series."
189 The final gathering of the Winchester manuscript has not survived; for a discussion of the reference to "Turkes" as a possible Caxton addition, see Roland, "Arthur and the Turks."
190 Near the conclusion of the Roman War tale, Malory refers once to Rome as "the cité of Syon that is Rome callyd" (187.27), a poetic but perhaps also geo-spiritual decentering of Jerusalem.
191 *The Alliterative Morte Arthure*, line 835.
192 Vinaver, 198.6–8 (Caxton version).
193 See Matthews, "A Question of Texts," and Field, "Caxton's Roman War" in *The Malory Debate* for a discussion of the Caxton text's addition of Flanders. Both scholars consider the addition in the context of whether Caxton was the reviser of the Roman War tale—and come to different conclusions.
194 Vinaver, 191.7–8 (Caxton version).
195 For a discussion of Caxton's insertion, see Field, "Caxton's Roman War," 135.
196 Matthews, "Where Was Siesia-Sessoyne?" 681; Vinaver, 227.5–6 (Caxton version).
197 Note that in Vinaver's edition, he provides the name of the city in brackets as [Urbyne], indicating that the text is an emendation from Caxton's edition, 242. Shepherd follows the same practice (148). The manuscript transcription (fol. 94r) is "Virvyn" and the transcription of Caxton's edition is "virbyn." Vinaver, 1404, n. 242.
198 Vinaver, 205 (Caxton version).
199 *The Alliterative Morte Arthure*, line 3114.
200 Robert William Ackerman, *An Index of the Arthurian Names in Middle English*, vol. 10 (Palo Alto, CA: Stanford University Press, 1952), 236.
201 Vinaver, 1404, n. 242. Vinaver also cites the possibility, suggested by William Matthews, that Malory's city name might refer to *Cressyn* (region of Quercy) in southeastern France, which seems highly tenuous geographically.
202 See Gautier-Dalché.
203 Vinaver, 246.15–16–247.1–2 (Caxton version).
204 Peta Mitchell, "Literary Geography and the Digital: The Emergence of Neogeography," in *The Routledge Handbook to Literature and Space*, ed. Robert T. Tally, Jr. (London: Routledge, 2016), 85–94.
205 Summit, 311.
206 McGann, *The Textual Condition*, 58.
207 Jerome McGann's critical assessment of Vinaver's edition formed a central example of McGann's rejection of the theoretical principles of "critical editions." Surprisingly, despite McGann's enormous influence in textual scholarship, his work is rarely cited by Malory scholars, with the exception of Thomas Crofts.

See Jerome J. McGann, *A Critique of Modern Textual Criticism* (Charlottesville, VA: University of Virginia Press, 1983, reprint, 1992), see especially, 81–83 and 138–139.

208 Summit, 311.13.

209 Matthew Sparke, *In the Space of Theory* (Minneapolis, MN: University of Minnesota Press, 2005), xvi.

210 Tomasch, 6.

211 McGann, *The Textual Condition*, 33.

212 Natalia Lozovsky, *The Earth Is Our Book: Geographic Knowledge in the Latin West, 400–1000* (Ann Arbor, MI: University of Michigan Press, 2000).

213 For a discussion of the impact of the Roman wall built around the city of London on the literary imagination of early modern England, see an essay written by my students "The Wall," for *The Map of Early Modern London*, available at: https://mapoflondon.uvic.ca/WALL2.htm.

214 Matthew Sparke, "Political Geographies of Globalization (3): Resistance," *Progress in Human Geography* 32:1 (2008): 1–18; 12.

215 Robert Hughes, *Rome: A Cultural, Visual, and Personal History* (New York: Vintage, 2012).

216 Cosgrove, xi.

217 Ibid., 47.

218 Bachelard, xxxvii.

219 Matthew Sparke, "Everywhere but Always Somewhere: Critical Geographies of the Global South," *The Global South* 1:1 (2007): 117–126; 117.

220 Neil MacGregor, *Shakespeare's Restless World* (London: Penguin Books, 2014). See especially his chapter, "London Becomes Rome," concerning the Roman-styled triumphal arches created for James I's coronation procession.

221 J.R.R. Tolkien, *The Fall of Arthur*, ed. Christopher Tolkien (Boston: Houghton Mifflin, 2013), lines 1–5.

222 Ibid., 85.

223 Ibid., 88–89.

224 Ibid., 110.

2 Cartographic Caxton

The Myrrour of the World and early English print

In the span of a lifetime, the world-view of an educated European—as represented by a map they might encounter in a book or hanging on a manor or church wall—underwent a visual reimagination from a Jerusalem-centered, sacred cosmology to one based on mathematical projection and exploration-based accounts. This process of the "rationalization of space"[1] was not instantaneous nor absolute, however: maps or globes based on the mathematical projections of Claudius Ptolemy, such as the Behaim globe (see Chapter 4), continued to be in dialogue with aspects of the sacred, marvelous, or exotic. As Ptolemaic-based maps began to be produced—both in manuscript and then in print—cartographic practices and sea-based expeditions rapidly developed. such that from the mid-fifteenth century to the early sixteenth century, multiple geographic models coexisted. As Matthew H. Edney has observed, it was a period that preceded the cartographic "semantic stability" of the seventeenth and eighteenth centuries.[2] The concept of the Earth as a sphere, as noted in the Introduction, was part of the scholarly inheritance from classical cosmography, including Greek and Arabic scholarship[3] and had been widely studied as part of European university and monastic scholarship. In texts such as Johannes de Sacro Bosco's *De Sphaera*, the *Polychronicon*, and Mandeville's *Travels*, the spherical properties of the Earth were widely represented. Even in schematic cartographic images such as "T-O" maps the spherical properties of the Earth were represented. (T-O maps, *orbis terrarum*, are named for their graphic representation of a "T" inscribed within an orb to represent the three known continents.)[4] Maps based on the integration of Claudius Ptolemy's second-century *Geography* into European intellectual practices, however, were part of a shift from the representation of the Christian, time-based world-view of medieval *mappaemundi* to a cartographic structure based on non-historicized, mathematically-constructed space. On Ptolemaic maps, Alessandro Scaffi explains, "space is homogenous, ordered by a geometrical grid."[5] These new maps typically oriented world maps to the north, no longer depicted Jerusalem as the sacred center, and represented spatially-proportional continents (within known limits), although as Edney argues, mathematically-proportional scale is itself a major fiction of the field of cartography.[6] The text of Ptolemy's *Geography* was translated into Latin from the Greek by 1410 by Jacopo Angeli and published, most likely, the following year.[7] The first print edition of Ptolemy's *Geography* was published in

DOI: 10.4324/9781003096245-3

Vincenza in 1475, followed by an edition with maps produced in Bologna in 1477, then, by 1550 editions printed in Rome, Ulm, Vienna, Strasbourg, Paris, and, Lyon.[8] Together these early print editions of Ptolemy's *Geography* constitute part of the Latinate revival of Greek scholarship activated in Europe through Italian humanistic translation and disseminated through merchant and scholarly networks of book production. Further, these maps did not hang on a church wall or in a grand hall: cartography was now a component of the printed book. Scaffi notes this important contextual shift: the Ptolemaic map "is spread over two pages in a bound, folio-size book."[9]

Once a producer of world maps, England ceded cartographic production during the fifteenth century to the advanced technology of continental map makers. As I have argued throughout this book, some form of a concurrent geographic transformation in terms of representations of the world took place in England but it was undocumented by English-produced cartographic artifacts, such as world maps or globes. This lack of production, however, has been overdetermined in the assessment of English geographic thought as removed from the great change in spatial thought that took place in Europe. Kathy Lavezzo has evaluated the production gap to make the claim that "England's 'backwardness' regarding map production suggests a construction of a 'geographic otherness' in terms of English national identity, a disregard for new images of the world, and a lingering national investment in the old medieval image of the *oikoumené*."[10] Lavezzo's argument regarding England's national sensibilities as an island nation on the edge of the former Roman Empire comes under pressure, however, in light of English humanism and the transcultural nature of English book culture.[11] Production alone does not signal all of a culture's engagement with an object or construct in a globalizing culture, and established courier routes connected northern European trading centers, including London and Bruges, to their Italian counterparts on regular routes.[12] That is not to say that maps were a common consumer item in late fifteenth century England—they were not—but mercantile networks between London and Italy, and between English printer William Caxton (c.1422–c.1491) and his European counterparts, were well established by the time that Caxton set up his print shop in Westminster in 1476.

In *Printers Without Borders*, A.E.B. Coldiron argues that the pan-European nature of English early print book culture is evident not only in terms of the mobility of books themselves through mercantile networks but also in terms of the process of "Englishing texts," the process of translating and printing foreign language vernacular texts by England's early printers.[13] Coldiron asserts that "what is asserted and promulgated as English literature was actually founded on and thoroughly permeated by the foreign."[14] Such "internationalized" literary texts indicate the way in which the dissemination and acculturation of Ptolemaic-based geographic thought in England moved through the channels of book culture, including translation and importation. While some English writers may have claimed the special status of an island nation, the new geography made its way into English reading culture in a pattern that follows the process of vernacular translation and acculturation via the book trade and early print.[15]

Maps drawn on the technical principles laid out in Claudius Ptolemy's *Geography* were included in most fifteenth-century manuscript and print editions of the work, exerting an enormous influence on European visual culture; initially part of the prestige culture of gift and exchange,[16] the *Geography* became a highly-valued text for scholars and bibliophiles alike. After its translation from Greek to Latin, its subsequent transmission coincided with the development of print technology but accompanied by beautifully colored, hand-painted maps. Although England had a vigorous history of map production in the medieval period, no maps based on Ptolemy's projection were produced in England during the Ptolemaic revival in Europe and, as discussed in the Introduction, the scholarly consensus has been that England was a backwater in terms of geography and cartography.

Such a view of English geographical thought during the fifteenth and sixteenth centuries is at odds, however, with the cosmopolitan nature of the English book trade (of which maps and geographical texts were a part), as well as the inter-connectedness of English humanists and printers with continental scholars and printers. The concurrent and interlocking cultural structures of European humanism, early print, and geographic writing in the Ptolemaic revival era sheds light on the nature of English engagement with Ptolemaic-based geography and, in particular, the role of William Caxton in this process.

Denis Cosgrove calls attention to the way that knowledge constructs are "inscribed in globes and world maps." That is, that cartographic objects represent a visual compression of cultural knowledge in material form. This process of geographical distillation, he points out, "frames and constructs poetic geographies, shaping meaning across the global surface."[17] Without English-made globes or world maps, however, how can we determine "English geographical thought" at the juncture of medieval and modern cartography and at the juncture of pre-modern and modern culture? It is the work of this chapter, and indeed of this entire book, to find means other than production that offer a glimpse into English geographic thought and its channels of transmission, looking instead to translation, importation, library holdings, English humanism, the popularization of scientific texts in early print, and book illustration. Informed by book history, this approach resists any teleological narrative of cartographic practices as leading inexorably to map-based representations and considers "mapping" and geographic thought broadly.[18] Geographical culture in England seems vacant only when viewed through the presumption of geographic representation as solely equated with cartography. Fifteenth-century geographic writing in England, in contrast, offers a hybrid, active articulation of geographic space, one not at odds with Ptolemy's own presumption of geographic information as intended to be continually added to and improved upon by his readers.[19] Tom Conley has proposed the writer as topographer of both the page and of national space, drawing attention to "cartographic writing" as an important conceptual reservoir of geographic expression in the early print period. According to Conley, "a new cartographic impulse" was linked to the relationship of self and space in early modern France.[20] In this regard, the cognitive processes associated with mapping precede and accompany the actual making of maps and, further, are evident in a variety of literary and popular texts of the period that are neither explicitly about mapping nor contain maps. A changing sense of the space constructs underlies and engages with the cartographic innovations of the fifteenth and sixteenth centuries.

In the absence of the production of cartographic objects—world maps, editions of Ptolemy's *Geography*, or globes—how was geographic knowledge disseminated in England during the European Ptolemaic revival? Though he is not known to have ever printed a Ptolemaic-based world or regional map, we must look to William Caxton, the English book trade, and English humanism as the context within which to trace the transmission of geographic thought in England. By investigating the forms of geographical culture that *were* present in late medieval and early modern England, what emerges is a hybrid knowledge construction highly integrated with literary forms and transnational exchanges. Geographic representation in text and image was still deeply influenced, both in England and on the Continent, by a poetic and religious conception of space. As such, the continental production of cartographic objects such as maps and globes did not exclude the potent continuity of classical and medieval geographic narratives nor the intermingling of scientific and poetic forms. As Kellie Robertson has noted, medieval scientific debates on the make-up of the human body "influenced late medieval vernacular poets in surprisingly specific ways."[21] In turn, that poetic traditions influenced geographical writing is evident in Florentine humanist and diplomat Francesco Berlinghieri's 1482 verse adaptation based on Ptolemy's *Geography*. In Berlinghieri's *The Seven Days of Geography*, the poet-narrator is at once a humanist scholar and geographical traveler with sensibilities that harken back to Macrobius' *Dream of Scipio* and Dante's *Divina Commedia*.[22] The division between scientific writing, including what has come to be the discipline of Geography, and humanistic or literary writing was far less sharply defined than modern disciplinary boundaries.[23] Humanists and cosmographers were not distinct categories of readers in the late fifteenth and early sixteenth centuries, and scholars such as Desiderius Erasmus and Gerardus Mercator, for example, were engaged in humanistic study and the production of cartographic editions. For English print culture, Coldiron points to Caxton as "the first of England's 'printers without borders,'"[24] a printer and bookseller deeply involved in the transformative process of translation and, I will argue, in the negotiation and presentation of new transnational literary and geographic influences in his editions of the *Description of Britain* and *The Myrrour of the World*. Caxton's apprentice and successor, Wynkyn de Worde, subsequently produced an extraordinary illustrated edition of Thomas Malory's *Le Morte Darthur* based on Caxton's edition that introduced world landscape imagery, influenced by early Netherlandish cartographic landscape painting, into early English print. Through the trajectory of narrative and non-cartographic visual geographic production by the print house of Caxton and de Worde, English readers actively engaged with the cognitive processes of the Ptolemaic re-organization of space.

William Caxton: humanism and geographical transmission in early English print

The key to the puzzle of the dissemination of geographic thought in England is at once transnational and local: first, geographical texts and artifacts were highly mobile and pan-European; and, second, Caxton is an unrecognized agent in the formation and dissemination of geographical thought in England. Printer of medieval romance, chronicle, poetry, and moral treatises, Caxton is not typically recognized for any role in the dissemination of new geographical knowledge in England or, for that matter, Renaissance humanism. Edmund Childs claims that Caxton's press "played no part at all in the spreading of Renaissance ideals in England."[25] Notwithstanding such comprehensive claims, two of Caxton's publications, the *Description of Britain* and *The Myrrour of the World*, [26] form a central core of an emerging geographic discourse in England, particularly for the vernacular reading audience Caxton cultivated. Caxton's 1480 edition of the *Description of Britain* is an adaptation of one section of John Trevisa's translation of Ranulf Higden's fourteenth-century *Polychronicon*. Caxton's 1481 and 1490 editions of *The Myrrour of the World* [27] provided a translation from a fifteenth-century French edition of the thirteenth-century *Image du Monde*, a work that included a world geography and universal history. The publication of these English-language narrative geographies by Caxton provides clear evidence of his estimation of his readers' interest in world geography. Caxton's *Myrrour* gives evidence of a process of augmentative and corrective geography, a spirit of knowledge-building that aligns with the spirit of Ptolemy's original scholarship in producing the *Geography*. Ptolemy approached scholarship as a process of knowledge-building and "expected people to go on refining" his geographic data.[28] Unlike an instrumentalist approach to geography, which portrays any given map as static, Caxton's narrative geographies are fluid, solicitous of emendation, and in dialogue with the reader in a pattern similar to that of his famed Preface to Malory's *Le Morte Darthur*, in which he invites readers to consider Arthur's historicity as they see fit, based on the conflicting evidence he has provided. Despite his dim view of Caxton's involvement with humanism, Childs does recognize the role that Caxton's continental connections in the book trade played in bringing humanist works to English readers. The international book trade in London, of which Caxton was a vital part, facilitated the dissemination of humanist-related books to an expanding reading community in England alongside the traditional medieval scholastic books sought by academic and religious libraries. Although university print shops met specific needs for their reading communities, the expanding public sphere of print in which Caxton operated did not quarantine the scholarly from the popular and contributed to the transnational exchange of books between England and Europe during the late fifteenth century.

William Kuskin argues that the logic of fifteenth-century literary production has been overlooked or misunderstood. It is a logic, he argues, of appropriation and reduction, a recursive dynamic rather than a static process of "material and intellectual consolidation."[29] Both the *Description of Britain* and *The Myrrour of the World* represent just such a dynamic process of appropriation and reduction, of geographic knowledge translated, or "reduced into English" as Caxton puts it, augmented with eyewitness accounts, and corrected based on new knowledge and hypotheses. While both texts ostensibly continue the medieval geographic narrative tradition, utilizing texts in circulation for well over a century, Caxton's selection of these particular texts, his translation and editorial framing, his illustrations, and his paratextual framework produce narrative geographies influenced by the new positional geography of the Ptolemaic revival.[30] Contextualizing Caxton's work in the Ptolemaic revival reframes the premise of English geographic isolationism by paying close attention to the reproduction and re-formation of geographical thought Caxton made available to English readers through his work as a vernacular humanist collaborator.

Humanism in England

Refuting an oft-repeated maxim that the Alps stood as a barrier to the dissemination of Italian-based humanism, David Rundle counters with his own maxim: "humanism traveled."[31] To disabuse the presumption of geographic containment, he argues that the cultivation of international reputations—through practices such as correspondence, intellectual friendship, book purchasers, and personal travel—was considered of central importance for aspiring Italian humanists, an aspiration which brought them or their work to cities across Europe, including England.[32] Sean Roberts examines the role of the beautifully illustrated 1482 edition of the *Geography* by Francesco Berlinghieri as part of a practice in which luxury books "came to materially connect geographically distant authors and readers."[33] Roberts traces the presentation of luxury editions of the Florence-printed *Geography* to two Ottoman princes, one in exile in Savoy, and therefore entailing a literal journey over the Alps, and the other in Constantinople. According to Roberts, through their use as diplomatic gifts, as prized books purchased by influential intellectuals, and their placement in renowned libraries, books such as the *Geography* "constituted diffuse and potent communities of readers."[34] Roberts also notes the "expanded economic exchange of visual culture" as part of the multinational material trade in books in the fifteenth century.[35] The *Geography* was a prestigious and important part of mercantile networks and exemplified the trade in "worldly goods."[36] The channels of production and circulation of cartographical and cosmographical materials were part of humanist and mercantile collaborations and reinforced a growing interest in geography and place-names. Thus, two strands of humanism—geographic writing and cartography—were inextricably (and literally) bound together.

In Roberto Weiss' long-influential work, *Humanism in England During the Fifteenth Century*, published in 1943, he suggests a perspective that became a commonplace assumption, that is, that English "geographical isolation and the conservative characteristics of its culture began to respond to the influence of humanism at a later stage than France."[37] Murray Newton Rothbard, however, identifies humanism as having arrived in England as early as 1435–1440 with Pietro del Monte, a collector of papal revenues in England and literary advisor to Duke Humphrey of Gloucester, "the first English patron of humanism."[38] Rundle argues that the opportunities for intellectual exchange with Italian humanists were more numerous than previously thought and that English readers interacted with Italian scholars via the lecture circuit of these scholars in England, via letters and literary works, and via travels to Italy to study with renowned humanists. In his reassessment of English humanism, Rundle provides the example of the Milanese humanist Stefano Surgioni who traveled between the university towns of Oxford, Louvain, and Cologne between the 1450s and 1470s "in all places presumably making his living by primarily teaching."[39] Cornelio Vitelli, the first professor of Greek at an English university, taught at New College, Oxford, in the 1470s,[40] and Duke Humphrey retained Tito Livio Frulovisi as his secretary.[41] Poggio Bracciolini, working in England a half-century earlier, from 1418 to 1423, famously lamented that in England there were "very few lovers of literature and those few barbarians, trained rather in trifling debates and in quibbling than in real learning."[42] But by the mid-fifteenth century, Bracciolini's bibliographically bleak years in England seem to have given way to a more vital exchange between English and Italian humanism. Kristian Jensen has noted that manuscripts of classical authors were prominent in the university libraries of Cambridge, including works by Cicero, Juvenal, Plutarch and Ptolemy.[43] In addition, Benjamin Weiss points out that in many instances, the *Geography* might not be evident as a self-standing text in a scholar's library, but that "fragments of the *Geography* or references to Ptolemaic geography show up with some regularity" in scholars' compendia, a phenomenon he compares to today's course reading packets.[44] Weiss argues that the structure of the *Geography* as a stand-alone text was unsuited to university curricula and that it was not until the popular print edition of Peter Apian's *Cosmographia liber* in the 1530s that the intellectual framework of Ptolemy's *Geography* became available in affordable print editions.[45]

While Rundle acknowledges that Italian humanists who traveled north of the Alps were a minority, he also argues that Italian humanists who did not physically travel were eager for their work to circulate to the "ends of the earth—that is England."[46] Pier Candido Decembrio, the Milanese humanist and bookseller, was eager to publicize his association with Duke Humphrey, and Flavio Biondo boasted that "highly embellished copies" of his work

Roma Triumphans were circulating in England, France, and Spain.[47] With his contacts in Italian humanist circles, Duke Humphrey directed Pier Candido Decembrio in 1451 "to acquire various classical Latin texts, including "Pomponium Melam et Ptolomei *cosmographium*."[48] That Ptolemy's *Geography* was in England is irrefutable, corroborated by references in library catalogs, probate records, and humanist exchanges.[49] Such evidence of international contacts led Rundle to conclude "that the *studia humanitatis* was less about the genius of place than mastery over space,"[50] suggesting that humanism, indeed, traveled.

Italian humanists, eager to cultivate their reputation and the international circulation of their work, contributed to the importation of books and new learning into England, including geographical texts. As European culture reimagined global spatial organization, humanist scholars were both cause and effect of this form of globalized thinking. Ecclesiastical councils were also an important venue for humanistic exchange, though such exchanges were not the primary purpose of such gatherings. The Council of Arras, for example, convened in 1435 to resolve the Hundred Years War and included scholars and scribes from a variety of European countries, including England.[51] By the mid-fifteenth century, as J.B. Trapp has argued, book importation for private and institutional libraries was augmented by connections made between Italian and English scholars with the result that "English professional scribes had learned to write competently and even elegantly, the new upright Italian humanist hand."[52] The 1577 edition of *Holinshed's Chronicles*, claimed a heritage of mid-fifteenth-century English humanists such as John Free and John Gunthorpe who "went into Italie, where hc heard that eloquent learned man Guarinus read in Ferrara."[53] The *Chronicles* also cites as evidence of England's learned tradition "Iohn Tiptoff, a noble man borne, a great traueller, excellentlie learned, and wrote diuerse treatises."[54] These references a century later indicate the lasting cultural impact of these fifteenth-century English scholars whose intellectual exchanges and "feet-on-the-ground" travel connected England's intellectual life with that of the Continent throughout the fifteenth century.

International commerce also played a critical role in the dissemination of humanistic texts and ideas. William Gray (1414–1478), later Bishop of Ely, traveled to Italy as a student in Padua and Ferrara and subsequently served as the English royal proctor in Rome.[55] Rundle speculates that Gray supported his lavish scholarly and diplomatic lifestyle via bank transfers through the Venetian banker Giovanni Marcanova, who had previously spent time in England. Merchants carried books between Italy and England and served as diplomatic and financial couriers, facilitating humanistic contacts.[56] Rundle cites English cleric William Sellyng's correspondence with the Italian merchant Umfredo Gentili, regarding a copy of Livy's *Decades* as yet another example of a merchant functioning as what Rundle terms a "humanist collaborator."[57] Such a definition provides the vocabulary for the role of London's preeminent printer, William Caxton. That Caxton facilitated contacts with printers and book buyers is undeniable; his own translation output, as Anne F. Sutton argues, was prodigious, and few English printers embraced "the new art" of printing as enthusiastically as did Caxton.[58] Printing of the *Geography* began first in Italy, then in German-speaking cities where Caxton had lived and established professional networks. Caxton's earlier stint in Cologne, where he learned the technical aspects of the print business, as well his considerable network from his years as Governor of the English Nation of Merchant Adventurers in Bruges no doubt seeded his awareness of geographic printing as editions of the *Geography* began to emerge from presses from Rome to Ulm during this time. Caxton was England's first printer and premier book importer, and his ties to European book culture were extensive. Jennifer Goodman presciently invokes a geographic metaphor for Caxton's role in England literary culture vis-à-vis the Continent when she argues that "William Caxton's Continent is still in many ways an unexplored territory."[59]

Continental Caxton: humanist collaborator

Caxton's reputation has waxed and waned over the decades, from noble literary hero to workmanlike clod. N.F. Blake suggests a dim view of Caxton resulted

> partly because he played such a small part in the intellectual and economic life of the country and partly because there has been a natural antipathy toward biographical enquiry toward people of his rank in the enquiry into wider social, constitutional, and political issues.[60]

Fortunately, less dismissive studies of Caxton's class and role in shaping English literary culture reject the once-common projection of Caxton as a "zone of corruption."[61] Despite mid-twentieth-century representations of Caxton as a printer of provincial taste with a crass focus on selling books, Goodman locates Caxton at the nexus of Anglo-European connections and "the most cosmopolitan of English literary figures."[62] As evidence, she cites Caxton's 32 years working on the Continent and the multilingual environment of Bruges as formative experiences that he incorporated into his printing, bookselling, and writing practices once back in London, a consummate "cultural amphibian."[63] Given Caxton's Burgundian connections,[64] it is also likely he directly saw a manuscript copy of Ptolemy's *Geography* in the renowned libraries of Louis de Bruges and of the Duke of Burgundy, no doubt a luxury edition.[65] Caxton's vernacular multilingualism (if limited Greek and Latin) and extensive network created the conditions for humanist collaboration that flourished outside the Latin-oriented textual culture of the universities and monasteries.

Caxton's prose romances draw from Belgian, Swiss, and Provençal source texts. In terms of geographical sweep, Caxton's six major prose romances take the reader from the English setting of King Arthur's court to the continental locales of Genoa, Venice, Prussia, and Norway (in *Blanchardyn and Eglantine* and *Paris and Vienne*) as well as referencing locations in the East, including Jerusalem, Babylon, and the mythical land of Prester John. As Goodman suggests, Caxton's prose romances are "superlative guides to the cultural geography of their day, supplying us with a copy of the writer's map of the world."[66] The geographic structure of such imaginative literature precedes the more instrumental "map consciousness" ascribed to military and governmental interest in strategic cartography during the reign of Henry VIII.[67] This attentiveness of Caxton's continental prose romances to geographical locales, his apparent plan to publish an edition of Mandeville's *Travels*, [68] and his editions of the *Description of Britain* and *The Myrrour of the World* suggest Caxton was both respondent to and agent of a growing interest in the new geographical framework of the Ptolemaic revival. Caxton's editions constitute an unrecognized populist strand of geographic writing in England, an expression of geographic curiosity across a range of imaginative, pseudo-scientific, descriptive, and philosophical texts which expanded the sense of "the world" far beyond the English island nation. Studies in book history and Renaissance humanism often draw a sharp line between chivalric reading communities and humanist reading communities, artificially delineating a fragmented book culture in contrast to the way in which humanist texts, prose romances, "Kalenders," and vernacular translations sat spine to spine in the vats and fardels that brought continental book culture to English readers.

Caxton was not a printer of scholarly, humanist texts, an endeavor that relied on a requisite professional knowledge of Latin. Such books met a relatively small academic market, the result of which was that many humanist printers on the Continent struggled financially. As a printer, there would have been little need to take such risks when learned book buyers in England could readily purchase Latin books printed on the Continent. Rundle points out that commercially-successful printing houses "favored works less learned and more popular than the *studia humanitatis* – as several early printers learned … only when the specter of bankruptcy overshadowed them."[69] Lienhart Holle, for example, went bankrupt after printing his 1482 deluxe edition of the *Geography*, complete with dazzlingly blue seas on the accompanying maps.[70] Over a century ago, an exasperated Gordon Duff, responding to the charge that Caxton's press failed to produce humanist works, responded, "Of what use would it have been to us, or profit to our printer, to reprint editions of the classics which were pouring forth from foreign presses, and even there, where most in demand, were becoming unsaleable?"[71] The foreign stationers in London readily provided Latin classics to interested London readers and could certainly have done so for a print edition of Ptolemy's *Geography* sought after by an English bibliophile. With its Latin text, the visuality of its maps, and the trans-lingual nature of its data tables, there was little need for Caxton to produce a text that needed no translation and that had a dismal record of financial success for its continental printers. The technical demand of printing copper-etched maps would have only added to Caxton's rationale for not printing an English version of Ptolemy's *Geography.* While print technology is acknowledged as central to the development of Renaissance cartography,[72] Caxton's narrative geographical editions have not been recognized as part of this symbiotic relationship, an omission this chapter seeks to redress. Caxton's selection and editing of geographic narratives provide the material embodiment of a "vibrant and hybrid" geographical culture as expressed in English popular print.[73] Roberts has argued that the intellectual history of the transmission of Ptolemaic-based geographic concepts is forged by the "movement of books."[74] And no one in late medieval England moved more, or translated, or printed, or sold more books that William Caxton.

Robert Wilson's determination that Caxton's contact with humanism "is small"[75] is reinforced by Raphael Lynn's lament that the first translation of Ovid that emerges "from the cradle of printing" in England "is not a humanist triumph of classical learning but William Caxton's verbatim version of a French redaction of the *Ovid Moralisé.*"[76] Thus, an imposed hierarchy of translation practices sequesters Caxton's output from the work of new learning. Crofts, however, points to Caxton's Preface to *Le Morte Darthur* as evidence of a highly conscious "rhetorical modulation" based on fifteenth-century practices in the genre of the *Historia.*[77] Though not a university-trained scholar, Caxton was a multilingual prose writer steeped in some of the same rhetorical practices as his humanist counterparts. Caxton was what we might today call a promoter of the public humanities, an endeavor still slightly suspect in academia. Caxton published a body of work related to humanists' interests, including *The Dicts and Sayings of the Philosophers* (a popular compendium of philosophical proverbs and maxims attributed to Homer, Socrates, Plato, Galen, and Ptolemy, among others),[78] Ovid's *Metamorphoses,* Laurentius' *Latin Treatise on Rhetoric,* Chaucer's translation of Boethius' *Consolatio Philosophiae,* Cato's *Dicts,* Cicero's *Dei Amicitia,* as well as a French/English vocabulary. In 1490, Caxton published *Eneydos,* his translation of the *Aeneid,* complete with his usual preface that sets forth his translation practices.[79] To this list we must add Caxton's translation and editing of *The Myrrour of the World,* which drew on classical authors such as Isidore of Seville, Alexander Neckam, and Ptolemy, as well as his edition of the *Description of Britain,* drawn from Higden's *Polychronicon,* a title meaning "many stories." These works do not classify Caxton, strictly speaking, as a humanist printer but they certainly constitute a significant engagement with humanist texts and translation.

Caxton cultivated connections with two well-known English bibliophiles: John Tiptoft, Earl of Worcester, and Anthony Woodville, Earl Rivers. Tiptoft was both a lover of Italian humanist books and a notoriously cruel political operative of Edward IV.[80] Tiptoft utilized the services of the Florentine bookseller Vespasiano du Bisticci to procure volumes and also translated Cicero's *De Amicitia,* a work printed by Caxton in 1481 in a volume entitled

Of Old Age, of Friendship, of Nobility.[81] Caxton maintained business connections and, apparently, a friendship with Tiptoft, authoring Tiptoft's eulogy. Woodville, also an accomplished translator and bibliophile, was closely connected with the court of Edward IV as well as with Caxton's former patroness in Burgundy, Margaret of York. Woodville was translator and patron of Caxton's edition of the *Dicts and Sayings of the Philosophers* and had himself traveled to Rome to meet the scholarly Pope Sixtus IV.[82] Despite his own facility in translating from French to English, Woodville requested that Caxton review his translation of the *Dicts*, a review that resulted in Caxton's restoration of the final misogynistic "Socrates" passage which Woodville had omitted. Coldiron suggests Caxton's addition stands as "an extreme example of the authority Caxton transfers to the reader,"[83] a propensity of Caxton's to provide multiple sources for the reader's analysis. These associations with English bibliophiles confirm that sequestering Caxton from the intellectual force of European humanism is off the mark; rather, the translation and publication practices of Caxton situate him as one of England's most prominent humanist collaborators during the height of the Ptolemaic revival.[84]

Ultimately, Crofts puts the question of Caxton's humanist "credentials" aside and argues that Caxton's prose in his prologues and epilogues serves as evidence of the "sophistication of Caxton's critical sense."[85] Within the framework of a mercantile humanist collaborator, we ought to consider Caxton's interaction with humanist texts, scholars, and patrons without a concomitant condemnation of his commercial practices. Caxton, we may agree, was no Desiderius Erasmus, but his cultural and intellectual role in the dissemination of geographic thought in England is irrefutable, particularly his editions of the *Description of Britain* and *The Myrrour of the World.* As yet, Caxton's role in the material and discursive practices of geographic writing has yet to be explored. Kathleen Tonry suggests that "Caxton offers an innovative historiographical practice," in his edition of the *Polychronicon*, one defined by his "juxtaposition of sources,"[86] a practice common in geographical humanism. Caxton's geographic editions utilize an amalgamation of geographic information from his base texts, additional textual sources, and individual eye-witness experiences, and Caxton presumes the reader's capacity to discern among multiple authorities as part of an active rather than passive reading practice. Caxton's geographical editions, closely linked as they are to his historiographical practices in his edition of the *Polychronicon*, rely "upon links drawn by the reader around the juxtapositions Caxton shapes."[87] Caxton's geographical editions exhibit both oscillation and accrual, attributes that distinguish the geographic aesthetic of the late fifteenth century and that frame an interpretive space for the reader. In Caxton's geographic editions, the medieval and the Ptolemaic mutually constitute a polyvalent expression of early modern geography, one of capaciousness rather than insistence on a single authority. Certainly, Ptolemy himself would have approved of this process, one that called upon readers to incrementally assess and add new knowledge. The foregrounding of readerly interpretation in fifteenth-century geographic culture is also part of Caxton's editorial practice and evident in his geographic editions of the *Description of Britain* and *The Myrrour of the World.*

Of books and maps: importation and transmission

As Paul Needham has argued, "The importation of early printed books into England was not an interesting sideline, but a primary factor in the history of the English book-trade."[88] He notes that while the scale of the book trade is "difficult to gauge,"[89] the best source of evidence is that of the customs rolls, kept for purposes of taxation by the Crown. Needham documents a fascinating history of book entry into and out of England in "chests, coxers, fardels, hogsheads, maunds, pipes, trusses and vats,"[90] not only as a result of William Caxton's prodigious import/export activities but also due to the number of foreign booksellers active in the London trade. While the book trade between England and the Continent was a significant commercial activity, Margaret Ford points out that the custom rolls indicate the volume but rarely detail the exact titles.[91] Italy, in particular Venice, dominated the early trade in printed books with England.[92] In Venice, where "nearly a quarter of all publications in Europe were produced," publishers such as the famed Aldine Press contributed to a wide range of humanist texts, including subjects such as geography and pilgrimage.[93] Geographical texts produced in Venice include the 1486 edition of Bernhard von Breydenbach's *Peregrinatio in terram sanctum* (an illustrated account of a pilgrimage from Venice to the Holy Land),[94] a 1477 edition of Periegetes Dionysius' *De situ orbis,* and a 1497 edition of Ponponius Mela's *Cosmographia.* Given the production of such texts and maps in Venice, the "vast distribution network" of Venetian booksellers across the continent,[95] and London's mercantile network with Venice, the importation of Venetian geographical books and maps into England seems a reasonable assumption, though impossible to itemize. By 1520, as a result of Reformation-Catholic tensions and the influence of the continental humanist Desiderius Erasmus in England, Basel surpassed Venice as the primary source of imported books into England.[96] A great friend and correspondent of English humanist Thomas More, Erasmus edited the first Greek-language edition of the *Geography,* printed in Basel in 1533. For Erasmus, as with humanist scholars generally, the connection between history and the identification of classical and contemporary place names was a major focus of humanist historiographical and geographical scholarship.[97] Time and place, ultimately, are also the basis of cartographic projection and formed the raw data of Ptolemy's locational tables, that is, the calculation of latitude and longitude coordinates based on the length of the day at various points along the equator. It is through the book trade that networks of humanist geographical thought were established between London and continental centers of book production, and, as discussed in Chapter 3, the study of geography became an integral part of early Tudor historical and moral education.

Needham estimates there were over 150 book importers and at least 600 cargoes of books in the almost 30-year period between 1462 and 1490 to English ports of call and, of interest to this study, an additional cargo that included dozens of "kalenders,"[98] no doubt the proto-geographical texts known as the *Kalender of Shepherds* (see Chapter 5). The market for these early "calendars," primarily from French printers, attests to an emerging interest in natural or prognosticative astronomy in early English book culture, a phenomenon that will flourish over the next century in the form of inexpensive print editions of the *Kalender of Shepherds* and *The Compost of Ptholomeus.* Needham concludes that "readings of astronomical, and subsequently geographical, material" created a trajectory of scientific reading[99] which, along with romance and chronicle, formed part of the transmission of Ptolemaic-influenced geographic culture in England.

Part of the difficulty in tracing the dissemination of new geographic thought during this period is that the early map trade was virtually synonymous with the early book trade, with the majority of extant maps included as accompaniments to books. Maps that were not part of books were subject to stronger deterioration pressures and for these we have scant documentary evidence. While customs records list the importation of barrels of books by booksellers such as Caxton, they typically do not specify titles, nor whether a book was illustrated or accompanied by maps. For example, the records of the London Subsidy Roll show that on 25 April 1488, Caxton imported 1,049 books, but the exact nature of those books is unknown. The London Subsidy Roll for the period of the 1470s through 1491 only survives for the Michaelmas year 1487–1488, evidence that the surviving records are slim, indeed.[100] Geographic objects such as continentally-produced globes are only cursorily documented in wills and library holdings and therefore difficult to trace in England during this period. But even such slim evidence points to the fact that the importation of books is key to unlocking the transmission of geographic thought in England during the Ptolemaic revival in a context of "literary globalism."[101] In addition to importation, vernacular translation provided a wide avenue for transmediation, all part of Caxton's effort "to bring foreign works to new English readers."[102]

In tandem with importation records, studies of library holdings also shed light on English engagement with Continental works through catalog listings of religious, scientific, philosophical, humanist, and literary books published on the Continent. *The Corpus of British Medieval Libraries* itemizes the holdings of individual academic and monastic libraries, including the major universities and religious institutions such as Dover Priory or Syon Abbey.[103] David Bell notes that the "vast majority" of books in religious houses came from the Continent and that while the "monk scholars preferred the works of authors who belonged to an earlier tradition: writers like Bede, Hrabanus Maurus, the Victorines and Stephen Langton,"[104] some monks studied a wider range of topics than the traditional liberal arts curriculum, and included "mathematics, astronomy, French, and occult pseudoscience texts."[105] It is clear from these records that Ptolemy's *Almagest* and *Tetrabiblos* (typically listed as the *Quadripartitum*) were among the Latinate book collections of university scholars[106] and that an interest in astronomy, especially for the purposes of the establishment of church calendar of feast days, is evident. Copies of Ptolemy's *Cosmographia* (the Latinate title of the *Geography*) are present, though less frequently than the *Almagest*. In Christ's College, for example, a copy of the 1482 Ulm edition of the *Cosmographia* is listed among the holdings as well as a second copy, possibly the 1511 Venice edition, both of which were published fully illustrated with maps.[107] A similar listing occurs in the records of Corpus Christi College. Based on such holdings, it is irrefutable that Ptolemy's *Geography*, with accompanying maps, made its way into learned English book culture, if only in small numbers.[108]

The holdings of British libraries during the period of the Ptolemaic revival show that foreign importations met the demands of a learned market "which extended beyond the universities,"[109] a market that would have included the customers of William Caxton. His edition of the *Description of Britain* and *The Myrrour of the World* played a foundational role in vernacular early print culture in England and met an increasing demand for geographic texts by a non-academic audience. Subsequent print editions of the *Kalender of Shepherds* by all the key English printers, including Caxton's apprentice, Wynkyn de Worde, became a primary trajectory for an even broader popular dissemination of mathematical astronomy or pseudo-science, one that, ultimately, traces popular geographic thought in England. Although the term pseudo-science may sound pejorative, it equates with what today is called popular science for a general reading audience. As part of this vernacular book culture of importation, translation, and print, Caxton's *Myrrour* provides a key to English geographic reading in the polyglot, internationalized trade of the Ptolemaic revival and to the cache of Ptolemy as a symbol of this new learning.

"the discripcion of this said Ile of Britayne with the commoditees of the same": Caxton's *Description of Britain*

Printed in August of 1480, Caxton's *Discripcion of Britayne* [110] has not received a great deal of critical attention, I suspect, because it seems to be a relatively minor printing of just one section of the *Polychronicon*, a text he subsequently printed in full, two years later.[111] The *Polychronicon*, composed in Latin in the 1340s by the English monk Ranulf Higden, was subsequently translated into English in 1387 by John Trevisa. It is Trevisa's translation that Caxton uses as the basis of his print editions of the *Description of Britain* and the *Polychronicon*. A chronicle and universal history in seven books, the *Polychronicon* includes a geographic description of the regions of the world, including Britain. The geographical context of the work includes the iconic *mappamundi* for which Higden's *Polychronicon* is renowned, accompanying 19 of the extant manuscripts (Figure 2.1).[112] From Book I, the geographical section of the *Polychronicon*, Caxton culled the text he subsequently entitled the *Description of Britain*.

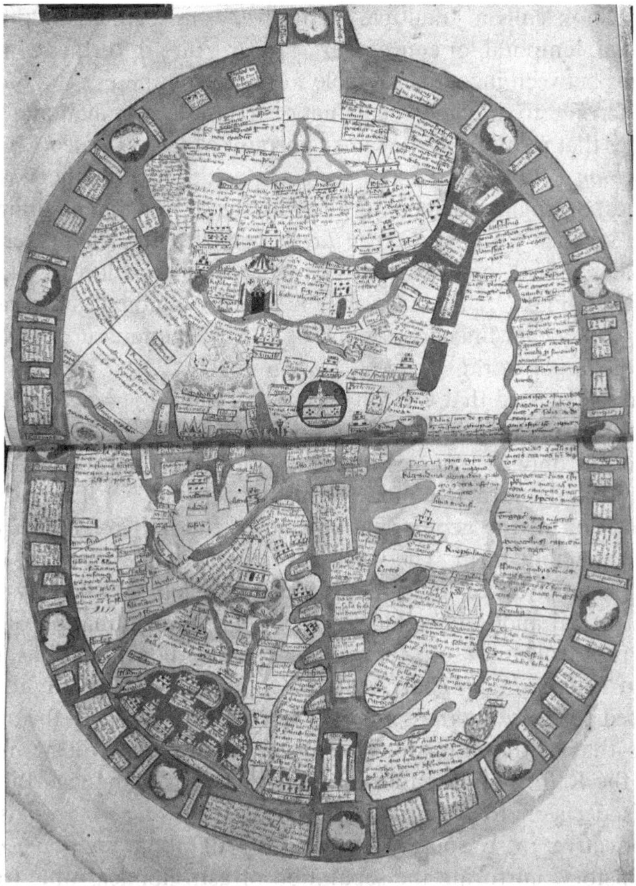

Figure 2.1 Ranulf Higden, *Polychronicon*, c. 1400
Source: © British Library Board, Royal 14 C. IX, ff.1v-2.

Lavezzo has argued that England's continued fascination with the *Polychronicon* and its sub-section, the *Description of Britain*, shows a deeply conservative geographic imagination, indicative of England's "investment in the old global perspective."[113] She argues that Caxton's 1480 adaptation "highlights and expands upon the traditional perception of England as an angle of the world or even otherworld."[114] The popularity of the *Polychronicon* is attested to by the seven editions from 1490 to 1528, including a 1510 edition by Richard Pynson, the King's royal printer, that includes both the full *Polychronicon* as well as Caxton's *Description of Britain*, indicating its status as a stand-alone text. Lavezzo suggests "the notion of English global marginality so stressed by Higden requires the adoption of a premodern perspective on the *entire* world."[115] However, like Caxton's "bestseller" the *Legenda Aurea*, a compilation of saints' lives and miracle narratives, the *Description of Britain* structures what

Catherine Sanok calls a "multipart narrative," one that links "differences in scale—spatial, temporal, or conceptual … to structure or further complicate the relationships between incommensurable forms of community."[116] Sanok is specifically referring to implied local, national, and "supranational" Christian communities that result from the saints' narratives in *The Golden Legend.* But in the *Description of Britain*, Caxton similarly seeks to link the wonders (and commodities) of English regional locales with the supranational global sphere of geographic wonder. Caxton's *Golden Legend* enabled English readers to "think about local, regional, and national communities in relation to supranational religious formations."[117] So, too, are the specific regional geographies in the *Description of Britain* linked as exemplum of the supranatural collective geography of the larger world.

Tonry has suggested that Caxton's edition of the *Polychronicon* functions as "a capacious form" one that "relies on an audience knowledgeable about and sensitive to not only the sources Caxton incorporates but to their amalgamation as a distinctly attractive mode of history writing."[118] The publication of Caxton's edition of the geographic portion of the full *Polychronicon* materializes this capacious form. The work participates in the aesthetic of multiple sources and reader-based interpretation that characterizes Caxton's approach to both historical and geographical content. The amalgamation of sources as a reading protocol draws out the "reader's active judgement," as Tonry suggests, in an "increasingly reader-centered context."[119] In this light, such multiply-sourced texts are quintessentially early modern, shifting the authority of the text to that of the individual reader, a practice that will be mirrored, of course, in the Reformation publication of the English-language Bible.

Caxton's decision to excerpt a section of the *Polychronicon* is, in itself, highly innovative: in so doing, he created a geographic text *independent* of a universal history, mirroring the separation of astronomical time from historical time in Ptolemaic maps.[120] The continued production of the *Polychronicon* shows the way in which universal history, to be sure, retained its traditional force in England, but the *Description* must also be seen as fundamentally *a new form* complementary to Ptolemaic maps. In selecting the relevant regional geographic chapters of the *Polychronicon* to form a new text, the *Description of Britain*, Caxton affects a separation between the long-standing classical and medieval tradition of pairing universal history and geography, a paring visually represented in *mappamundi.* Emily Steiner has argued that in writing a universal history comprised of a geography (Book I) and a history (Books II–VII), Higden linked the genres of geography and history through "the traditional analogy between the natural world and the body of man."[121] The act of separating out Book I of the *Polychronicon,* however, announces Caxton's sense that the genre of geographic writing could be distinct from a Christian universal history in the same way as the maps of Ptolemy's *Geography.*

Caxton's geography-focused adaptation of the *Polychronicon* does not trade on the marginality of England—far from it. Trevisa's translation of Book I of the *Polychronicon* narrates the geography of the known world concluding with a narrative of English geography. However, in the *Description of Britain*, the pairing of England and Rome is now the starting point rather than the conclusion of the geography section. Whereas the *Polychronicon* begins with an account of Julius Caesar, Caxton begins with the etymology of "Albion" along with a condensed version of the legendary Roman origins of British settlement. In so doing, Caxton's *Description* enacts a geo-textual transference of power from Rome to England as he sets forth a newly reordered "discripcion of this said Ile of Britayne." Caxton's foregrounding of the unique geography of Britain positions England as both a new Eden and a new Rome. Far from positing a marginalized nation, Caxton's edition is aimed at a cosmopolitan English readership attentive to the forging of a geographical and imperial possibility for a globalizing English nation. With the publication of the *Description*, the textual representation of the English nation now functions as stand-alone spatial and vernacular geographic narrative, one that claims a British nation inclusive of Wales, Scotland, and Ireland. If, as Tom Conley argues, "'description' is taken in the figural sense of topography or chorography," then Caxton is England's first printer-topographer, defining national space for fifteenth-century English readers.[122]

Perhaps not coincidentally, Ptolemy's *Geography* also begins its narrative description of the known world with *Britannia*, rather than Rome. Citing his choice of the best projection for making maps, Ptolemy argues that it is best to work from northern latitudes to southern, and from western to eastern "since to the eye of the writer or reader the northern localities appear in the upper part."[123] Consequently, Ptolemy begins with *Britannia:* "let us begin our particular descriptions with the western part of Europe according to its provinces or prefectures."[124] Ptolemy begins with the north coast of Wales and, as is his custom, lists coastal cities first, from north to south, and then lists interior cities in the same north to south pattern. It is possible that Ptolemy's north-to-south positional description of the known world inspired Caxton to set aside earlier depictions of England as "an angle" of the world and to create, instead, an excerpt of the *Polychronicon* that follows the practice of Ptolemy's *Geography.*

In the *Description of Britain*, Caxton also inverts the order of the geographical passages about the British Isles as found in Trevisa's text. In Caxton's *Description*, the chapters about England are relocated, instead, to the premier position, followed by the sections on Wales, Scotland, and Ireland, now as English territories. A.S.G. Edwards identifies Caxton's reordering as indicative of his astute awareness of his book-buying audience: the foregrounding of England as the prime meridian, so to speak, placed "the priority where he felt his London customers would want it."[125] In his Epilogue to the *Description*, Caxton explains his rationale for altering the order of his source text: "bicause Irlonde is vnder the reule of englond & of olde tyme it hath so continued therfore I haue sette the descripcion of the same after the said brituyne."[126] Geography, he makes clear, is political space. In producing the

Description of Britain, Caxton redefines British national space and elides or mystifies any former sovereignty of the Irish, Welsh, and Scots, constructing a new territorial space. Caxton also smooths out Higden's chunky citational practice, updating Trevisa's fourteenth-century translation into fifteenth-century prose as well as modernizing the lexicon, syntax, and orthography.

In contrast to Higden and Trevisa, Caxton's *Description* links geography not with universal history but with trade. In his Epilogue, Caxton proposes that his new edition will be of interest to English readers "bicause it is necessarie to alle englisshmen to knowe the propretees co[m]moditees & meruailes of them / therefor I haue sette them in enprinte according to the translacion of Treuisa."[127] This interest in commodities forms the basis for Caxton's production of national identity, and is suggestive of his long experience in international trade: as a mercer, as Governor of the English Nation in Bruges, and as a bookseller. Caxton makes a similar connection between geography and commodities in his brief prologue to the work: "Therfor I entende to sette in this booke the discripcion of this said Ile of Britayne with the commoditees of the same."[128] Caxton offers a rationale for his publishing of the geographic excerpt in terms of the lack of geographic description generally available:

> Hit is so that in many and diuerse places the comyn croniclez of englond ben had and also now late enprinted at westmynstre And for as moche as the descripcion of this londe whiche of olde tyme was named albyon / and afar Britayne is not descriued ne comynly had / ne the noblenesse and worthynesse of the same is not knowen.[129]

Caxton printed his edition of the *Chronicles of England* in the same year as the *Description* (1480) so he was well positioned to articulate a potential market for a "descripcion of this londe" that an interested readership, up to then, had not "comynly had." As Caxton well knew, chronicles were commonly had among the widening circle of London book buyers, including merchants and other non-academic readers,[130] but these chronicles did not include a comprehensive geographic description of Britain.[131] Caxton's instinct that English readers would find this new form of interest was well-founded: in addition to his edition of the *Description* and his full edition of the *Polychronicon* in 1482, Wynkyn de Worde brought forth an edition of the *Description* in 1498, attesting to the formative role of Caxton's press in shaping English geographic discourse. Like Arthur's symbolic vanquishing of Rome in Malory's *Le Morte Darthur*, Caxton's publication and re-chaptering of the geography section of the *Polychronicon* for his edition of the *Description of Britain* challenge the premise of England's sense of geographic periphery and suggest the cultural rupture that will come to define Reformation-era Anglo-Roman ecclesiastical relations. The publication of the geographical matter from the *Polychronicon* as a separate text and Caxton's paratextual shaping of that text produce a new native-based geographic expression that asserts England's centrality, not its marginality, and claims the spatial metaphor of imperial first nation in the emerging genre of secular geographic narrative.

The Myrrour of the World

Caxton's 1481 publication of the geographic encyclopedia, *The Myrrour of the World,* [132] reprinted in 1490, exemplifies the process of readerly construction engendered by Ptolemy's *Geography* during this period of intense redefinition of the known world. Kuskin points out that while early print culture may stage "a series of disabling paradoxes," English literary history "is constructed by a variety of readers, each consolidating previous readers' efforts in different material conditions."[133] This view of English literary history, corroborated by Tonry's analysis of Caxton's historiographical praxis, emphasizes a process of consolidation and addition rather than replacement and sets the stage for a reappraisal of Caxton's edition of *The Myrrour of the World* as part of a dynamic and hybrid geographical framework rather than a negative exemplum of an outdated pre-modern world-view. Although notable as the first printed book in England to include images, the reputation of Caxton's *Myrrour* has suffered from critical judgment of the work as a retrograde medieval publication. Citing the *Myrrour* as evidence of Caxton's intellectual indiscrimination, H.S. Bennett states that Caxton made "no attempt to bring the *Myrrour* up to date."[134] George Kimble similarly criticizes Caxton's *Myrrour*, arguing that he "printed one hundred year old information as if it were current."[135] This conception of a distinction between "old" and "new" scientific information, however, fails to recognize the aesthetic context in which Ptolemy's *Geography* came into print. Roberts describes fifteenth-century geography as "a wide-ranging discipline" that included geography, ethnography, history, myth, morality, and ethics.[136] As part of this milieu, Caxton's *Myrrour* is a comprehensible project, one in which the combination of old and new information was a vital enterprise, one "understood as geo-*graphic*—that both inscribed and pictured the earth."[137]

To set Caxton's editorial methodology for the *Myrrour* into a broader context, Johann Muller (more widely known as Regiomontanus) printed a pared-down version of Ptolemy's *Almagest* in 1462 in Venice that also integrated segments from other astronomers and ancient scholars from the Middle East.[138] While Regiomontanus was a humanist scholar, his methodology of utilizing an ancient text with interpolations from other sources provides insight into Caxton's vernacular approach. Moller characterizes the work of the early printers of astronomical and scientific texts as a process of "assimilation and correction,"[139] the very methodology which Caxton employs for his edition of the *Myrrour.* As Roberts observes about Francesco Berlinghieri's 1482 Florentine edition of the *Geography,* written in *terza rima*, geographic culture embraced the "combination of the novel and the archaic."[140] The popularity of the *Myrrour* and, later, the *Kalender of*

Shepherds belies the premise of a phlegmatic English geographic reading culture and indicates a growing lay readership interested in and curious about astronomical, astrological, and geographical material. As such, Caxton's *Myrrour* deserves a reappraisal as a hybrid and discursive geographical text—one that links English readers' interest in traditional works of universal history with texts that integrated positional geography, inspired by the cultural cache of Ptolemy's *Geography.* The second printing of the *Myrrour* speaks to continued interest in the subject matter of Caxton's edition by English readers and a willingness to tackle the interpretive process he requested.

There is a significant interconnection between humanist works and astrological/geographical texts, as Gautier-Dalché has argued, constituting a body of work of which the *Myrrour* must be considered a part. Gautier-Dalché stresses that Ptolemy was not read solely as a scientific writer in the fifteenth century, but simultaneously as an astrologer, astronomer, and moral sage both in England and throughout European humanist circles.[141] The first dated book printed in England is Caxton's edition of *Dicts and Sayings of the Philosophers,* a compendium which included quotations ascribed to Ptolemy's *Tetrabiblos,* his treatise on natural astrology.[142] Although Ptolemy's reputation is now established primarily via the *Geography,* his most widely published work available in England was his *Almagest* in which, as Charles Burnett points out, "the teaching and study of the exact science of mathematical astronomy [have] a moral purpose."[143] Such a moral grounding for the study of astronomy is the shared context of Caxton's *Myrrour,* a milieu of moralized geographical reading all but inscrutable to later scientific rationalism and yet an integral part of cartographic writing in the European Ptolemaic revival.

Caxton's edition of *The Myrrour of the World* is based on *Image du Monde,* translated into French from Latin in 1245 either by Gossouin, or possibly Gauthier de Metz.[144] Caxton, it seems, came across a French prose translation of the text while living abroad in Bruges. Royal MS 19A IX, dated "Bruges 1464" and now held by the British Library, is considered "beyond a doubt" to be Caxton's exemplar.[145] Caxton states that he translated the text from French into English at the behest of Hugh Bryce, a fellow Mercer and holder of various governmental offices in London, including Keeper of the King's Exchange and Governor of the King's Mint in the Tower.[146] *Image du Monde* is a compendium of classical geographic thought drawn from a variety of sources, including Isidore of Seville's *Etymologiae,* Alexander Neckam's *De Naturis Rerum, De Laudibus Divinae Sapientiae* and Ptolemy's *Almagest.* Organized into three parts, an organization Caxton follows in his edition, *Image du Monde* is comprised of a sacralized history of the creation of the world (including a description of the seven liberal arts), a geographical section (including the standard marvels of medieval travel narratives), and a third section on astronomy drawn from Ptolemy's renowned astronomical work, the *Almagest. Image du Monde* was a popular compendium on matters ranging from the excruciating experience of souls in hell to the existence of elephants in "Inde;" from calculations of the circumference of the Earth to a discourse on why lightning is seen before thunder is heard. The thirteenth-century version of *Image du Monde*, it is worth noting, was a poetic text: a lengthy, rhymed-verse rendition of geography and universal history. While Caxton's *Myrrour* is an English prose translation, it functions as a genre similar to that described by Cosgrove as the legacy of the classical tradition, a "hybrid genre of whole-earth description at once scientific, poetic, and geographical."[147]

To say that Caxton "merely" translated *Image du Monde* is not only incorrect, but also misses the textual nuances of his edition as well as the hybrid context of the Ptolemaic revival. Caxton does more than merely produce a word-for-word translation of the *Myrror*: he provides an expanded preface that is twice as long as that of his source[148] and makes strategic editorial interventions. W.J.B. Crotch argues that Caxton's "enthusiasm for the work was unbounded" and that, via his additions to the prologue, Caxton urges the work be read carefully, not in haste, so that the reader may "vnderstonde it right well."[149] In the opening to Ptolemy's *Geography*, Ptolemy states that it is "the prerogative of Geography to show the known habitable earth as a unit in itself, how it is situated and what is its nature."[150] Similarly, in his prologue to the *Myrrour*, Caxton urges the usefulness of the book for readers wanting to understand "the creacion of this world, the gretnes of the firmament and lytilnes of therthe in regard of heuen, how the vii sciences were ffounden and what they bee." Ptolemy credits the "exquisite accomplishment of mathematics to show all these things to the human intelligence," but also considers it "fitting at the outset to put forth that which is the first essential, namely a reference to the history of travel, and the great store of knowledge obtained from the reports of those who have diligently explored certain regions."[151] Caxton sets forth a similar intention of learning for his readers, that they "may lerne, and that lerned to reteyne, and that reteyned so teche."[152] Caxton's pedagogical enthusiasm speaks to a growing lay literacy and interest in learning as well as a heightened interest in geography as a branch of knowledge. Three times in his continuation of the prologue, Caxton references the "fourme and condicion of this worlde," that is to say, geography. Oliver H. Prior, editor of the 1913 edition of Caxton's *Myrrour*, observed that while Caxton for the most part "keeps strictly to his text," the places where he intervenes as editor are precisely at passages in which Caxton feels compelled to express "surprise at some geographical eccentricity of the French work."[153] Suggestions that Caxton was merely repackaging dated material are at odds with the passion Caxton expresses for the subject matter of the book when he presses his readers to "rede this said volume treatably, auisedly, and ordynatly" and his hope that the reader will "lerne grete partie of the fourme and condicion of this worlde."[154] For Caxton, such an understanding of geography will lead to the "parfyght scyence and knowleche of God."[155] Sceptics may attribute Caxton's plea as a mere marketing ploy to further book sales, but Caxton's earnestness, I would argue, deserves credence as a marker of his seriousness of purpose in publishing historio-geographical writing for English readers in the era of Ptolemy's *Geography.*

Beyond his expansion of the prologue, a close study of the text and images in comparison with his exemplum shows that Caxton, in contrast to Bennett's critique, does make subtle changes and most particularly in the matter of geography. In so doing, Caxton signals a critical engagement with his source text via three processes: augmentation, juxtaposition, and correction. Caxton's editorial intervention through augmentation is marked by small additions to his source text. Each time the University of Paris is mentioned, for example, Caxton augments the text by including, in a nod to English readers and scholarship, the universities of "Oxenford and Cambrige"[156] as among the great European centers of learning: "fountayns where men may drawe out most science."[157] In the chapter on European geography, Caxton draws on his own life experience to augment the list of towns along the Rhine River. Ever attentive to cities and place names, Caxton adds the city of "Coleyn" (the city where he first learned the craft of printing), to that of "Basyle, Strawsburgh, Magounce, and Covelence" as well as a more detailed description of the flow of the "Riin."[158] An intriguing connection to Malory's Roman War account, Caxton identifies a "montayne" as the source of the Rhine and Rhone rivers, a possible reference to the mountain range which figures prominently in Arthur's campaign across the Alps to Rome in Malory's *Le Morte Darthur* (see Chapter 1). Caxton also clarifies the time it would take a hypothetical Adam to walk from Earth to heaven. The source text states a daily rate of 25 miles to which Caxton adds: "xxv myles of Fraunce, which is • L • English myle".[159] Caxton, like Fra Mauro and other fifteenth-century geographers, participates in the Ptolemaic conviction that "geographical knowledge is constantly accumulating."[160]

Caxton's second editorial strategy is that of juxtaposition: when his source text places Greece, Tuscany, Spain, Portugal and other southern European locales in the African continent, Caxton attempts to negotiate an alternative geographic framework: "And how be it that the Auctour of this book saye that thise contrees ben in Affryke, yet, as I vnderstonde alle thise ben within the lymytes and bou*n*des of Europe."[161] Caxton does not obliterate the assertions of his source, a process Tonry identifies in Caxton's editing of the *Polychronicon*; rather, he adds contrasting information and leaves the critical judgment to the reader. When *Image du Monde* similarly locates Jerusalem in Africa, Caxton inserts the equivocal comment: "After thoppynyon of somme is that this holy londe longeth to Asye."[162] In these two examples, Caxton provides the original text, then offers an addendum which negotiates—and asks his readers to negotiate—the alternative geographic frameworks coexistent in the late fifteenth century. A similar example of such a geographic juxtaposition is evident on the Behaim globe (see Chapter 3) in which Prester John is depicted in two locations, both India and Africa, acknowledging the conflicting yet authoritative textual traditions then in circulation. Such multivalent citations concede a field of knowledge in flux, a fifteenth-century version of the role of multiple subjectivities now known as the "Rashomon effect."

Caxton employs his third strategy, that of correction, sparingly, preferring to leave judgment to the reader. As such, even when he offers a correction, the original account is typically presented, then contested. One such example is Caxton's refutation of the account of the Legend of St. Patrick's Purgatory, a pilgrimage site in Ireland in which after a night spent in grief, torment, in a penitential cave, one's sins were cleansed. As befits the times, it was a popular pilgrimage site and, moreover, the legend's transmission in Latin and vernacular versions had what Stephen Greenblatt categorized as "stunning mobility," in its textual proliferation.[163] Although the account seems to recount an earthly "event," the legend constitutes a form of geographical romance that "derives from the tours of Hell familiar from the second-century Apocalypse of Saint Peter … and other ancient texts in the Christian tradition."[164] Caxton's edition challenges the textual tradition, however, with the authentication of an eyewitness account and intervenes into the text in the matter of earthly geography. True to his propensity for multiple sources of authority, Caxton provides the text as it appears in his manuscript source, but then adds a lengthy interpolation, a form of a correction based on eyewitness authentication and his own research:

> hit may wel be that of auncyent tyme it hath ben thus as a fore is wreton, as the storye of Tu*n*dale & other witnesse, but I haue spoken with dyuerse men that haue ben therin. And that one of them was an hye chanon of Waterford whiche told me that he had ben therin v or vi tymes. And he sawe ne suffred no suche thynges … & other thyng sawe he not. And in lyke wyse tolde to me a worshipful knyght of Bruggis named sir John de Banste that he had ben therin in lyke wyse and see none other thyng but as afore is sayd.[165]

Thus, when the textual tradition of a geographically-based phenomenon, loosely defined, is contested by eyewitness account, Caxton provides a narrative structure of inquiry and a corroboration of eye-witness accounts that suggests to the reader the need for critical analysis. Clearly, the weight of such testimony puts the original text into doubt, forcing the reader to make a critical judgment as part of the act of reading. Tellingly, these textual junctures in the *Myrrour* are primarily focused around geographical contestations. Coldiron identified a similar propensity for "interpretive pluralism" in Caxton's editions of the *Recuyell of the Hystoryes of Troy* and in *Dicts and Sayings of the Philosophers,* about which she notes that "the discordant elements in Caxton's productions seem to ask readers to decide for themselves."[166] Such discordant elements in the *Recuyell* open "wide interpretive spaces" within a stabilizing frame.[167]

Caxton appears, uncharacteristically, to silently emend certain of the mathematical calculations in the *Myrrour* related to the "fourme and condicion" of the

Earth. He alters the distance from the Earth to the moon from 12 times the circumference of the Earth, as found in his source text, to 15 times the circumference of the Earth. Next, he alters the size of the Earth to 29 times that of the circumference of the moon rather than the figure of 39 in his source text.[168] A possible source of Caxton's emendations has yet to be identified. One possible source is a different manuscript edition of *Image du Monde*, [169] suggesting that Caxton consulted multiple texts. Whatever the origin, the "correction" of these calculations is a clear indication of Caxton's active engagement with the content rather than a slavish adherence to his source text and the increasing influence of the Ptolemaic-influenced practices.

Caxton's augmentations, juxtaposes, and corrections form a convincing portrayal of Caxton's keen awareness of geographic information as a dynamic process of tradition and re-constitution. In addition, Caxton's prologue invites his readers to explicitly enter into this process of knowledge construction. While offering the traditional trope of blaming any errors on his copy, Caxton also proposes a reader-based process of correction and emendation. As editor, he positions himself as:

> humbly requyryng alle them that shal fynde faulte, to correcte and amende where as they shal ony fynde, and of suche so founden that they repute not the blame on me, but on my copie whiche i am charged to folowe as nyghe as God wil gyue me grace; whom i most humbly beseche to gyue me scyence, connyng and lyf tacccomplysshe and wel to fynysshe it, etc.[170]

Here, Caxton encourages readers to reinscribe their own copy of the book with augmentation and corrections, a lay equivalent of humanistic scholarship. Caxton repeats this request for correction and emendation in his Epilogue, invoking a model of geographical authority as contingent and under pressure, a restructuring of classical geography with incremental new information validated through new structures of authority: that is, alternate sources, personal experience, and eye-witness accounts. Ptolemy would have lauded this kind of constructivist approach as in keeping with his own scientific practice. In his introduction to the *Geography*, Ptolemy states:

> We consider it necessary therefore for us to pay more attention to the newer records of our time, weighing, however, in our description these new records and those of former times and deciding what is credible and what is incredible.[171]

Caxton, too, weighs the credible and incredible and invites his readers to do so as well.

In his Epilogue, Caxton asserts that he rendered his translation "so playn that euery man reasonable may vnderstonde it"[172] and returns to an issue that seems to have vexed him: the accuracy of the mathematical calculations of the

Earth and moon's relative size, as noted above. "And yf ther be faulte in mesuryng of the firmament, Sonne, Mone, or of therthe, or in ony other meruaylles herin conteyned, I beseche you not tarette the defaulte in me but in hym that made my copye."[173] This abrogation of responsibility rings hollow, of course, and counters Caxton's actual practice of intervening into and augmenting the text. In blaming possibly erroneous calculations on his scribe or copy text, Caxton reveals an anxiety about the potential accuracy of the equations and his awareness that these measurements were under greater scrutiny in the age of the Ptolemaic revival. As printer, Caxton is both "charged to follow" the authority of his source while simultaneously offering "an extended, unpredictable engagement with the open interpretive field of the reader,"[174] one that is inclusive of multiple or competing accounts and a field of information in flux. His correction of the calculations in tandem with his simultaneous disavowal of the validity of the calculations speaks to Caxton's fraught engagement with the increasingly mathematical models of the world within a work of universal history. In the age of the Ptolemaic revival, Caxton actively engages in the mathematical rationalization of space that is underway, but within the structure of a universal history.

Visualizing the world: Caxton's cartographic images

Despite the dismissive attention given to the two "T-O maps" reproduced in Caxton's *Myrrour*, there are, in fact, *multiple* images that represent a global Earth in the text. Two are the progeny of the long-standing Greco-Roman tradition: cartographical sketches of the known world as represented by a "T" inscribed in a circle, an abstract configuration of the three known continents. Following his source text, Caxton reproduces these schematic images of a round Earth to accompany the sections of text that explain the Earth's spherical properties (Figure 2.2). But there are added illustrations in Caxton's *Myrrour*: an image that demonstrates the roundness of the Earth and two of scholars in contemplation of a globe, discussed below.[175] These images have been noted by book historians but have been dismissed as incidental to England's later cartographic culture. While globes were not produced in England until the late sixteenth century, at least none that are extant, it is not a corollary that globes were not present in England. As I argue in Chapter 4, a stage-prop globe was made for John Rastell's 1518 production of *A New Interlude and a Mery, Of the Nature of the Foure Elements*, earning it the previously unrecognized distinction as the first known globe made in England. In addition, academic and private libraries list globes in their inventories.[176] But globes are also represented—and quite early on in the Ptolemaic revival—by images in books and, specifically, in Caxton's *Myrrour of the World*. As Cosgrove observes, "representations of the globe have exercised an especially powerfully grasp on the Western imagination,"[177] and the two illustrations that include a globe, both original additions by Caxton into the illustrative scheme, merit careful attention. In each of these two images, a solitary scholar gazes upon a three-dimensional globe, indicative of Caxton's

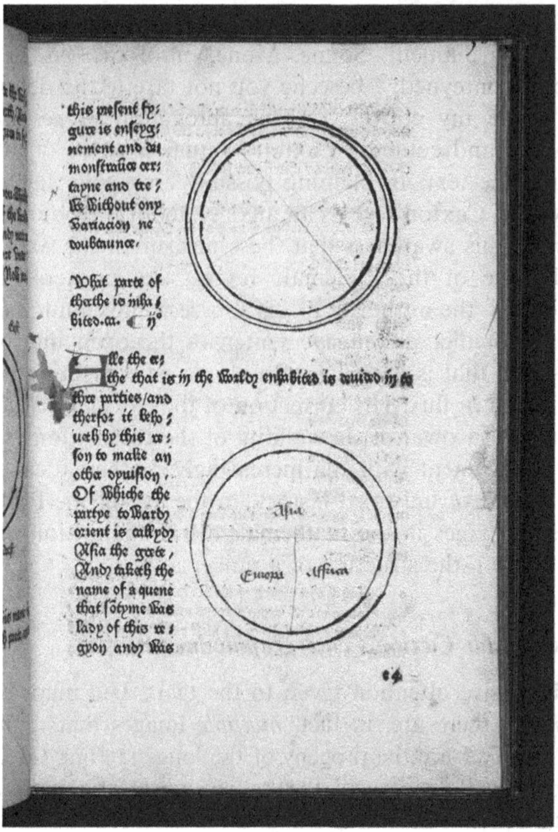

Figure 2.2 T-O map, William Caxton, *The Myrrour of the World*, 1481 (STC 24762)
Source: British Library/Creative Commons.

response to the increasing interest in maps of the whole world and globes during the Ptolemaic revival. In addition to Caxton's textual interpolations, the visual geographic representations in the *Myrrour*, including T-O maps, images of three-dimensional globes, round Earth illustrations, and the figure of God holding a Roman-style imperial orb, suggest in their totality that Caxton's *Myrrour* should be read as a native English production of the hybrid geographical practices of the late fifteenth century and as part of the early transmission of Ptolemaic geographic thought in England.

In his Prefaces, interpolations, and concluding "recapitulacion," Caxton does not specifically mention Ptolemy's *Geography*, but he does interpolate the term "Mappa Mundi" into the text twice: once to refer either to the book in the reader's hand or the geographical excursus about to commence ("this Mappa mundi") and a second time in his recapitulation when he states, "In the seconde partye is declared to yow whiche parte of therthe is inhabyted, & of the dyuysion of mappa mundi."[178] Blake has argued that the phrase is not a reference to an actual map but to the narrative of the whole world; in addition, however, the phrase "dyvysion of mappa mundi" signals both the traditional tripartite *mappamundi* as well as the great Ptolemaic project: that is, the division of the world into visually and narratively-described regions *within the form of the book*. In his paratextual organization of the book, Caxton maintains the tripartite division of *Image du Monde*: the first part on theology and the seven liberal arts, the second part on geography, the third part on astronomy, the measurable Earth, and philosophy. In his introduction to the second part, Caxton reinforces geography as a field of study ("this Mappa mundi") and the continued link between narrative and visual geography. In summarizing the third part of the book in his recapitulation, Caxton recalls the text's discussion of "the firmament & of the sterres; & how the world was measured, & the heven and therthe; of the kynge Tholomeus & of his prudence."[179] As Catherine Delano-Smith and Roger Kain note, Ptolemy's *Almagest* was "circulating widely in the West from the twelfth century onwards,"[180] and while Caxton's reference in this case refers primarily to the *Myrrour*'s reliance on the *Almagest*, this reference also captures a core principle of Ptolemy's work in both the *Almagest* and the *Geography*: the mathematical measurability of the Earth.

While Caxton's name is rarely associated with the history of English cartography, except in the most dismissive of references, his editing of the *Myrrour* contains the first printed maps in England, albeit rudimentary woodblock images. It seems that both the geographical content and the lovely illustrative scheme of comets and planetary relational order in his exemplar, Royal MS 19A IX, inspired Caxton to attempt the first printed text with images in England. In *Image du Monde*, the text repeatedly calls attention to the importance of the images for the explanation of the astronomical concepts, undoubtedly influencing Caxton's decision to provide illustrations in his edition: "in lyke wyse maye ye see alle the manere and fashion by these two fygyres the whiche ben here to you represented and shewde all entirely."[181] Thus, the textual and visual components of the book are highly integrated. Caxton takes up the enormous technical challenge of providing an illustrative scheme in the new medium of print, remediating the interplay of text and image which explicates the action of comets, astrological phenomenon, and the Earth's sphericity found in the manuscript of *Image du Monde*.[182] Erhard Ratdolt, a German humanist printer living in Venice, is renowned for his innovative 1482 edition of Euclid's *The Elements*, in which he added geometrical diagrams to the text, using composite woodcuts for the illustrations;[183] Caxton had tackled this technical problem two years earlier.

Caxton's two T-O maps are basic woodblock images (see Figure 2.2) that suffer in comparison to the sophisticated Ptolemaic-based maps being printed on the Continent, either as copper-etched or woodblock images. The 1482 edition of Ptolemy's *Geography* printed in Ulm, for example, has hand-colored woodblock maps with fine detail (Figure 2.3). In contrast to these continental maps, Gordon Duff describes the images in the *Myrrour* as,

Figure 2.3 Claudius Ptolemy, *Cosmographia*, Leinhart Holle, Ulm, 1482
Source: Boston Public Library.

"Rudely designed and coarsely cut."[184] Hodnett is in agreement: "The first native hand made the cuts for the *Myrrour of the worlde* and no more—let us hope because Caxton was disgusted by such hacking."[185] Hodnett adds, "England stumbles on to the book-illustration stage with some of the poorest cuts ever inserted between covers."[186] He then heaps further disparagement: "The blocks, *all* by one hand, are miserably executed. The outlook for English illustration could hardly be worse."[187] It seems reasonable to note in this context, however, that the Ulm edition was largely responsible for the brief duration of Leinhart Holle's print venture. Holle produced a beautifully illustrated woodblock edition of the *Geography*, the execution of which was expensive and, ultimately, contributed to the demise of his fledgling print business.[188] The 26 or so maps required to produce an illustrated *Geography* required considerable craftmanship, investment, and risk. In Caxton's case, his shrewd calculation, though less aesthetically and cartographically impressive, led to a long, successful press.

The two maps in Caxton's edition are abstract schemata that gesture to the traditional "T-O" maps, a geographic abstraction of the Earth, not intended to convey specific positional geographical information, typically oriented with East at the top. Following the medieval *mappamundi* tradition, the Caxton maps are comprised of the three major continents: Africa, Europe, and Asia. Of the two T-O maps, one illustrates the more common tripartite world while the other illustrates the relatively rarer representation of a quadripartite world, which includes a fourth "uninhabitable" region in the south. In both images the continent names were written in by hand. Royal MS 19 A IX, Caxton's direct source, does not include T-O maps but Harley 334, another manuscript version of *Image du Monde* from the fifteenth century, does, providing evidence that Caxton consulted at least two versions of his source text. These maps indicate Caxton's decision to incorporate multiple sources as well as his process of amalgamation, including both traditional geographic information and slight emendations to accommodate the new.

In addition to the T-O maps, Caxton's edition adapts an illustration from his source of a popular anecdote describing how a man could "goo aboute therthe … lyke as a flye goth round aboute a round apple."[189] Another image illustrates an anecdote similar to one found in Mandeville's *Travels*: two travelers depart from each other, "one went alle way to ward the eest and that other to ward the weste." Eventually the two "shold mete agayn in the opposite place fro where as they departed,"[190] revealing to both the travelers and to the reader that the Earth is a sphere (Figure 2.4). Contrary to the popular modern conception of pre-Columbus "flat Earth" theories, the Earth had long been acknowledged as a spherical body in learned texts such as Johannes Sacrobosco's *Da Sphaera Mundi*, c. 1230. Caxton's translation follows his exemplar in referring to the Earth as an apple, "Syth we haue vnderstande how the erthe is rounde on all partes as an apple."[191] Written approximately 100 years prior to the earliest Mandeville manuscript, *Image du Monde* attests to the popularity of the apple metaphor and may be a

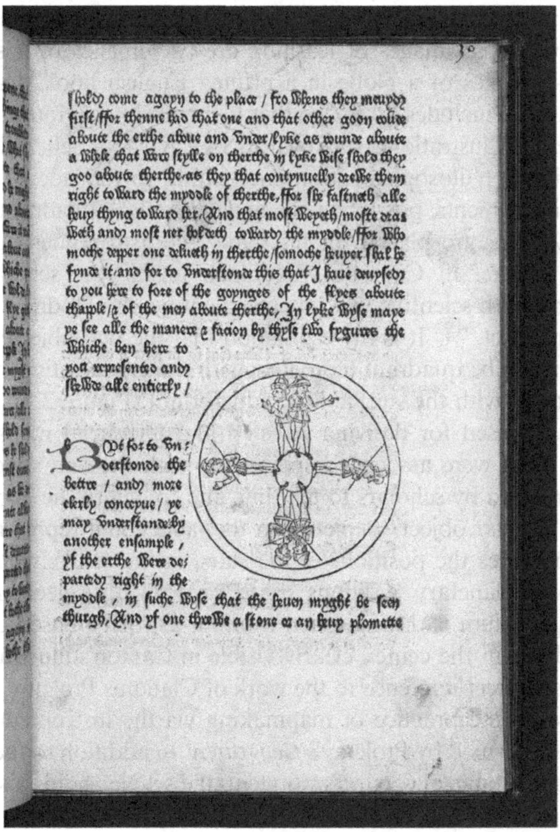

sholde come agayn to the place / fro whens they meuyd first/ffor thenne had that one and that other goon rounde aboute the erthe aboue and vnder/lyke as rounde aboute a whele that were stylle on therthe in lyke wise shold they goo aboute therthe/as they that contynuelly drawe them right toward the myddle of therthe/ffor she fastneth alle heuy thyng toward her/And that most weyeth/moste draweth and most nere holdeth toward the myddle/ffor who moche depper one deluyth in therthe/somoche heuyer shal he fynde it/and for to vnderstonde this that I haue deuysed to you here to fore of the goynges of the flyes aboute thapple/& of the men aboute therthe/In lyke wyse maye ye see alle the manere & facion by thyse two fygures the whiche ben here to you represented and shewde alle entierly /

But for to vnderstonde the bettre / and more clerly conceyue / ye may vnderstande by another ensample / yf the erthe were departed right in the myddle / in suche wyse that the heuen myght be seen thurgh/And yf one threwe a stone or an heuy plomette

Figure 2.4 William Caxton, *The Myrrour of the World*, 1481 (STC 24762)
Source: British Library/Creative Commons.

source for Mandeville's well-known "Earth-apple" metaphor, a metaphor subsequently utilized by Martin Behaim and incorporated into his "Introduction" to his "Erdapfel" of 1492 (see Chapter 3). The anecdote of the Earth as an apple exemplifies the way in which literary texts disseminated frameworks of geographic thought in anticipation of and contemporaneously with the publication of Ptolemaic-based maps, maps often pointed to as denoting a convenient but false demarcation of a "cartographic break" between the medieval "flat world" and early modern cartography. Caxton's text continues the tradition of non-scholarly works such as Mandeville's *Travels* and *Image du Monde* in promoting the principle of a spherical Earth for a broad reading audience, a task taken up by John Rastell in his early modern theatrical production.

Printing the globe

The printed reproduction of the two T-O maps has been used to categorize Caxton's *Myrrour* as a retrograde medieval geographic text rather than one printed on the cusp of the early modern period. However, Caxton's edition also includes a second group of original woodcuts, not in *Image du Monde*, comprised of various scenes of a master with his pupils as well as individuals representing the liberal arts (gramaire, logyke, rethorique, arsmetrike, geometrye, musique, astronomye). Included among these images of learning are two images of a scholar with a globe, the first images of a globe in a printed English book. Caxton's source, Royal MS 19A IX, includes an image that is the likely inspiration for one of the two Caxton globe illustrations. In Royal MS 19 A IX, a delicately-colored illumination depicts a philosopher/astronomer sitting to the side of a table where astronomical instruments, possibly a triangular astrolabe quadrant and an astrolabe or equatorium, are placed. In his hand, the astronomer holds a stylized onion-shaped sphere. In Caxton's woodblock image, the scholar is similarly seated at a table with scientific instruments, identified by Hodnett as "a segment and a circular object."[192] It is my estimation that the two objects are a realistic and detailed astrolabe quadrant (complete with the two sighting holes for aligning the instrument with the sun) and an equatorium (Figure 2.5). Both scientific instruments were used for deriving celestial and terrestrial measurements and locations and both were associated specifically with Ptolemy's work. Astrolabe quadrants were used by scholars to tell time and to determine geographical latitude.[193] The "circular object" depicted on the table could represent a full astrolabe, which measures the positions of the stars, or, more likely, an equatorium, which calculates planetary positions based on Ptolemy's geocentric planetary theory. An equatorium is distinguished from an astrolabe based on the circular epicycles, or disks, in the center, clearly visible in Caxton's illustration. Caxton's image, then, is a direct reference to the work of Claudius Ptolemy, as a historical figure, but also to the practice of mapmaking via the instrumentation of measured space, as informed by Ptolemy's *Geography.* In addition to the astronomical instruments on the table in Caxton's woodcut, the scholar holds a spherical globe

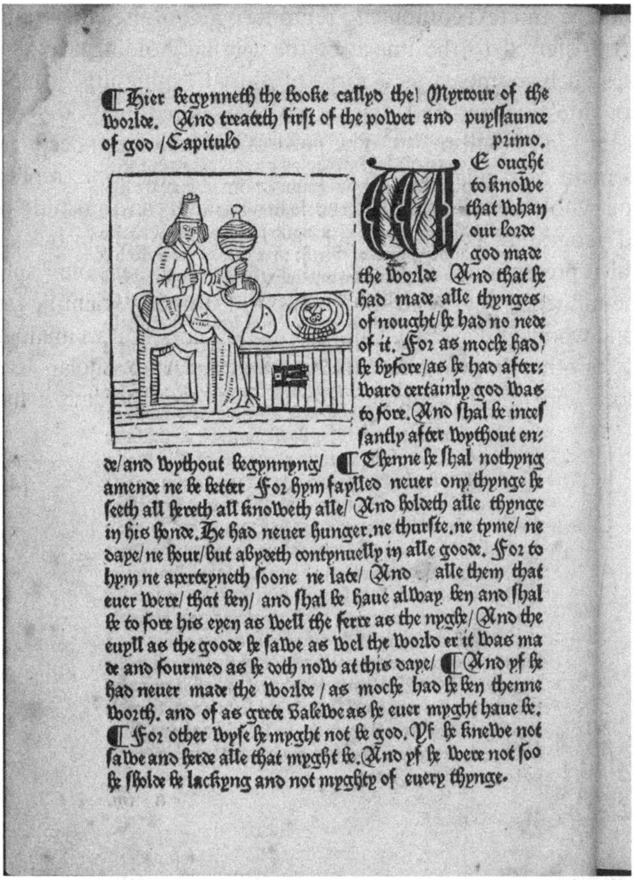
¶ Hier begynneth the booke callyd the Myrrour of the worlde. And treateth first of the power and puyssaunce of god / Capitulo primo.

We ought to knowe that whan our lorde god made the worlde And that he had made alle thynges of nought / he had no nede of it. For as moche had he byfore / as he had afterward certainly god was to fore. And shal be incessantly after wythout ende / and wythout begynnyng / ¶ Thenne he shal nothyng amende ne be better For hym faylled neuer ony thynge he seeth all hereth all knoweth alle / And holdeth alle thynge in his honde. He had neuer hunger. ne thurste. ne tyme / ne daye / ne hour / but abydeth contynuelly in alle goode. For to hym ne apperteyneth soone ne late / And alle them that euer were / that ben / and shal be haue alway ben and shal be to fore his eyen as well the ferre as the nyghe / And the euyll as the goode he sawe as wel the worlde er it was made and fourmed as he doth now at this daye / ¶ And yf he had neuer made the worlde / as moche had he ben thenne worth. and of as grete valewe as he euer myght haue be. ¶ For other wyse he myght not be god. Yf he knewe not sawe and herde alle that myght be. And yf he were not soo he sholde be lackyng and not myghty of euery thynge.

Figure 2.5 William Caxton, *The Myrrour of the World*, 1481 (STC 24762)
Source: By permission of the University of Glasgow Library, Archives, and Special Collections.

set upon a stand. The globe does not depict specific continents but has wavy zonal or latitudinal demarcations along with a darker equatorial line. Like its exemplar, the illustration suggests the scholarly study of the globe but is less ethereal and may indicate the rotation of a globe on a stand.

The image of the astronomer and globe decorates the beginning of the *Myrrour*, adjacent to the opening lines of the first chapter:

> Ye ought to knowe that whan Our Lord God made the world and that he had made alle thinges of nought, he had no nede of it; ffor as moche had he bifore as he had afterward. Certainly God was to fore, and shal be incessantly after, without ende and withoute begynnyng. Thenne he shal nothyng amende ne be better, ffor hym faylled neuer ony thynge. He seeth all, hereth all, knoweth alle, and holdeth alle thynge in his honde.[194]

The visual image and text combine to reproduce an omniscient and divine global gaze, now transferred to the image of the scholar holding a globe. Cosgrove characterizes such moments as a form of global "sublimity" which "prompts reflection on life's origins and destiny, so that eschatology and teleology are shaping themes in globalism."[195] The gaze of the scholar, indeed, constitutes a sublime moment of contemplation at the outset of the book, representative of the text as a whole: a beholding of the whole Earth via the study of universal history and geography as seen through a Christian paradigm but also through the material representation of the world as a three-dimensional sphere, not to mention the influence of Arabic scholarship in European scientific treatises.

The second woodcut depicting a globe does not have a direct analogue in Royal MS 19 A IX or other known source. In this image, an Arab scholar—depicted with pointed shoes and fez—is standing near a citrus tree, holding a mathematical

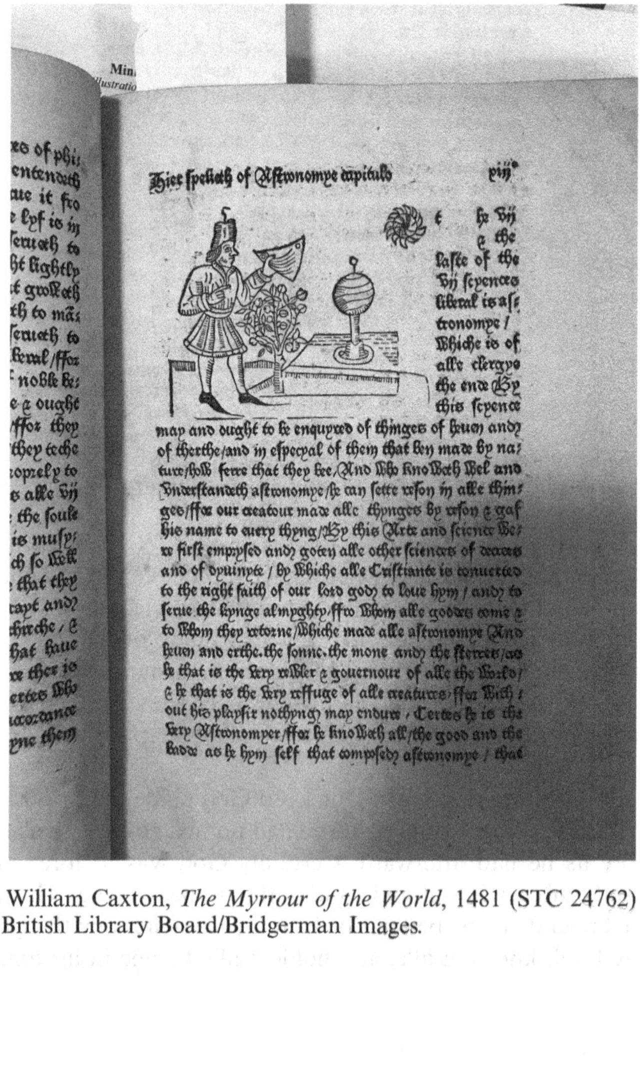
Hier spekith of Astronomye capitulo xiij

t he vij & the laste of the vij scyences liberal is astronomye / Whiche is of alle clergye the ende By this scyence may and ought to be enquyred of thinges of heuen and of therthe/and in especyal of them that ben made by nature/how ferre that they bee/And who knoweth wel and Vnderstandeth astronomye/he can sette reson in alle thinges/ffor our creatour made alle thynges by reson & gaf his name to euery thyng/By this Arte and science were first empryseda and goten alle other sciences of clerkes and of dyuinyte / by whiche alle Cristiante is conuerted to the right faith of our lord god to loue hym / and to serue the kynge almyghty/ffro whom alle goodes come & to whom they retorne/Whiche made alle astronomye And heuen and erthe. the sonne. the mone and the sterres/as he that is the very reuler & gouernour of alle the world/ & he that is the very refuge of alle creatures/ffor with out his playsir nothyng may endure/Certes he is the very Astronomyer/ffor he knoweth all/the good and the badde as he hym self that composed astronomye / that

Figure 2.6 William Caxton, *The Myrrour of the World*, 1481 (STC 24762)
Source: © British Library Board/Bridgerman Images.

triangle (Figure 2.6). Placed on the table nearby is a similar globe to that of the first image: it rests upon a pillar and base, suggesting that it can revolve and is similarly marked by wavy zonal delineations. Caxton's woodcut suggests a representation of Ptolemy, referred to as "kynge Tholomeus" throughout the text, "a kynge moche subtil in Astronomye," an erroneous but common misidentification on the part of medieval writers of the royal Ptolemaic family of Egypt as the Alexandrian scholar.[196] In the Harley 334 manuscript of *Image du Monde*, Ptolemy is illustrated as an astronomer-king, wearing a crown (Figure 2.7), and in Caxton's text, the image of the Arab scholar and globe fittingly accompanies the chapter devoted to "Astronomye." Caxton's representation of Ptolemy is suggestive of the hybrid aesthetic of fifteenth-century geographic thought, a synthesis that affirmed medieval tradition and was attentive to the long-standing respect for Ptolemy's *Almagest*, but that was also alert to the humanist revival of Ptolemaic geography. In Caxton's

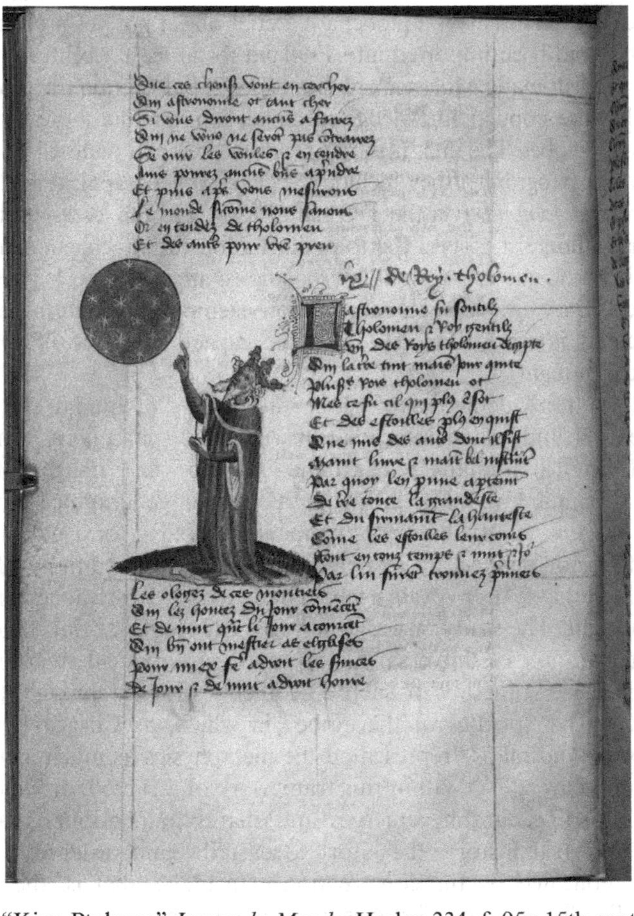

Figure 2.7 "King Ptolemy," *Image du Monde*, Harley 334, f. 95v 15th century
Source: © British Library Board.

Myrrour, the text adjacent to the image of Ptolemy reads: "By this scyence may and ought to be enquyred of thinges of heuen and of therthe."[197] In this pairing of an image of a three-dimensional globe with the image of an Arab scholar, Caxton visually signals the transmission of Ptolemaic geographic culture in English early print. The images of the globes in Caxton's *Myrrour* evoke a connection to humanist study: the images of masters and students, the lively representations of the liberal arts, and the scholarly study of geography provide a visual popularization of humanistic scholarship for a non-academic reading community. In Caxton's text, the medieval genre of the "mirror" intersects with the ascendance of the visual, shifting from the sense of a manual of how the world "should" be, to one which establishes the potential for images to reflect the world as it seems to be. Like the broader genre of "speculum" literature, the Latin origin of "mirare" encompasses both a sense of wonder as well as the sense of looking upon something extraordinary.

It is significant that the images in Caxton's text include an increasingly prized geographical object—that of a globe, a material object not yet produced in England. Globes had been imported into England for at least a century prior to the renowned English-made Molyneux globes of 1592, but Caxton's *Myrrour* provide evidence that the popular knowledge of globes long pre-dates the production of the Molyneux globes. The images in the *Myrrour* present a multivalent moment in the history of cartography: they include the classical geographical tradition in the form of the *mappamundi* (via the T-O maps) as well as the iconography of three-dimensional, humanist globes. Caxton's illustrations exert a phenomenon similar to that of his prologue and textual interpolations: an oscillation between multiple geographic frameworks. This amalgamation exemplifies Caxton's penchant for reader-oriented interpretation, but also the capaciousness of fifteenth-century geographical thought.

Beyond Caxton's depiction of globes as objects of humanistic study, his various textual and visual interpolations reveal a symbolic aspiration for a culture of English humanistic learning. The text and images of Caxton's *Myrrour* are indicative of a growing interest in geographic knowledge in England, one that subtly augments its medieval source for a new audience of print readers not defined by ownership of the luxury editions of the *Geography* which have been exclusively associated with advancement in geographic thought. The study of geography in early English print, then, includes a populist tradition of universal history and astronomical treatise. Caxton's text exemplifies a mingling of sacred geometry with a representation of the globe to create a "poetics of the globe" in which, as Cosgrove points out, global models and maps "represented the metaphysics as much as the physics of earth and heaven."[198] Within this framework of a sacred, ordered universe, the discipline of geography emerges and shares foundational questions. In works of universal history, the effort to describe and order the Earth and cosmos is replicated by the construction of a globe, that is, the "dream of order."[199] The first images of globes in England appear in a text that presents both the cosmological framework of the past and an indication of the emerging discipline of geography and scientific cartography. Gautier-Dalché, whose study of Ptolemy's *Geography* as part of European humanistic circles emphasizes the connection of geographical and astrological works within a broader intellectual and religious context, observes that for such geographical texts, "the overall framework of knowledge is more significant than the actual contents."[200] Thus, the accuracy of Caxton's calculations of the Earth's circumference, while erroneous, must be set within the geo-moral context of the work as a whole. Gautier-Dalché argues that any analysis that

> focuses solely on the 'medieval' aspects of these texts, contrasting them with the supposedly 'modern' and underlining the absence of 'new input,' would entirely miss the interest in and meaning of these works, which reflected a very particular intellectual climate.[201]

In his 1620 *Novum Organon*, Francis Bacon argues that it would be a "disgrace to mankind" if the geographic discoveries of the sixteenth and seventeenth centuries "were confined to the discoveries and narrow limits of the ancients,"[202] but in the late fifteenth century, classical authors, religious thought, and early modern geography had yet to be quarantined from each other. Caxton's *Myrrour* tells a story of one stream of the constitutive process of geographic thought in England in the early print period, one that integrated the practices of the medieval *mappamundi* with the prestige of Ptolemy as inspiration for new mathematical and visual representations of a global Earth.

If print production actively reproduced prior traditions in dynamic cultural reinterpretations, as Kuskin argues, the ways in which the medieval book was an active agent in the culture of the early modern period bear further scrutiny.[203] Indicative of such a continuing cultural agency, *The Myrrour of the World* met with enough interest to merit a reprinting by Caxton in 1490[204] and again by Andrew Lawrence in 1527. Caxton's editions of the *Myrrour* indicate the kind of historical and geographic knowledge Caxton felt lay readers in England would seek out and, as such, provides an important source for the study of geographic thought via vernacular texts in England. Caxton's *Myrrour* and the multiple editions of the *Kalender of Shepherds* trace incremental changes in geographic thought over the *longue durée* and thus collectively record what Bernard Kline has termed "a cartographic transaction," that is, "the mental and material renegotiation of the lived space of experience,"[205] a renegotiation in full force during the Ptolemaic revival. During a critical period of changing spatial conceptions of the world, Caxton's editions of the *Description of Britain* and *The Myrrour of the World* renegotiated for English readers the conception of English identity in the context of a new global geography. The English scholar gazing upon the globe in Caxton's *Myrrour* reflects the medieval interest in a divinely-ordered world, but it also represents an emerging spatialized and scientific study of the Earth in which English readers could now take part.

'Caxton me fieri fecit:' mapping the text via rubrication and textual space

A feature for which Caxton is well known among book historians and bibliographers has not been considered in the context of emerging spatialized practices: the rubrication of the text. Caxton's famed phrase at the conclusion of his second edition of the "*Myrrour, Caxton me fieri fecit,*" roughly translates as "Caxton had me made."[206] And indeed, Caxton's editorial and paratextual work defines the making and the spaces of the text. In Caxton's print practice, the introduction of rubrication, or what we now call a table of contents, coincided with his production of the *Description*, the *Chronicles of England*, and the *Myrrour.*[207] In 1480, Caxton not only produced his geographic texts but also simultaneously introduced the spatializing practice of front matter rubrics into English books.[208] Previously, Caxton provided prologues or epilogues for his publications as well as chapter rubrics that followed the chapter headings *within* the body of the text. But with the printing of the *Chronicles of England* in June 1480, the *Description of Britain* in August 1480, and the translation of the *Myrrour* in that same year, Caxton began the editorial practice of "rubricating" his prose texts—that is, not only denoting chapter breaks within the text but providing a list of chapters in the front matter of the book as a guide to the reader.

According to Blake, the *Chronicles of England* was the first book Caxton printed with page signatures, alphanumeric references on the bottom right-hand corner of the page, noting that "Caxton seems to have experimented" with different pagination systems.[209] By 1481, the year after his geographical publications, signatures as a form of organizing the properties of a book became "a standard feature" of Caxton's books.[210] The content of his 1480 print productions may have spurred, or at least coincided with, a system that was becoming part of the spatial structure of print books as we have come to know them. Caxton did not, as a rule, provide rubrics for poetic works, but when prose works had rubrics in his manuscript copy text, Caxton would reproduce it in his print work as well. Blake does not see this as any innovation on Caxton's part, dryly commenting that Caxton's rubrics do not "imply any editorial initiative on the part of the printer."[211] However, for his 1480 edition of the *Description* and his 1482 edition of the *Polychronicon*, Caxton has, in fact, taken editorial "initiative:" he supplies the rubrics, or table of contents, for the excerpted matter in *Description* as well as for the long prose content of the *Polychronicon* where one did not previously exist in his exemplum. It is here that the landscape of book organization comes into a new form. Of importance in the context of the present study is that Caxton's practices—simultaneous with his geographical publications—spatialize the book. Through his rubrics he provides nothing less than a map of the book: the locational coordinates—the textual latitude and longitude—by which a reader may navigate the text.

In reshaping the geographic portion of the *Polychronicon* for print publication, Caxton expanded the chapter divisions, creating 33 chapters from the previous 19. In Caxton's *Description of Britain*, the very space of Britain becomes larger and more detailed than his exemplum. Tellingly, Caxton compressed Higden's six chapters on English bishoprics into a single chapter while expanding Higden's six chapters on Welsh and Irish geography and commodities into 13 chapters. Fundamentally, the criteria for the new divisions are framed as primarily geographic rather than ethnographic or historic. For example, Higden's chapter entitled "Of Ireland," becomes four chapters in Caxton's text: "Of the discripcion of Irlonde," "Of the boundying of Irlonde," "Of the gretnes & qualite of that londe," and "Of the defautes of that londe." Caxton similarly adds a chapter heading, "Of the comoditees of the londe of Walys" not in his exemplum. Caxton does not significantly alter the verbal landscape of the chapters' content, but rather provides a new geographical framework through his "enchapterying" and emphasis on geographical description. Caxton focuses the reader's attention on the physical boundaries of Britain's territories by providing rubrics in the front matter of the book which created, in effect, a rationalization of space akin to the concomitant process in Ptolemaic-based maps. In Caxton's *Description*, the rubrication shapes the order in which the reader encounters the geography of England, Scotland, Wales, and Ireland and, in so doing, provides a map of fifteenth-century British geographic identity.

In his various editions, Caxton's rubrication often provides a new organizational framework superimposed upon a previously continuous narrative. For example, in his edition of *Le Morte Darthur*, Caxton divides the prose text into 21 "books" containing 507 chapters, a new organization of the text distinct from that of the Winchester manuscript which, as is generally accepted, represents an earlier format of eight lengthy tales. According to Blake, this division of Malory's text, provided "a reasonable system of reference … in what was after all a very long book."[212] In fact, Caxton's chapter divisions and rubrication create a new bird's-eye view of *Le Morte Darthur*, so to speak, providing the reader, at the outset of the book, with an overview of the narrative landscape to come. Much like a map, Caxton's rubrication of the *Polychronicon*, the *Description of Britain, The Myrrour of the World*, and *Le Morte Darthur* articulates an unspoken claim that a particular location in the text may be sought out and read, allowing for a reading practice that is atemporal and geolocational, in much the same way that spatialized Ptolemaic maps replaced the narrative-based *mappaemundi*.

While it is true that cartographic production *per se* is not a significant feature of early English print production, Caxton's narrative geographies in the form of the *Myrrour* and the *Description*, along with the practice of spatializing books through tables of contents, signatures, and indexes serve as evidence of the polystream that carried the conceptual framework of Ptolemaic-based mapping in late fifteenth-century England book culture. Caxton's geographic print productions, along with his edition of *Le Morte Darthur*, reordered the textual geography of his exemplars. We can say with justification that despite the fact that William Caxton never printed a map based on Ptolemy's *Geography*, it would be an omission to segregate him or his works from the development of cartographic culture in England. Caxton's geographic publications from the early 1480s established a new cartographic context for his editing and publishing of Malory's "grete book" in 1485, a work Helen Cooper identifies as "geographically huge,"[213] and for which in posterity Caxton is best known.

Coda: the cartographic landscape of Wynkyn de Worde's *Le Morte Darthur*

After Caxton's death in 1491, his long-time journeyman Wynkyn de Worde took over the Westminster print shop at the sign of the Red Pale, but not first without a legal wrangle with Caxton's daughter, Elizabeth, that brought production to a near standstill through 1493.[214] Having emerged as the sole legal proprietor of the successful publication house by 1495, de Worde completed projects thought to be in process at the time of Caxton's death, such as an edition of Mandeville's *Travels* (1499), and issued new editions of previous Caxton publications such as the slightly retitled *Le Morte d'Arthur* (1498) and the *Descrypcyon of Englonde* (1498).[215] In addition to the editions published in the late 1490s, de Worde went on to publish another edition of the *Morte* in 1529 as well as two more editions of *Mandeville's Travels* in 1503 and 1510.[216] De Worde's newly illustrated edition of *Le Morte d'Arthur* reinforces the assertion of Chapter 1 that Malory's romance must be read as a cultural agent in the transmission of Ptolemaic-based geography in the medium of popular print. Through de Worde's *Morte*, the next generation of readers engaged in an increasingly cartographic, illustrated text of Malory's work. While ostensibly visually narrating the iconic scenes of chivalric deeds, courtly life, and combat, the illustrations of de Worde's 1498 *Le Morte d'Arthur* reveal a startlingly close and previously unrecognized engagement with Flemish landscape painting and the influence of Ptolemaic cartography in visual culture.

Although Mandeville's *Travels* and Malory's *Morte* presumably shared workspace in de Worde's print shop in the late 1490s, the two texts do not exhibit "shared geographical stratagems."[217] The major popular medieval work to assert the sphericity of the world, Mandeville's *Travels* posits a "new geography," as Karma Lochrie argues, and a "geographical decentering."[218] And yet, the iconographic program of the two texts is markedly distinct. De Worde's edition of the *Travels* includes some images specific to the text, but is largely illustrated by generic woodcuts, familiar across his large output of illustrated books. The illustrations of de Worde's *Le Morte d'Arthur*, however, were commissioned images uniquely and specifically created for this work. Of interest to the present study, the continentally-influenced designs of the *Morte*'s illustrations directly participated in the emerging visual language of early Netherlandish landscape painting which itself was in dialogue with the mathematical depiction of space in Ptolemaic-influenced cartography.

The woodcut images of de Worde's edition of Mandeville *Waye to the Holy Lande* (known now as Mandeville's *Travels*) appear to interpret the text as primarily concerned with the marvels of the East. That is, Mandeville's text is illustrated, not as an inherently geographical text, but one of curiosities and spectacle, showcasing exotic animals and one-footed men, many of which will make appearances in other of de Worde's texts, such as the *Kalender of Shepherds.* Such images include, for example, an image of the hand of doubting Thomas reaching out from a coffin in front of a blank visual field. The illustrations of de Worde's *Mandeville* are copies of a Lyon edition that, in turn, had been copied from a German edition by Anton Sorg and printed in Augsburg in 1481.[219] Sorg frequently used images of the naked human body, and these images are replicated in de Worde's edition. In the opening illustration, Mandeville is depicted against a blank background embarking on his journey to Constantinople, naked and on horseback—surprising attire for a long-distance journey, but in keeping with Sorg's *œuvre.*

De Worde was an enthusiastic adopter of illustration, producing 450 illustrated editions; Caxton, in contrast, produced just 20 illustrated editions.[220] De Worde was constantly experimenting and innovating, without any one "house style," and he typically used stock and composite images in his illustrative schemes to create a general rather than specific correspondence to the text. They include, for example, what Martha Driver has called the "Everyman and Everywoman" of early print, so-called for their ubiquitous appearances as all manner of literary, biblical, or historical characters.[221] De Worde's print house included hundreds of other stock images used to illustrate his editions. In contrast, the woodblock images for his edition of Malory's *Le Morte Darthur* are unique, textually specific, and attentive to a new visual language of landscape painting and cartography. As Edward Hodnett notes, de Worde's Malory is the only English edition of a literary work prior to 1535 that has "an original series of interpretive illustrations" made in England.[222] For Hodnett, the woodcuts "create for the modern reader a true world of the imagination," and these images augment the geographic scale of the text. De Worde's woodcuts for the *Morte* capture both a sense of the geographical restlessness implicit in the *Morte* and reveal a shared iconography, not with his other illustrated books, but with Ptolemaic-influenced maps, engravings, and landscape paintings. Whereas de Worde's other illustrated books do not deploy the attributes of early Netherlandish landscape painting, this influence is notably evident in the illustrations for *Le Morte Darthur.*

Though I have referred to the images as the de Worde's woodcuts, the images in de Worde's *Morte* must be credited to the anonymous woodcut artist de Worde commissioned, an artisan whose reading of the *Morte* and artistic aesthetic is acutely attuned to spatial representation, including linearity, depth, and the volumetric articulation of relational, measured space. Certain aspects of the woodcuts—for example, the large circular eyes of the individuals—have led print scholars such as Hodnett to speculate that the artist was either from, or trained in the tradition of, the Low Countries,[223] an association that speaks to the international flux of books and artisans in early print culture, in this case, the network of artisans and mercantile exchange between London, Bruges, and Antwerp.

Kevin Whetter has argued that "the early texts of the *Morte Darthur* were Caxton texts."[224] This is irrefutable, of course, in terms of the verbal text; however, de Worde's addition of illustrations utterly defines *Le Morte Darthur* for a new generation of readers increasingly attentive to visual culture, including landscape paintings and maps of the world. And, indeed, these iconic images have continued to visually define the work over the centuries. The illustrations of the *Morte* draw our attention, first, to what the artisan/reader saw as the defining episode of each book, but also to what the artisan saw as the fitting genre—that of *landskip* or landscape prints—for the matter of the book.[225] The cartographically-informed landscape images of de Worde's *Morte* were certainly influenced by early Netherlandish landscape painting, a genre stirred by the work of the Flemish painter Jan Van Eyck (1390–1441). Just 20 years before Caxton lived and worked in Bruges, and where Caxton undoubtedly met Wynkyn de Worde, Van Eyck spent his last years there painting commissioned works for Philip the Good, Duke of Burgundy, including his most famous work, the Ghent altarpiece of 1432, a work which famously set biblical scenes in naturalistic landscape settings.[226] In the case of the de Worde illustrations, Malory's text was innovatively linked to the early Flemish landscape movement and cartographically-influenced landscape prints, generating a new and creative visual experience of Malory's text.[227] Combining text and image, de Worde's *Le Morte d'Arthur* accumulated an increasing identification with early modern cartographic visual language.

For his edition of Malory's romance, de Worde used Caxton's front-end rubrics but also co-located them at the beginning of each of the 21 books and at the chapter divisions. As Tsuhyoki Mukai has noted, de Worde did so in an attempt to clarify and disambiguate Caxton's book and chapter headings.[228] In addition, de Worde added a woodblock image for each of the 21 books that articulates and anticipates an episode, as James Wade has pointed out, from the first or second chapter of each book.[229] The images compound the reader's foreknowledge of the well-known Arthurian tales and visually reinforce these episodes in the imaginative landscape of the text. Hodnett calls the de Worde illustrations "wild and wonderful"—and they really are, especially when seen as a thematic group, rather than the one-off often reproduced for an essay or to illustrate a web page. The opening image of the work illustrates the seduction of Igraine by Uther Pendragon, an episode which sets in motion Arthur's conception and kingship (Figure 2.8).[230] Through the conventions of distant city and foreground of Netherlandish landscape images, the illustration conveys the tension between Tintagel and the court of Uther—the geography of power that forms the central core of the work. The most reproduced of the de Worde images is that from Book Two depicting Balyn unsheathing the sword of the damosel sent by the Lady of the Lake to Camelot (Figure 2.9). Here, as in the previous image, the figures are set outside the formal structure of the castle amid the raised horizon of a landscape that evokes distance and a skyscape suggestive of a sense of flux. The tension between the formal power structures of Camelot and the intimacy of the

The fyrste boke.

Here begynneth the fyrst bo/ke of the noble kyng . Kyng Arthur . somtyme kynge of Englonde and of his noble actes and feates of armes of chyualrye & his noble knyghtes & table rou͂de . and is deuyded in to . xxi . bokes.

¶ How Utherpendragon sent for the duke of Cornewayll and Igrayne his wyf/and of theyr departynge sodaynly agayne. Capl'm Primu͂.

HIt befell in the dayes of Utherpendragon / whan he was kyng of all Englonde/and so regned that there was a myghty duke in Cornewayll that helde warre ayenst hym longe tyme . And the duke was called the duke of Tyntagyll / & so by meanes kynge Uther sende for this duke/ chargynge hym to brynge his wyfe wt hym/for she was called a fayre lady/ & a passynge wyse/and hyr name was called Igrayne . So whan the duke & his wyf were comen vnto the kyng by meanes of grete lordes they were accorded both/the kyng lyked & loued this lady well/& he made them grete there out of mesure / & desyred to ; aue lyen

a i

Figure 2.8 Book 1, *Le Morte d'Arthur*, Wynkyn de Worde, 1498 (STC 802)
Source: Copyright of the University of Manchester.

adventures the court will set in motion are captured in the depiction of both individuals, city, and landscape. All of these chapter illustrations convey a sense of a vast landscape, amplifying the setting in which the individual characters must negotiate the power structures within and the challenges to Arthur's kingdom.

Of the 22 images in the edition, 19 are bespoke illustrations that directly recount an episode in Malory's text.[231] The images speak to the role of *Le Morte Darthur* in the transfer of emerging carto-landscape images into early English print culture, synthesizing iconic episodes of Arthurian legend, such as the Grail quest or the Knight of the Cart with Netherlandish landscape imagery. For the woodcut artist, the fictional Arthurian landscape inspired a connection to qualities of Netherlandish landscape painting, indicating an already established dialogue between Renaissance cartography, landscape, and literary work.[232] Seth Lerer suggests that de Worde's woodcuts often have "a

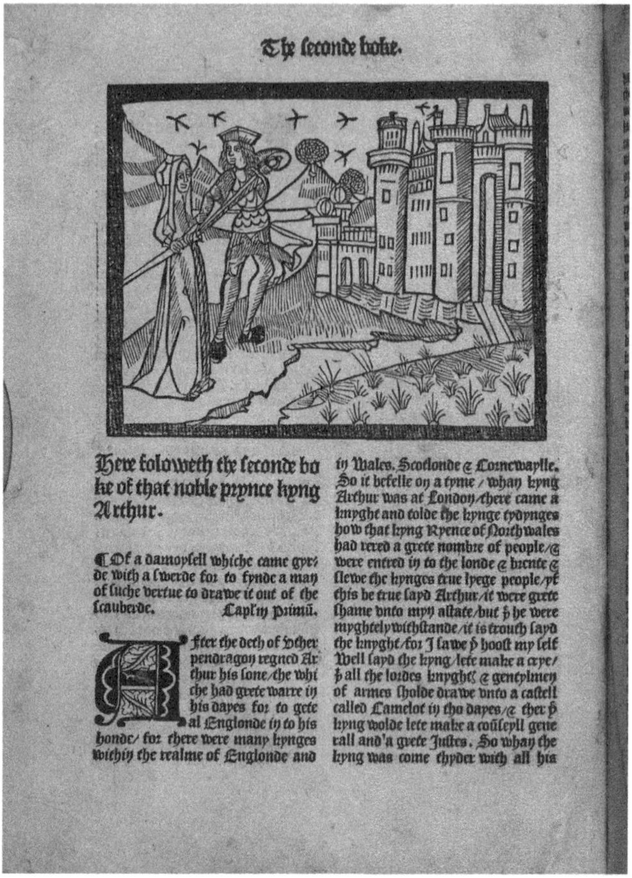

The seconde boke.

Here foloweth the seconde boke of that noble prynce kyng Arthur.

¶ Of a damoysell whiche came gyrde with a swerde for to fynde a man of suche vertue to drawe it out of the scauberde. Capl'm primū.

After the deth of Uther pendragon regned Arthur his sone/the whiche had grete warre in his dayes for to gete al Englonde in to his honde/ for there were many kynges within the realme of Englonde and in Wales. Scotlonde & Cornewaylle. So it befelle on a tyme / whan kyng Arthur was at London/there came a knyght and tolde the kynge tydynges how that kyng Ryence of Northwales had rered a grete nombre of people/& were entred in to the londe & brente & slewe the kynges true lyege people/yf this be true sayd Arthur/it were grete shame unto myn astate/but þ he were myghtely withstande/it is trouth sayd the knyght/for I sawe þ hoost my self Well sayd the kyng/lete make a crye/ þ all the lordes knyghtes & gentylmen of armes sholde drawe unto a castell called Camelot in tho dayes/& ther þ kyng wolde lete make a coūseyll generall and a grete Justes. So whan the kyng was come thyder with all his

Figure 2.9 Book 2, *Le Morte d'Arthur*, Wynkyn de Worde, 1498 (STC 802)
Source: Copyright of the University of Manchester

precise function" in his editions and that the visual appearance and layout of de Worde's texts reveal "aspects of his critical approach as well as the cultural importance of the texts he chose to print."[233] The fusing of England's great chivalric tale with the iconography of landscape is a new form of English hybrid cultural geography. Further, the unique imagery of *Le Morte d'Arthur* speaks to de Worde's or his artisan's sense of the work's inherent geographic capaciousness. Whether de Worde felt the woodcuts for Malory's work deserved to be uniquely associated with the text or whether he cast aside these cuts as unsuitable for any future work remains a mystery.

In the late fifteenth century, Flemish, Belgian, Netherlandish, and German artists began to define a genre that would have incalculable impact on the field of visual art. The associated German word for this new genre, *Weltlandschaft*, translates as *orbis terrarum* in Latin or, in English, world landscape. Influenced by cartography and the representation of space, the painter most closely associated with the development of landscape painting as a genre is the Flemish painter Joachim Patinir (1485–1524). Although Patinir, considered the first landscape painter, overlaps with de Worde's later output in the early sixteenth century, emerging landscape painting was already in development in the late fifteenth century, influenced by Van Eyck and Rogier van der Weyden, in Brussels. Drawing upon the technique of linear perspective, these early Netherlandish landscape paintings depicted distant townscapes, dramatic outcroppings, aerial views, and the suggestion of birds scattered across the sky to create an atmospheric perspective and cartographic view. Artist and engraver Albrecht Durer, a contemporary of de Worde and friend of Patinir, utilized modified linear perspective to depict background townscapes and soaring birds to evoke a sense of distance, space, and movement within the pictorial view of his engravings and prints. Many painters augmented their work in the fine arts by supplying woodcut designs, and one such artist created the woodcuts for de Worde's *Le Morte Darthur*, distinguishing the book as part of the early landscape movement.

In early Netherlandish landscape compositions, a dramatic, panoramic landscape is depicted from an elevated or bird's eye view. In Joachim Patinir's *Landscape with St. Jerome* (1516), for example, Patinir raises the line of the horizon, leading to an increase in the space devoted to the foreground and landscape (Figure 2.10). He also creates volumetric space through the

Figure 2.10 Joachim Patinir, *Landscape with St. Jerome*, c.1516–1517
Source: © Museo Nacional del Prado.

placement of smaller-scale scenes and cityscapes in the background. These early landscape paintings invite the viewer into the world of the image, breaking the barrier between the pictorial space and that of the viewer. Creating expansive terrain by elevating the horizon, depicting distant townscapes, and including dramatic outcroppings, these landscapes co-developed with the mathematical, relational principles and God's-eye gaze of cartography as well as the increasingly spatial-realism of narrative geographies. Italian artists were intrigued by the Flemish technique of creating landscapes from "both close by and far away,"[234] a dual pictorial organization present in the woodcuts of de Worde's *Morte* in which the narrative act is foregrounded, but with a distant landscape, implying that the action of the romance occurs both in private and in public spaces, a visual re-creation of the mixed-scale Malory used in his narrative (see Chapter 1). As Manfred Sellink explains, the public's growing interest in geography and cosmography was an "important stimulus to the rapid growth of the landscape as an autonomous genre in the sixteenth-century Netherlands."[235] To augment the illusion of a sweeping, panoramic view and to evoke, symbolically, a world in motion, landscape artists included birds in flight, despite the sky's relative smaller field in the painting. The effect, reproduced more simply in the de Worde woodcuts, is one of a long, cartographic prospect: a map-like view. While the representation of distant townscapes and birds in motion is not unique to the Malory cuts, it is so much a part of the iconography of de Worde's *Morte d'Arthur* as to be part of the defining imagery of the text, as in the image of Nyneve and Merlin (Figure 2.11). Contextualized in the cartographic milieu of the Ptolemaic revival, de Worde's iconographic program for Malory's *Morte* is replete with the visual language of spatial representations of maps and Netherlandish landscape painting, reinforcing the link between cartographic thinking and Malory's narrative.

Netherlandish landscape engravings and prints, or *landskip*, a term that came into use in late sixteenth-century England, were part of an Anglo-Dutch network that "facilitated the circulation of prints, exchange of ideas, and mutual influence between English and Dutch engravers and printmakers."[236] The de Worde artisan should be recognized as a very early innovator in this circulation of Anglo-Dutch visual print culture. Martha Driver has evocatively demonstrated that woodblock artists took images from paintings or manuscript illumination and rendered them into the artform of woodblock carving.[237] While their adaptation to print required a simplification in form, David Areford argues that:

> between the conceptions of the medieval and the modern, fifteenth-century prints are best explored as they are: experimental, multimedia, and dirtied by the hands of viewers who activated the power and potential of each image for practical and religious goals.[238]

Early Flemish landscape paintings emphasized a horizontal field, an orientation which complemented the double-column image size of the folio pages of

The fourth boke.

Here foloweth the fourth boke of that noble prynce kyng Arthur.

¶ How Merlyn was aſſoted & dooted on one of the ladyes of the Lake / and how he was ſhette in a roche vnder a ſtone & there deyed. Capl'm Primū

SO after thyſe queſtes of ſyr Gawayne / ſyr Tor / and kyng Pellynore. It felle ſoo that Merlyn felle in a dotage on the damoyſell that kyng Pellynore brought to court and ſhe was one of the damoyſels of the Lake þ hyght Nyneue. But Merlyn wolde lete haue her noo reſte but alwayes he wolde be with her. And euer ſhe made Merlyn good chere tyll ſhe had lerned of hym all maner thynge that ſhe deſyred & he was aſſoted vpon her that he myght not be from her. So on a tyme he tolde kyng Arthur that he ſholde not dure longe / but for all his craftes he ſholde be put in the erthe quycke / & ſo he tolde þ kyng many thynges that ſholde befalle / but alwayes he warned the kyng to kepe well his ſwerde & the ſcauberde / for he tolde hym how the ſwerde & the ſcauberde ſholde be ſtolen by a woman frō hym that he mooſt truſted. Alſo he tolde kyng Arthur that he ſholde myſſe hym / yet had ye leuer than all your

Figure 2.11 Book 2, *Le Morte d'Arthur,* Wynkyn de Worde, 1498 (STC 802)
Source: Copyright of the University of Manchester.

de Worde's *Morte,* a book form de Worde later relinquished in favor of the more affordable quarto-size, perhaps explaining why these images are not used in any of his subsequent publications, despite his practice of liberally reusing woodcuts. The townscapes, high horizons, rocky outcroppings, and implied movement of birds form a visual interpretation of de Worde's *Morte* that amplifies the already-activated participation of the work with a world imaginary of geographic specificity and expansion. The de Worde woodcut artist visually reinforces the new Ptolemaic principles of time and space, highlighting Malory's great innovation of the de-entrelacement, or narrative linearization, of his source text's episodic romances. De Worde's illustrator intuitively grasped the cartographic aesthetic of Malory's text and used it as a leitmotif to co-create a new version of Malory's romance as an early modern illustrated book. It is irrefutable that the de Worde woodcut artisan selected the technique of world landscape as a fitting visual accompaniment for the deeply human, large-scale emotionality of Malory's literary space and for the text's evocation of geo-spatiality. In so doing, the artisan and printer of the second edition of Malory's Arthurian epic created a visual metaphor of the figurative and cartographically-aligned territory of the text. Sadly, by the time de Worde's 1529 edition was printed, some of the original woodcuts had been lost or damaged, such that the second edition is a less coherent blend that includes some of the original *landskip*-inspired woodcuts as well as generic woodcuts from his storehouse of images, losing some of the potency of the original cartographic aesthetic.

The majority of de Worde's hundreds of woodcuts are familiar to any reader of early modern print editions, many with low horizons and blank backgrounds which emphasize the figure in the foreground. However, a few stylistically-related woodcuts can be found in the end-leaf of de Worde's edition of *Dives & Pauper* (1496), on his title-page of *The crafte to lyue well and to dye well* (1505) and, notably, in an image repeated across three other texts he printed, each of which figures in the narrative transmission of geographic thought in England: *De proprietatibus rerum*, the encyclopedic compendium which, as Michael Twomey has pointed out, included a geography that is largely inflected by its classical sources rather than any correspondence to English geography;[239] the *Chronicles of England*; and the *Description of England* (de Worde's re-titling of Caxton's *Description of Britain*).[240] The image depicts a panoramic landscape, complete with the elements that define early Netherlandish landscape painting: an elevated horizon, dramatic, rocky outcroppings, distant townscapes, and birds soaring across the sky (Figure 2.12). The townscapes are depicted with winding paths which seemingly lead from the viewer to these distant towns, each with a church tower rising from its center, evoking a sense of a journey about to commence and suggesting that urban spaces and a wider world await the reader/viewer's exploration.

The visual aesthetic of the 1498 *Morte d'Arthur* seems to have eventually had an influence on the visual strategy of the 1515 edition of Mandeville's *Travels*.[241] Though not as finely shaded or rendered as the Malory woodcuts, de Worde's 1515 edition eschews the earlier imagery of Sorg's naked Mandeville set against a blank background. Instead, Mandeville's *Travels* now enters into dialog with the cartographic aesthetic of landscape painting and the images of de Worde's edition of Malory. On the title page, a fully clothed Mandeville with staff in hand strikes out across the visual depth of an English landscape, leaving behind a turreted castle, bound for Constantinople.

Explicit

Caxton's and de Worde's print works were critical in forming English geographic identity, but that identity was diffuse and not exclusively "English." Geographic identity formation in England was influenced by Continental humanist book practices, personal expatriate experiences (in Caxton's case, in Bruges; in de Worde's case, in London), diplomatic and ecclesiastic envoys,

Figure 2.12 de proprietatibus rerum, Wynkyn de Worde, c. 1495, fol F.iii (v)
Source: © British Library Board.

military campaigns as part of the Hundred Years War and Crusades, pilgrimages to the Holy Land, trade networks, map dissemination in the form of Ptolemy's *Geography*, the translation of Arabic and Latinate scholarship, early landscape painting, and the movement of artists and books between Italy, France, the Low Countries, and England. To identify Caxton as a geographic innovator, highly attuned to Continental trends, is to recognize his critical role in the "bringing into being of a new, Tudor, print culture"[242] and, as part of that culture, the dissemination of elements of Ptolemaic geography. The artisan of de Worde's *Le Morte d'Arthur* similarly synthesized the cartographic visual culture of landscape artists into early Tudor print culture, reimaging the landscape of a centuries-old English-French hybrid text for English readers within a Ptolemaic milieu. An essential though overlooked contribution to early Tudor print culture and the dissemination of Ptolemaic-influenced geographic principles include the formative geographical editions of William Caxton's and the singular illustrative scheme of Wynkyn de Worde's *Le Morte d'Arthur*.

Notes

1 David Woodward, "Maps and the Rationalization of Geographic Space," in *Circa 1492: Art in the Age of Exploration*, ed. Jay A. Levenson (New Haven, CT: Yale University Press, 1991), 83–88.

2 Matthew Edney, *Cartography: The Ideal and its History* (Chicago: University of Chicago Press, 2019), 3.

3 Violet Moller, *The Map of Knowledge: A Thousand-Year History of How Classical Ideas Were Lost and Found* (New York: Doubleday, 2019).

4 An image of a T-O map is available at www.princeton.edu/~his291/T-O_Map.html

5 Alessandro Scaffi, "After Ptolemy: The Mapping of Eden," in *Ptolemy's Geography in the Renaissance*, eds. Zur Shalev and Charles Burnett (London and Turin: The Warburg Institute – Nino Aragno Editore, 2011), 187–206; 188.

6 Edney, 167. See, also, the discussion of scale in his Chapter Two.

7 J. Lennart Berggren and Alexander Jones, *Ptolemy's "Geography": An Annotated Translation of the Theoretical Chapters* (Princeton, NJ: Princeton University Press, 2002).

8 Benjamin Weiss, "The Geography in Print: 1475–1530," in *Ptolemy's Geography in the Renaissance*, eds. Zur Shalev and Charles Burnett (London and Turin: The Warburg Institute – Nino Aragno Editore, 2011), 91–120.

9 Scaffi, 188.

10 Kathy Lavezzo, *Angels on the Edge of the World: Geography, Literature, and English Community, 1000–1534* (Ithaca, NY: Cornell University Press, 2006), 119.

11 See, for example, David Rundle, "Humanism Across Europe: The Structure of Contact," in *Humanism in Fifteenth-Century Europe*, ed. David Rundle (Oxford: Society for the Study of Medieval Languages and Literature, 2012) and A.E.B. Coldiron, *Printers Without Borders: Translation and Textuality in the Renaissance* (Cambridge: Cambridge University Press, 2015).

12 Peter Spufford, *Power and Profit: The Merchant in Medieval Europe* (New York: Thames and Hudson, 2002). See especially 22–26.

13 Coldiron, 1.

14 Ibid., 104; 2.

15 For the debate regarding early print and its impact, see Elizabeth Eisenstein, *The Printing Press as an Age of Change: Communications and Cultural Transformations in Early-Modern Europe* (Cambridge: Cambridge University Press, 1979), Adrian Johns, *The Nature of the Book: Print, and Knowledge in the Making* (Chicago: University of Chicago Press, 2000), and D. McKitterick, *Print, Manuscript and the Search for Order, 1450–1830* (Cambridge: Cambridge University Press, 2003). Robert Darnton proposes a communications circuit model in "What Is the History of Books?" in *The Book History Reader*, eds. David Finkelstein and Alastair McCleery (London: Routledge, 2002), 65–83. Works on textual theory that inform this context include D.F. McKenzie, *Bibliography and the Sociology of Texts* (Cambridge: Cambridge University Press, 1999) and Roger Chartier, *Inscription and Erasure: Literature and Written Culture from the Eleventh to the Eighteenth Century*, trans. Arthur Goldhammer (Philadelphia, PA: University of Pennsylvania Press, 2007).

16 Sean Roberts, *Printing a Mediterranean World: Florence, Constantinople, and the Renaissance of Geography* (Cambridge, MA: Harvard University Press, 2013).

17 Denis Cosgrove, *Apollo's Eye: A Geographic Genealogy of the Earth in the Western Imagination* (Baltimore, MD: The Johns Hopkins University Press, 2001), 12.

18 For a discussion of pre-cartographic mapping, see Nicholas Howe, *Writing the Map of Anglo-Saxon England: Essays in Cultural Geography* (New Haven, CT:

Yale University Press, 2007). See also Keith D. Lilley, *City and Cosmos: The Medieval World in Urban Form* (London: Reaktion Books, 2009).

19 Alexander Jones, "Ptolemy's *Geography*: A Reform that Failed," in *Ptolemy's Geography in the Renaissance*, eds. Zur Shalev and Charles Burnett (London and Turin: The Warburg Institute – Nino Aragno Editore, 2011), 15–30; 17.

20 Tom Conley, *The Self-Made Map: Cartographic Writing in Early Modern France* (Minneapolis, MN: University of Minneapolis Press, 1996), 10.

21 Kellie Robertson, "Materiality and the Hylomorphic Imagination," in *Medieval Literature: Criticism and Debate*, eds. Holly A. Crocker and D. Vance Smith (New York: Routledge, 2014), 367–375; 367.

22 See Roberts.

23 For a discussion of scientific and literary writings that share similarities in form, style, and rhetorical mode, see Matthew Boyd Goldie, *Scribes of Space: Place in Middle English Literature and Late Medieval Science* (Ithaca, NY: Cornell University Press, 2019), 4.

24 Coldiron, 36.

25 Edmund Childs, *William Caxton: A Portrait in a Background* (London: Northwood Publications, 1976), 125.

26 *The Myrrour of the World* (Westminster: William Caxton, 1481, 1490), STC 24762, 24763; Ranulf Higden, *the discripcion of Britayne* (Westminster: William Caxton, 1480), STC 13440a; and *Polychronicon* (Westminster: William Caxton, 1482), STC 13438.

27 Dating on the reprint varies from 1489 (Glasgow University Library), to 1490 (British Library), to 1491 in *Oxford Handbook to Tudor Literature, 1485–1603*, eds. Mike Pincombe and Kathy Shranks (Oxford: Oxford University Press, 2009).

28 Jones, 21.

29 See William Kuskin, *Symbolic Caxton: Literary Culture and Print Capitalism* (Notre Dame, IN: University of Notre Dame Press, 2008), 16. See also Kuskin's other works on Caxton and early print: *Recursive Origins: Writing at the Transition to Modernity* (Notre Dame, IN: University of Notre Dame, 2013) and *Caxton's Trace: Studies in the History of English Print* (Notre Dame, IN: University of Notre Dame Press, 2006). See also Clare Ingram, "Losing French: Translation, Nation, and Caxton's English Statutes," in *Caxton's Trace*, 275–298.

30 Indeed, Trevisa refers to his narrative geography as a "*mappamundi.*"

31 David Rundle, "Humanism Across Europe: The Structure of Contact," in *Humanism in Fifteenth-Century Europe*, ed. David Rundle (Oxford: Society for the Study of Medieval Languages and Literature, 2012), 307–336; 307.

32 Ibid., 309.

33 Roberts, 11.

34 Ibid., 11.

35 Ibid., 13.

36 Lisa Jardine, *Worldly Goods: A New History of the Renaissance* (New York: Nan A. Talese, 1996).

37 Roberto Weiss, *Humanism in England During the Fifteenth Century*, edited and prepared for electronic publication by David Rundle and Anthony John Lappin (Oxford: Society for the Study of Medieval Languages and Literature, 2009), 10.

38 Murray Newton Rothbard, *Economic Thought Before Adam Smith: An Austrian Perspective on the History of Economic Thought* (Auburn, AL: Ludwig von Mises Institute, 1995), 195.

39 Rundle, 311.

40 Rothbard, 195.

41 Rundle, 313.

42 Poggio Bracciolini, quoted in Stephen Greenblatt, *The Swerve: How the World Became Modern* (New York: W.W. Norton, 2012), 207.
43 Kristian Jensen, "Universities and Colleges," in *The Cambridge History of Libraries in Britain and Ireland,* (No. 1), eds. Elisabeth Leedham-Green and Teresa Webber (Cambridge: Cambridge University Press, 2006), 345–362; 350.
44 Benjamin Weiss, 107, 113.
45 Ibid., 116.
46 Rundle, 314–315.
47 Ibid., 315.
48 Patrick Gautier-Dalché, "The Reception of Ptolemy's *Geography* (End of the Fourteenth to Beginning of the Sixteenth Century)," in *The History of Cartography,* vol. 3: *Cartography in the European Renaissance,* Part 1, ed. David Woodward (Chicago: University of Chicago Press, 2007), 285–364; 319.
49 For example, see Elizabeth Leedham-Greene, *Private Libraries in Renaissance England: A Collection and Catalogue of Tudor and Early Stuart Book-Lists,* vols. 1–4 (Binghamton, NY: Medieval and Renaissance Texts and Studies,1993). In volume 3, *Oxford Inventories Mostly for the Purpose of Probate,* one William Brown, Scholar (died 1558) is listed as owning: "*Ptolomei geographia libri 8 greece,*" and "Claudius Ptolemy. *Geographia.*" The latter is determined to have been printed on the Continent: date indeterminable. Leedham-Green's reference number is 67.189. Such references to Ptolemy, however, are relatively few throughout the four-volume study.
50 Rundle, 309.
51 Ibid., 324.
52 J.B. Trapp, "The Humanist Book," in *The Cambridge History of the Book in Britain,* eds. Lotte Hellinga and J.B. Trapp (Cambridge: Cambridge University Press, 1999), 283–325; 287. As one example, Rundle notes that Thomas Candon worked in Italy for a time as a scribe. See Rundle, 319–321.
53 Raphael Holinshed, *Holinshed's Chronicles,* 1587 edition, vol. 6, 710. *The Holinshed Project.* Available at: http://english.nsms.ox.ac.uk/holinshed/.
54 Ibid., 710.
55 Rundle, 326.
56 Ibid., 326.
57 Ibid., 327.
58 Anne F. Sutton, "Merchants," in *A Companion to the Early Printed Book in Britain, 1476–1558,* eds. Vincent Gillespie and Susan Powell (Cambridge: D.S. Brewer, 2014), 127–133; 127–128.
59 Jennifer Goodman, "Caxton's Continent," in *Caxton's Trace: Studies in the History of English Printing,* ed. William Kuskin (Notre Dame, IN: University of Notre Dame Press, 2006), 101–123, 101.
60 N.F. Blake, *William Caxton and English Literary Culture* (London; The Hambledon Press, 1991), 19.
61 Thomas Crofts, *Malory's Contemporary Audience: The Social Reading of Romance in Late Medieval England,* Arthurian Studies LXVI (Cambridge: D.S. Brewer, 2006), 42.
62 Goodman, 101.
63 See A.E.B. Coldiron, "Cultural Amphibians," *Yearbook of Comparative and General Literature* 51 (2003–2004): 43–58.
64 Alexandra Gillespie, "'Folowynge the Trace of Mayster Caxton:' Some Histories of Fifteenth-century Printed Books," in *Caxton's Trace: Studies in the History of English Printing,* ed. William Kuskin (Notre Dame, IN: University of Notre Dame Press, 2006), 167–195.
65 George Doutrepont, *Inventaire de la "Librarie" de Phillipe le Bon 1420* (Geneva: Slatkine Reprints, 1977).

66 Goodman, 119.

67 Peter Barber, "England I: Pageantry, Defense, and Government: Maps at Court to 1550," in *Monarchs, Ministers, and Maps: The Emergence of Cartography as a Tool of Government in Early Modern Europe*, ed. David Buisseret (Chicago: University of Chicago Press, 1992), 22–56; 26.

68 Wynkyn de Worde published an edition of Mandeville's *Travels* in 1498. Caxton may have intended to publish an edition: a manuscript copy of the text appears to have been in his print shop. Prior to de Worde's print edition, manuscript copies of the text known as the Defective Version circulated among English readers as "the dominant form of the book in England." See M.C. Seymour ed. *The Defective Version of Mandeville's Travels* (Oxford: Oxford University Press for the Early English Text Society, 2002). Seymour estimates that the Defective Version was translated into English by the late 1300s.

69 Rundle, 330.

70 Benjamin Weiss, 91.

71 E. Gordon Duff, *William Caxton* (New York: Burt Franklin, 1970. Originally printed 1905), 14.

72 Robert J. Mayhew, "Geography, Print Culture and the Renaissance: 'The Road Less Travelled By.''" *The History of European Ideas* 27:4 (2001): 349–369; 352.

73 Roberts, 2.

74 Ibid., 179.

75 R.H. Wilson, "Malory and Caxton," in *A Manual of the Writing of Middle English, 1050–1500*, vol. 3., ed. A.E. Hartung (New Haven, CT: Connecticut Academy of Arts and Sciences, 1972), 757–807. Wilson's and Crofts' discussion of Caxton's Preface to the *Historical Libra*ry by Diodorus Siculus (translated from the Greek by Poggio Bracciolini and subsequently translated by Caxton into English possibly via a French intermediary text) may be worth re-examination in light of recent research into English humanism.

76 Raphael Lyne, *Ovid's Changing Worlds: English Metamorphoses, 1567–1632* (Oxford: Oxford University Press, 2001), 27.

77 Crofts, 32.

78 Lotte Hellinga, *William Caxton and Early Printing in England* (London: British Library, 2010). See also Henry Lathrop, *Translations from the Classics into English, from Caxton to Chapman, 1477–1620* (New York: Octagon Books, 1967), although Lathrop considers the list "a short one," 16.

79 For a discussion of Caxton's Preface and translation practices, see Allesandra Petrina, "The *Aeneid* of the North: William Caxton's *Eneydos* and Gavin Douglas's *Eneados*," *The Mediaeval Journal* 7:1 (January 2017): 145–162.

80 Lathrop emphasizes Tiptoft's "Italian ideals" in terms of book-buying and international relations, 16, 24.

81 See ibid., 24.

82 Lotte Hellinga, *Caxton in Focus: The Beginnings of Printing in England* (London: British Library, 1982), 85.

83 Coldiron, 65–69; 67.

84 In the context of the *Chronicles of England*, Lister M. Matheson came to a similar conclusion: that Caxton had been unappreciated for "the breadh of his reading in history and his skill as a compiler." See Lister M. Matheson, "Printer and Scribe: Caxton, the *Polychronicon*, and the *Brut*," *Speculum* 60:3 (July 1985): 593–614, 610.

85 Thomas Crofts, "Thynges Foresayd Aledged," in *Reviewing Le Morte Darthur: Texts and Contexts, Characters and Themes*, eds. Kevin Sean Whetter and Raluca L. Radulescu (Cambridge: D.S. Brewer, 2005), 54.

86 Kathleen Tonry, "Reading History in Caxton's *Polychronicon*," *JEGP* 111:2 (2012): 169–98; 171, 197.

87 Ibid., 195.

88 Paul Needham, "The Customs Rolls as Documents for the Printed-book Trade in England," in *The Cambridge History of the Book in Britain*, vol. 3, *1400–1557*, eds. Lotte Hellinga and J.B. Trapp (Cambridge: Cambridge University Press, 1999), 148–163; 148.

89 Ibid., 148.

90 Ibid., 160.

91 Margaret Lane Ford, "Importation of Printed Books into England and Scotland," in *The Cambridge History of the Book in Britain*, vol. 3: *1400–1557*, eds. Lotte Hellinga and J.B. Trapp (Cambridge: Cambridge University Press, 1999), 179–202.

92 Ibid., 183.

93 See Thomas F. Madden, *Venice: A New History* (New York: Viking Penguin, 2014), 319.

94 See Elizabeth Ross, *Picturing Experience in the Early Printed Book: Breyenbach's Peregrinatio from Venice to Jerusalem* (University Park, PA: Penn State University Press, 2014) and Bronwen Wilson, *The World in Venice: Print, the City, and Early Modern Printed Books* (Toronto: Toronto University Press, 2005).

95 Moller, 253.

96 Ford, 186.

97 R.J. Shoeck, "The Geography of Erasmus," in *Northern Humanism in European Context: 1469–1625: From the 'Adwert Academy' to Ubbo Emmius*, eds. Fokke Akkerman, Arie Johan Vanderjagt, and A.H. Van Der Laan (Leiden: Brill, 1999), 201.

98 Needham, 153.

99 Ibid., 153.

100 Needham, 154.

101 Coldiron, 5.

102 Ibid., 31.

103 See, for example, Peter D. Clarke, ed. *The University and College Libraries of Cambridge*, Corpus of British Medieval Library Catalogues, vol. 10 (London: The British Library in association with the British Academy, 2002).

104 David N. Bell, "The Libraries of Religious Houses in the Late Middle Ages," in *The Cambridge History of Libraries in Britain and Ireland*, eds. Elisabeth Leedham-Green and Teresa Webber (Cambridge: Cambridge University Press, 2006), 126–151; 127. See also James G. Clark, "University Monks in Late Medieval England," in *Medieval Monastic Education*, eds. George Ferzoco and Carolyn Muessig (London: Leicester University Press, 2000), 56–71.

105 Bell, 127.

106 Roger Lovatt, "Introduction," in *The University and College Libraries of Cambridge*, Corpus of British Medieval Library Catalogues, vol. 10, ed. Peter D. Clarke (London: The British Library in association with the British Academy, 2002), xxvii–xcii; lxxvi.

107 Clarke. See listing for Christ College, item 33 (p. 116) and Corpus Christi College, item 105 (p. 251).

108 For a convenient list of editions, see "Printed Editions of Ptolemy Containing Maps," *Map Forum*. Available at: www.mapforum.com/02/ptolemy.htm.

109 Ford, 193.

110 Ranulf Higden, *The Description of Britain*, The English Experience, No. 386, published in facsimile of the printed edition by William Caxton 1480, trans., John Trevisa (Amsterdam: Theatrum Orbis Terrarum and DaCapo Press, 1971).

111 Wikipedia incorrectly lists the date of publication of Caxton's edition of the *Polychronicon* as 1480. In fact, Caxton printed the *Description of Britain* in 1480 and the full version of the *Polychronicon* in 1482. William Blades suggests,

however, that Caxton's plan had been to print the *Description* as an "addendum" after completion of the *Polychronicon*; see *The Life and Typography of William Caxton, England's First Printer*, vol. 2 (London, 1863, reprinted Cambridge: Cambridge University Press, 2014), 112.

112 See, for example, Royal MS 14.C.IX, fol. 2v. For a discussion of Higden's text and the accompanying maps, see Evelyn Edson, *The World Map, 1300–1492: The Persistence of Tradition and Transformation* (Baltimore, MD: Johns Hopkins University Press, 2007).

113 Lavezzo, 119.

114 Ibid., 119.

115 Ibid., 119.

116 Catherine Sanok, *New Legends of England: Forms of Community in Late Medieval Saints Lives* (Philadelphia, PA: University of Pennsylvania Press, 2018), 3.

117 Ibid., 134.

118 Tonry, 187.

119 Ibid., 187.

120 Alessandro Scaffi, "After Ptolemy: The Mapping of Eden," in *Ptolemy's Geography in the Renaissance*, eds. Zur Shalev and Charles Burnett (London and Turin: The Warburg Institute – Nino Aragno Editore, 2011), 187–206; 189.

121 Emily Steiner, "Compendious Genres: Higden, Trevisa, and the Medieval Encyclopedia," *Exemplaria* 27: 1–2 (Spring/Summer 2015), 73–92.

122 Tom Conley proposes the early modern writer as topographer, defining national space, in "Early Modern Literature and Cartography: An Overview," in *The History of Cartography*, vol. 3: *Cartography in the European Renaissance*, Part 1, ed. David Woodward (Chicago: University of Chicago Press, 2007), 402–411; 405.

123 Claudius Ptolemy, *Geography of Claudius Ptolemy*, trans. Edward Luther Stevenson (New York: Cosimo Classics, 2011), 47.

124 Ibid., 48.

125 A.S.G. Edwards, "Geography and Illustration and Higden's *Polychronicon*," in *Art into Life: Collected Papers from the Kresge Art Museum Medieval Symposium* (Grand Rapids, MI: Michigan State University, 1995), 95–113; 96.

126 Ranulf Higden, *Discripcion of Britayne*, trans. John Trevisa, ed. William Caxton (Westminster: 1480), STC 13440a.

127 Ibid,, ca. xviv.

128 Ibid,, Preface.

129 Ibid., Preface.

130 Caroline M. Barron, "What Did Medieval London Merchants Read?" in *Medieval Merchants and Money: Essays in Honour of James L. Bolton*, eds. Martin Allen and Matthew Davies (London: University of London Press, 2016), 43–70. See also Kathleen. L. Scott, "Past Ownership: Evidence of Book Ownership by English Merchants in the Later Middle Ages," in *Makers and Users of Medieval Books: Essays in Honour of A.S.G. Edwards*, eds. C. Meale and D. Pearsall (Woodbridge: Boydell and Brewer, 2014), 150–177; and Anne F. Sutton, "The Acquisition and Disposal of Books for Worship and Pleasure by Mercers of London in the Later Middle Ages," in *Manuscripts and Printed Books in Europe, 1350–1550*, eds. E. Cayley and S. Powell (Liverpool: Liverpool University Press, 2013), 95–114.

131 Readers may enjoy a charming modern-English version, illustrated with photographs; see Marie Collins, *Caxton: The Description of Britain, A Modern Rendering* (New York: Weidenfeld & Nicolson, 1988).

132 The title is alternatively spelled *Mirrour of the World.* Caxton uses both spellings to refer to the book. I follow, here, the Catalogue of the British Library.

133 Kuskin, *Symbolic Caxton*, 18.

134 H.S. Bennett, *English Books and Readers*, vol. 1: *1475–1557* (Cambridge: Cambridge University Press, 1969), 110.
135 George Kimble, *Geography in the Middle Ages* (New York: Russell and Russell, 1938; reissued 1968), 200.
136 Roberts, 10.
137 Ibid., 11.
138 Moller, 229–230.
139 Ibid., 254.
140 Roberts, 2.
141 Gautier-Dalché. See also F.E. Robbins, trans. and ed. *Tetrabiblos* by Claudius Ptolemy (Cambridge, MA: Harvard University Press, 1940).
142 Robbins, *Tetrabiblos.*
143 Charles Burnett, "Why Study Ptolemy's *Almagest*? The Evidence of MS Melbourne, State Library of Victoria, Sinclair 224," *The LaTrobe Journal* 81 (Autumn 2008): 126–143; 126.
144 Oliver H. Prior, ed. *Caxton's Mirrour of the World* (Berlin: A. Asher & Co. for the Early English Text Society, Extra Series CX, 1913), ix.
145 Ibid., vii.
146 Ibid., v (n. 1).
147 Cosgrove, 49.
148 W.J.B. Crotch, *The Prologues and Epilogues of William Caxton* (London: Early English Text Society, Original Series, No. 176, 1928), cxiii.
149 Prior, 7.
150 Ptolemy, 25.
151 Ibid., 26.
152 Prior, 8.
153 Ibid., vii.
154 Ibid., 7.
155 Ibid., 8.
156 Prior, 27; at 31 it is spelled "Cambryge."
157 Ibid., 31.
158 Ibid., 92–99.
159 Ibid., 171.
160 Pietro Falchetta, *Fra Mauro's World Map* (Turnhout, Belgium: Brepols, 2006), 54.
161 Prior, 94.
162 Ibid,, 94.
163 Stephen Greenblatt, *Hamlet in Purgatory* (Princeton, NJ: Princeton University Press, 2013), 75.
164 Ibid., 79.
165 Prior, 99; italics mine.
166 Coldiron, *Printers Without Borders*, 37; 52.
167 Ibid., 45.
168 Prior, 124 and 170. See also Prior's discussion, xxii–xxiii. Caxton's "corrections" are not derived from Ptolemy. Ptolemy's calculations, which underestimated the circumference of the Earth, are from his *Almagest* and *Planetary Hypotheses.* For a technical discussion of Ptolemy's mathematical process, see Christián C. Carman, "Rounding Numbers: Ptolemy's Calculation of the Earth-Sun Distance," *Archive for History of Exact Sciences* 63:2 (March 2009): 205–242.
169 Prior, xxii.
170 Ibid., 7.
171 Ptolemy, 29.
172 Prior, 184.
173 Ibid., 184–185.

174 Tonry, 172.
175 Gadi Algazi argues that the scholar's study was a new architectural space in late medieval Europe in "At the Study: Notes on the Production of the Scholarly Self," in *Space and Self in Early Modern European Cultures*, eds. David Warren Sabean and Malina Stefanovska (Toronto: University of Toronto Press, 2012), 17–50; 38.
176 See Leedham-Greene.
177 Cosgrove, xii.
178 Prior, 68, 182.
179 Ibid., 182.
180 Catherine Delano Smith and Roger J.P. Kain, *English Maps: A History* (Toronto: University of Toronto Press, 1999), 41.
181 Prior, 53.
182 For the concept of remediation, the way in which new media refashions earlier forms, see J. David Bolter and Richard A. Grusin, *Remediation: Understanding New Media* (Cambridge, MA: MIT Press 1999).
183 Moller, 243.
184 Duff, 51.
185 Edward Hodnett, *English Woodcuts, 1480–1535* (Oxford: Oxford University Press, 1973), 1.
186 Ibid., 1.
187 Ibid., 2.
188 Benjamin Weiss, 91.
189 Prior, 51–52.
190 Ibid., 52.
191 Ibid., 61.
192 Hodnett, 111.
193 In 2005, an English-made brass quadrant astrolabe from 1388 was discovered at an archeological dig at the Agnes House B and B in Canterbury. For an image of the Merton College equatorium, see Derek J. Price, ed. *The Equatorie of the Planetis, Edited from Peterhouse MS. 75 I* (Cambridge: Cambridge University Press, 1955), frontispiece. This manuscript is possibly in Chaucer's hand.
194 Prior, 8–9.
195 Cosgrove, 7.
196 Prior, 150.
197 Ibid., 40.
198 Cosgove, 28.
199 Ibid., 13.
200 Gautier-Dalché, 342 (note 398).
201 Ibid., 350.
202 Francis Bacon, *The New Organon*, trans. Lisa Jardine and Michael Silverthorne (Cambridge: Cambridge University Press, 2000), 69, (Book I, Aphorism *LXXXIV*).
203 Kuskin, *Symbolic Caxton*, 3–5.
204 Of the first edition, 33 copies are extant; of the second Caxton edition, 19 copies are extant. Prior, vi.
205 Bernard Klein, *Maps and the Writing of Space in Early Modern England and Ireland* (Houndsmills: Palgrave Macmillan, 2001), 5.
206 A recent discussion among Malorians Michael Twomey, Thomas Crofts, Thomas D. Hanks, and Karen Cherewatuk teased out the various interpretations of this phrase: the most literal rendering is "Caxton made me to be made," or "Caxton caused me to be made," suggesting Caxton's self-conception as directing and shaping the making of books.

207 Caxton states that he translated the *Myrrour* in 1480 but does not specifically state that the book was printed in that year. The date of publication is presumed to be 1480 or 1481.

208 Certain classical texts did provide a model for such book organization: Pliny the Elder's *historia naturalis* included a table of contents and earlier Quintus Valerius Solanis apparently had a table of contents (though only a single couplet of his work survives). See John Henderson, "Knowing Someone Through Their Books: Pliny on Uncle Pliny (*Epistles* 3.5)," *Classical Philology* 97 (2002): 256–284.

209 N.F. Blake, *Caxton: England's First Publisher* (London: Osprey Publishing, 1976), 68.

210 Ibid., 68. See also N. F. Blake, *William Caxton and English Literary Culture* (London: The Hambledon Press, 1991), 101–103.

211 Blake, *Caxton: England's First Publisher*, 116.

212 Ibid., 119.

213 Helen Cooper, "Arthur in Transition: Malory's *Morte Darthur*," in *Romance and History: Imagining Time from the Medieval to the Early Modern Period*, ed. Jon Whitman (Cambridge: Cambridge University Press, 2015), 120–133; 123.

214 Henry R. Plomer, *Wynkyn de Worde and His Contemporaries: From the Death of Caxton to 1535* (London: Grafton & Co., 1925), 45.

215 *[Here endyth the boke of Iohn Maunduyle*] (London: Wynkyn de Worde 1499), STC 17247; *Le Morte d'Arthur* (Westminster: Wynkyn de Worde, 1498), STC 802; *descrypcyon of Englonde* (Westminster: Wynkyn de Worde, 1498), STC 13440b. See also Siân Echard, "Malory in Print," in *The New Companion to Malory*, eds. Megan G. Leitch and Cory James Rushton (Cambridge: Boydell and Brewer, 2019), 96–121.

216 *Le Morte d'Arthur* (London: Wynkyn de Worde, 1529), STC 803; *The way to the holy lande* (Westminster: Wynkyn de Worde, 1503), STC 17249.

217 Karma Lochrie, *Nowhere in the Middle Ages* (Philadelphia, PA: University of Pennsylvania Press, 2015).

218 Ibid., 90.

219 Josephine Waters Bennett, "The Woodcut Illustrations in English Editions of 'Mandeville's *Travels*,'" *The Papers of the Bibliographic Society of America* 47:1 (First Quarter, 1953): 59–69.

220 Edward Hodnett, *Five Centuries of English Book Illustration* (Aldershot: Scolar Press, 1998), 15, 11.

221 Martha Driver, *The Image in Print: Book Illustration in Late Medieval England and Its Sources* (London: The British Library, 2004), 75. Driver details connections between continental artists in Delft, and Antwerp with de Worde's editions. See also Martha Driver, "The Illustrated de Worde: An Overview," *Studies in Iconography* 17 (1996): 349–403.

222 James Wade, "The Chapter Headings of the *Morte Darthur*: Caxton and de Worde," *Modern Philology* 111:4 (May 2014): 665–667; 663.

223 Hodnett, *Five Centuries*, 16.

224 Kevin S. Whetter, *The Manuscript and Meaning of Le Morte Darthur: Rubrication, Commemoration, Memorialization* (Cambridge: D. S. Brewer, 2017), 23.

225 For the influence of the *landskip* tradition in early modern England, see Sara Trevisan, "The Impact of the Netherlandish Landscape Tradition on Poetry and Painting in Early Modern England," *Renaissance Quarterly* 66:3 (Fall 2013): 866–903.

226 Hodnett notes that Bruges was an international center for illumination, *Five Centuries*, 11.

227 For the intersection of the physicality of books and the creativity of book making, see Kathleen Tonry, *Agency and Intention in English Print, 1476–1526* (Turnhout, Belgium: Brepols, 2016), 2.

228 Tsuyoshi Mukai, "De Worde's 1498 *Morte Darthur* and Caxton's Copy Text," *The Review of English Studies* 51:201 (February 2000): 24–40.
229 See Wade.
230 Special thanks to the John Rylands Library for providing digital files of the de Worde edition, the only extant copy.
231 The composite illustrations for Books 10 and 11, in contrast, are made by a different artist and are typical of the generic woodcuts known to be part of de Worde's large collection, a collection he recycled and adapted to suit various editions.
232 For a discussion of the dialogue between Renaissance rhetoric and visual culture, see Larry Silver, "Arts and Minds: Scholarship on Early Modern Art History (Northern Europe)," *Renaissance Quarterly* 59:2 (Summer 2006): 351–373.
233 Seth Lerer, "The Wiles of a Woodcut: Wynkyn de Worde and the Early Tudor Reader," *Huntington Quarterly* 59:4 (1996): 381–403; 390, 381–382.
234 Manfred Sellink, "A New Look on the World, the Invention of Landscape," in *The Age of Van Eyck: The Mediterranean World and Early Netherlandish Painting, 1430–1530*, ed. Till-Holger Borchert (London: Thames and Hudson, 2002), 212–219; 215.
235 Ibid., 213.
236 Trevisan, 875. Trevisan focuses on Michael Drayton's *Poly-Obion* and the influence of Dutch Netherlandish landscape on sixteenth-century English literature, including Spenser, but the influence, I argue, is evident earlier in the de Worde woodcuts.
237 Driver, 57.
238 David S. Areford, "Print Trouble: Notes on a Medium in Between," in *From Minor to Major: The Minor Arts in Medieval Art History*, ed. Colum Hourihane (Philadelphia, PA: Penn State University Press, 2012), 254.
239 Michael W. Twomey, "The Exemplary Environment of Bartholomaeus Anglicus," in *Reading the Natural World in the Middle Ages and Renaissance*, ed. Thomas Willard (Turnhout, Belgium: Brepols, 2020), 71–88.
240 *Dives & Pauper* (Westminster: Wynkyn de Worde, 1496), STC 19213; *de proprietatibus rerum* (Westminster: Wynkyn de Worde, 1495), STC 1536; *Chronicles of England* (Westminster: Wynkyn de Worde, 1497), STC 9996; *the descrypcyon of Englonde* (Westminster: Wynkyn de Worde, 1498) STC 13440b.
241 *The way to the holy lande* (London: Wynkyn de Word, 1515), STC 14082.
242 Alexandra Gillespie, "Caxton and the Invention of Printing," in *The Oxford Handbook of Tudor Literature, 1485–1603*, eds. Mike Pincombe and Kathy Shrank (Oxford: Oxford University Press, 2009), 21–36; 28.

3 The equipollent Earth-apple

Mandeville's *Travels*, the Behaim globe, and globes in Tudor England

The geographic imagination of fifteenth-century Europe was deeply poetic and perhaps no other text shaped the geo-poetic imagination of European readers more than *The Book of John Mandeville*, a text that described a unique "poetics of global space,"[1] a whole-Earth adventure tale. Iain Macleod Higgins has called Mandeville's book "an unprecedented fusion and personal interpretation of both learned and popular traditions"[2]—part fictional itinerary, part ethnographic encyclopedia, part devotional pilgrimage itinerary, part theoretical geographical treatise. Imaginatively, at least, England was the purported home of John Mandeville, knight and "author" of this "most widely circulated" medieval text.[3] While the sources and texts that make up the tradition of Mandeville's *Travels*, as the text has come to be commonly titled, are transnational, the Mandeville-narrator and subsequent readers of the text participated in the claim of its origin and composition in England.

Between 1351 and 1357, an ostensible English knight returned from 34 years of traveling to write the story of his journey to Jerusalem, by way of Constantinople. But this account of an English traveler went well beyond a single wayfinding pilgrimage route to Jerusalem and tells, in addition, the stories of many travels that continued on to Ethiopia, India, and the Far East before he returned home. Just where that "home" was has been a mystery ever since. Despite evidence to the contrary, a plaque in the cathedral in St Albans still celebrates its professed famous citizen and world traveler. "Sir John Mandeville," the narrator of this heteroglossia of cultural encounters and *mirabilia*, tells us that he departed from his hometown of St Albans on Michaelmas [September 29] in 1322 and, upon his return home, dedicated himself to writing "for as moch as men desireth to hyre speke of the Holy Lond and they have therof greet solas, spoort, and comfort, I shal telle somwhat that I have sey."[4] While the *Travels* begins with an intended audience "of hem that wole and beth in purpos to visite the holy cité of Jerusalem,"[5] Mandeville recounts a multi-threaded journey, asserting that rather than one route, a traveler may go by "many waies after the contré that he cometh fro, and many of hem cometh to oon eynde."[6] In other words, it's more about the journey than the destination which, he asserts, a traveler will reach by and by.

DOI: 10.4324/9781003096245-4

In his text, Mandeville asserted what scholars had known since classical times—the Earth was a sphere, and this "English knight" championed the rotundity of the Earth for a wide swath of European readers. However, over a hundred years later, the popularity of Mandeville's *Travels* notwithstanding, globe and world map production were slow to take root in Mandeville's purported home country of England. According to Mandeville, he "wende so longe by londe and by see seynge aboute the worlde,"[7] a world he describes as various, delightful, faith-inspiring, and possible to circumambulate, "for all the erthe is round."[8] Mandeville's emphasis of the sphericity of the Earth in the *Travels*, however, is not evenly distributed across the extant manuscripts or early print editions that circulated in England. Indeed, a popular strand of the text in Middle English as well as the early print editions in England lack a significant chapter that provides the theoretical basis of the round Earth theory.[9] Despite the absence of this chapter in approximately one-third of the extant manuscripts known to have circulated in England (in Latin, French, and English), Mandeville's *Travels* exerted a strong influence on the English geographic imagination, including the Earth's sphericity. In addition, the *Travels* seeded a "cartographic transaction,"[10] in which the narrative was reinstantiated as a three-dimensional "edition" of the text: the Behaim globe of 1492. As a cartographic object, the Behaim globe synthesizes the marvels of Mandeville's *Travels* and the Jerusalem-centered *mappamundi* with the equipollent[11] grid of Ptolemaic geography.

The representation of the Earth as a globe is known to have been well-theorized and widely written about in antiquity by cosmographers and philosophers, including Aristotle, Pythagoras, and Anaximander of Miletus.[12] The works and theories of the classical philosophers, mathematicians, and cosmographers were, for the most part, well known to European scholars, if not laypersons, throughout the Middle Ages. The English scholar Roger Bacon (c. 1219–1292) described the use of longitude and latitude to identify location, and John of Sacrobosco (perhaps an Englishman himself) wrote *Tractatus de Sphaera* around 1230, an influential cosmographical text that draws on Ptolemy's *Almagest* for the understanding of the Earth as a sphere in a geocentric cosmos, a text that was an integral part of the European scholarly tradition. In European geographic culture, the earliest physical globes were produced in the mid-thirteenth century, but the oldest surviving globe, the Behaim globe, dates from 1492, on the cusp of Columbus's voyage. In our contemporary popular imagination, Columbus's journey is typically, though erroneously, associated as the inaugural factor in changing the conception of the world from flat to round. Indeed, Columbus and his men are portrayed as brave adventurers who overcame the common fear of sailing off the edge of the flat Earth. This popular myth of medieval geographic thought has been thoroughly discounted by scholars, but its cultural circulation has been persistent.[13] For vernacular readers in Europe, the Earth's sphericity may well have been introduced to them by Mandeville's mid-fourteenth-century text, in which he recounts his pilgrimage to Rome, Jerusalem, and exotic countries of the East. Mandeville's travel narrative is a vital link in the popularization of the Earth's sphericity, both in England and throughout Europe. As Moseley notes, the "explosion of geographical knowledge" in Europe in the late Middle Ages is in no small measure due to Mandeville and his "unusual insistence that the world was everywhere traversable."[14]

As M.C. Seymour points out, Mandeville was "probably not an Englishman"[15] but since its earliest manuscript production, Mandeville's *Travels* has been associated with "English" geographic thought. The production of globes, like the production of maps, was not a primary activity in England during the Ptolemaic revival but, with its descriptions of crocodile tears and umbrella-footed Africans (also known as *sciapods*),[16] Mandeville's *Travels* "popularized many of the facts and fictions of our classical inheritance,"[17] that reached back to Pliny the Elder's *Natural History* and earlier Greek mythology. Part of this tradition also included the conception of the Earth as a sphere, and, via Mandeville's *Travels*, this construct was disseminated in England through book culture. Denis Cosgrove calls attention to "the historical depth and cultural complexity of the earth's deceptively simple form:"[18] this chapter traces how that "simple form" manifested in culturally complex transnational geographical practices, including a pseudo-English travel narrative probably written in France, a German globe based on that narrative, the importation of continental globes into England, and, ultimately, the production of the first "English-made" globes by Emery Molyneux, the gores of which were printed by a Dutch engraver living in London. With its "provocative syncretism" of Latin, European, and Arabic texts,[19] the notion of Mandeville's *Travels* as a singular expression of any one nation's geographic thought is a moot point.

Over a hundred years after its inscription, the *Travels* prompted a kind of "cross-species transmission" to animate the geographical world-view of the 1492 Behaim globe of Nuremberg, Germany. Partially based upon the *Travels*, the Behaim globe embodies the synthesis of medieval cosmology and Ptolemaic cartography that is the hallmark of late fifteenth-century geographical thought. The inclusion of traditional marvels and quoted passages from Mandeville on the globe sustains the possibility of a world that harbors wonders and diversities and situates the globe in the textual tradition of Mandeville's work. In addition to functioning as a geographic object, the Behaim globe is also a redaction or epitome of Mandeville's *Travels*. As such, it also functions as a *literary text,* one which exemplifies the blending of narrative and cartographic practices of the late fifteenth century and, though a Continental production, draws upon narrative geographic thought that claimed to be English in origin. Like Mandeville's *Travels*, the globe is collaboratively created and multi-vocal, a three-dimensional icon of poetic cartography. To date, however, the Behaim globe has yet to be recognized as a part of the Mandeville textual tradition, as a material version of the *Travels*, one that articulates the imaginative impact of the purported English traveler within the poetic cartographies of the Ptolemaic revival in Europe.

This chapter provides an original study of the "Defective Version" of Mandeville's *Travels* in relation to the dissemination of a Ptolemaic, geographic framework in England and of the Behaim globe as a literary text.[20] As the earliest surviving globe, the Behaim globe has been studied almost exclusively by cartographic historians, yet it provides key insights into the influence of literary culture on the Ptolemaic world-view of the late fifteenth century. This chapter also briefly explores the importation of globes into England during the sixteenth century and the relatively late but influential production of the Molyneux globes in England at the end of the century. The production of geographical knowledge did not adhere to national borders in late medieval Europe and any cartographic or literary history must rely not on the "nation-based frameworks" of the nineteenth century, but on an "interconnected reckoning" of the Global Middle Ages.[21]

"knyght of Ingelond" and the "roundnesse" of the Earth

The *Travels*' peripatetic, tangential journeys led Stephen Greenblatt to deem the work a "hymn to mobility."[22] With its claim of an English author, the text characterizes the English as prone to "moche movnge," suggesting an apparent "national reputation for restlessness."[23] This English travel bug, according to Mandeville, is due to the influence of the moon:

> And in oure countre is a contrary for we ar in a clymate that is of the mone and of lyght styrynge and that is the planet of way. and therfore it gyveth us wyll to moche movynge and styringe and go in to dyvers countres of the worlde. for it gothe aboute the worlde more lightly than another planet doth.[24]

This sense of restless "styringe," stands in contrast to modern scholarly perceptions of the English of the late medieval era as isolationist or uninterested in world geography. As Karma Lochrie notes, Mandeville's account depicts "a restless, mobile, and unsettling geographic fretwork that eludes the univocities of place and Western perspective."[25] For Lochrie, this speaks to Mandeville's work as part "of an emergent medieval cosmopolitanism."[26] Mandeville's *Travels* encompasses many books, many peoples of the world, many possible "wais" one might travel, many perspectives on faith, and, in addition, the very form and condition of that "dyvers" world. In sum, Mandeville's *Travels* attests to a globe that is endlessly discoverable, one without ends or edges.

Despite its popularity, Mandeville's *Travels* remains a textual enigma, a "multi-text" of authors, languages, nationalities, and genres.[27] Tamarah Kohanski notes that the "general stigma of falsehood and imposture" that has surrounded the work is "compounded by a generic complexity that makes it difficult to define or categorize."[28] Its dissemination in various languages (French, English, Latin, Italian, and a handful of other European languages) and in various forms (prose, verse, abridgements, excerpts, interpolations) led Higgins to characterize the various instantiations of the text as "isotopes," each related to the larger "work," but each distinct in form, language, national identity, recension, order of interpolation, and physical dissemination.[29] There is not even consensus on how the book should be titled, with scholars referring to the work alternatively as *The Book of John Mandeville, Mandeville's Travels,* or simply the *Travels.*[30] While various versions circulated in England, the Middle English Defective Version was a key constituent of the transmission of the work both in manuscript and early print and therefore of critical interest in terms of the dissemination of geographical thought in England during the Ptolemaic revival.

Author(s) and text(s)

The global journey of Mandeville begins with a statement of localized English identity:

> Here bygynneth *The Book of John Maundevile*, knyght of Ingelond, that was y-bore in the toun of Seynt Albons and travelide aboute in the worlde in many diverse contreis to se mervailes and customes of contreis and diversiteis of folkys and diverse shap of men and of beistis.[31]

The manuscript thought to be the earliest of the Mandeville texts, however, is written in French, and seems to have originated in France, leading to suspicions about the identity claim of this supposed knight from the town of St Albans. Just over 20 miles northwest of London, St. Albans was once a Roman city (Verulamium) along Watling Street, the major Roman artery that ran from the eastern Kentish seaports through Canterbury, London, St Albans, and on to the northwest, terminating at modern Wroxeter, near the Welsh border. Watling Street was also specified in Malory's text as the most expedient route to Sandwich, one of the primary ports of departure for any journey to the Continent (Chapter 1). With its location along this ancient thoroughfare, St Albans seems a fitting start for a pilgrimage from England to Jerusalem and beyond to the marvels of the East. It is uncertain whether the author, sometimes referred to as a compiler, was actually from St Albans but writing in France toward the end of his life. As Higgins notes,

> an English knight, or English cleric claiming to be a knight, could have composed an Anglo-Norman book on the Continent, while conversely a French knight or cleric claiming to be an English knight could have composed the work in Continental French in England.[32]

From the earliest French manuscript, c. 1357, the text fanned out across the Continent and to England.[33] Seymour speculates that the author was French, while John Larner suggests the author was Flemish.[34] Whatever the author's nationality, references to the *Travels* in the English poem *Cleanness* indicate that by the 1370s, the *Travels* had arrived in England and achieved "a wide circulation."[35] Seymour speculates that the popularity of the *Travels* in England was "helped no doubt as much by its claim to be a genuine relation of a genuine English knight born and bred at St Albans as by its exotic description of the Holy Land and the fabulous East."[36] During the fourteenth and fifteenth centuries, English readers received the text as the work of an Englishman and assumed a shared national identity.

And what an identity! The Mandeville author booms off the page, telling rip-roaring stories of his visits to fabled cities such as Troy, Constantinople, Jerusalem, and Cathay, his *tête-à-têtes* with Saracen kings and the great Khan, as well as accounts of the lands of the Amazon women, Ynde and the far East. He tells his readers at the outset that he has made multiple journeys with good company:

> Y have many tymes y-passid and ryden to Jerusalem in company of greet lordes and other good companye.[37]

He has seen for himself, Mandeville asserts, how Ethiopia is "a gret lond" where there is:

> a welle, that on the day the water is so colde that no man may drynk hit, and on the nyght hit is so hoot that no man may suffre hit.[38]

In "Ynde" he has seen diamonds cultivated from a hillside where they can grow and even procreate:

> And Y have many tymes asayd that if a man kepe hem with a lytel of the roch, which he groweth uppon, and wete hym with Mayis dewe, they shal growe ech yer, and the smale shal wexe greet.[39]

The genial innkeeper of Chaucer's *Canterbury Tales* could not have been a better host for this ostensible pilgrimage and world adventure. Though he expresses harsh judgment of the Jews, Mandeville displays a surprisingly open curiosity, an oft-remarked-upon characteristic, about Saracens and Chinese. His observations include religious practices, as well as food and drink. Through the *Travels*, in fact, he frequently remarks upon the quality or absence of wine. On the way from Constantinople to Jerusalem, for example, Mandeville makes note of Marcha, observing that "ther groweth ryght good wyn and strong, that men callen wyn of Marcha."[40] In the Cotton manuscript version, our author notes with satisfaction that "Many gode cytees þere ben in þat contree & men han gret plentee & gret chep of all wynes & vitailles."[41] That he partook liberally can be the only conclusion from his story's end, in which he confides: "And now I am comen hom mawgree myself to reste for gowtes."[42] To conclude the tale of his 34 years of traveling, Mandeville asserts his eye-witness account:

> Y, John Maundevyle, knyght, that wente out of my contré and passide the see the yer of Our Lord 1332 — and Y have y-passed thorgh many londes, contrees, and iles and am now come to reste — Y have compyled this book.[43]

In so doing, Mandeville provides a frame narrative in which we begin and end with an identifiable authorial voice, one who recounts his grand tale, if with a bit of pain in his feet.

In England, versions in Anglo-French, Latin, and English were copied and imported with differing audiences and ownership patterns. The *Travels* is one of the earliest extended prose works in the English vernacular, exclusive of sermons, and exerted "a major influence on subsequent English writing."[44] The best-known and most oft-cited Middle English manuscripts of the work include the early Anglo-Norman Harley 4383 as well as the Cotton, Egerton, Royal, Metrical, and Defective Version translations.[45] In addition, the text circulated in Latin and French manuscripts. Of the 64 extant manuscripts known to have circulated in England, 15 are in French, 11 in Latin, and 38 in English.[46] As Seymour notes, "Every major abbey possessed a copy of one of these translations" as did the universities at Oxford and Cambridge.[47] Of the Middle English translations, a subgroup called the Defective Version[48] exhibits a copying error known as the "Egypt Gap," presumably due to a missing quire and resulting in a lacuna in the passage about Egypt.[49] This version becomes one of the primary means of the work's transmission in the vernacular in England. Ralph Hanna argues that of all the manuscript versions circulating in England, the Defective Version distinctively "has any real claims to be the English *Mandeville*"[50] because of the number of extant manuscripts. The Defective Version was selected as the basis of the first printed text of Mandeville's *Travels*, a 1496 edition by Richard Pynson. Pynson's text, in turn, was reprinted in the widely circulated illustrated editions by Wynkyn de Worde in 1498, 1503, and 1510.[51]

Sources, manuscripts, print

Ostensibly the story of one man's itinerary to Jerusalem and the Far East, Mandeville's *Travels* is, instead, a pastiche of geographic, cultural, and travel texts from multiple sources. The brilliance of the text is certainly not its veracity as one man's coherent experience but, instead, the way it distills a dizzying array of sources into a first-person account that is narratively fresh, witty, and alive with a sense of authorial persona. Any writer of a scholarly text, including the present author, must genuinely admire the Mandeville author's skill in synthesizing scholarly sources while bringing a lively authorial voice to the whole. As Kohanski and Benson note, the *Travels* draws from at least 12 major sources, including Odoric of Pordenone's *Itinerarius,* William of Boldensele's *Itinerarius,* and Vincent of Beauvais's *Speculum Naturale* and *Speculum Historiale,* but also from over two dozen minor sources, including Sacrobosco's *De Sphaera* for the chapter on the Earth's sphericity.[52] The text is salted with frequent biblical citations; for example, in describing the Holy Temple, Mandeville observes:

> And withynne the north dore is a poule of water, of which Holy Wryt spekith and seith thus: *Vidi aquam egredientem de templo.* That is to say: "I saw water comyng out of the Temple."[53]

These Latinate interpolations have led scholars to suggest that the Mandeville author was a cleric. A syncretic process, the *Travels* also distilled the geographic tradition of antiquity and the Middle Ages into a single potent text, but below the surface, the text shimmers as an assemblage of cultural and geographic observation and voices, authorially elusive and indeterminant.[54]

The popularity of the *Travels* is well-attested by the number of surviving manuscripts and the longevity of its circulation in English reading culture. Whereas the *Travels* survives in over three hundred manuscripts, Marco Polo's *Description of the World*, by way of comparison, survives in an estimated 70–100 manuscripts and in fewer languages.[55] Seymour notes that the *Travels* was read by gentry and clergy, by both "educated and informed readers" as well as by "a less particular audience who embraced the abridgements in prose and verse."[56] Whether the work's vernacular readers were "less particular" is open to debate, but the impact of the *Travels* on the geographic imagination of a wide spectrum of English readers is clear. Part of the work's appeal, no doubt, is its description of marvels amplified by Mandeville's attestation of his being an eyewitness to this geography of amazing sights.

The jocularity and tall tales that led to its long success have also served to sequester the work from consideration as part of the learned geographical tradition. Other than Caxton's *The Myrrour of the World* and de Worde's editions of the *Travels*, Mosely argues that English readers had to rely on Latin academic treatises for geographic information which limited readership and resulted in a "somewhat surprising lag" of English geographical interest; by 1500, Mosely concludes, "there is little interest in geography in England."[57] However, as the previous chapters substantiate, importation of geographical material along with Thomas Malory's romance, Caxton's geographical editions, de Worde's illustrations, and the circulation of Mandeville's *Travels* in England all comprise a body of evidence for vernacular, non-elite English engagement in Ptolemaic-based geography. As we shall see in Chapters 4 and 5, John Rastell's 1518 geographical play and the popular almanac the *Kalender of Shepherds* contributed to a geo-poetic culture in England. The underlying principles of a changing geographic framework were produced, reproduced, and circulated in English literary culture along with the importation of explicitly geographical texts or maps. Interest in geography was not limited to academic treatises on instrumentation and measurement.

For over a hundred and fifty years, Mandeville's *Travels* lit up the geographic imagination of a broad spectrum of European readers, both in Latin and in the vernacular and is, in part, a precursor for the reception of Ptolemy in Europe. Surprisingly, Mandeville's narrative both reinscribes the medieval *mappamundi* but opens room for a conception of the world in which Jerusalem is no longer the center of an ideologically hierarchical world-view.[58] His journey, after all, is not solely a journey to Jerusalem and return, but one of many journeys and many ways. Although Mandeville affirms the centrality of Jerusalem,[59] his text also offers a new world that keeps expanding to the East, one of equal fascination and emotional weight. Like the portolan maps in which ports of call were drawn with equipollent values across the map, Mandeville's work similarly created an alternative image of the world, one that obliquely decentered Jerusalem, that registered a captivation with other cultures, and that opened imaginative space for the mathematically-derived Ptolemaic maps of the Renaissance.[60] In Ptolemy's *Geography*, no city sits in an ideological center; rather he uses a system of mathematical coordinates, noting that geography "looks at the position rather than the quality, noting the relation of distances everywhere."[61] Of course, his text is not without its own ideology, but it is not the Jerusalem-centered world of the medieval *mappamundi*. In *Nowhere in the Middle Ages*, Karma Lochrie suggests a principle of literary history in which texts are contextualized as "imaginative projects whose circulation resists our tendency to assign them to moments of composition and publication."[62] Beyond its compositional moment, the various textual dissemination streams of Mandeville's multi-text, including its round-Earth anecdotes, prepared Europeans intellectually for the visualization of a non-hierarchical, traversable sphere. Mandeville's "globalizing perspective" interrogated the Christian ideological hierarchy that posited the centrality of Jerusalem on the *mappamundi*. The result, Lochrie argues, is that Mandeville's *Travels* disrupts "the centers and peripheries of medieval theological and geographic thinking."[63] Drawing on the work of Christiane Deluz, Lochrie affirms that Mandeville's popular narrative introduced "a new geography," one constructed from a "strategic provincializing" of Christianity.[64] With its emphasis on the Earth's navigability, the *Travels* anticipated an equipollent valence of places and an "oceanic hermeneutic"[65] that would mark the transition from the time-based *mappamundi* to the measured space of Renaissance maps.

The introduction of maps based on Ptolemy's *Geography* followed the initial publication of Mandeville's *Travels* by approximately 50 years and did not create a new vision wholesale of a measurable, mappable Earth; instead, the concept of a global Earth was already active in the cultural imagination when the *Geography* was reintroduced to the Latin West. During the Renaissance, as Zur Shalev observes, "the theory and practice of linear perspective helped readers to understand Ptolemy's methods rather than *vice versa*."[66] It is no surprise, then, that the oldest extant globe, the 1492 Behaim globe of Nuremberg, visually integrates the geography of Mandeville's *Travels* with the new, Ptolemaic-based image of the world. The Behaim globe has been characterized at times as an incoherent curiosity, a late instantiation of medieval map-making "which tended to happily juxtapose unaltered medieval descriptions of the world with maps which made a nonsense of them."[67] But this juxtaposition is more synthesis than chaos.[68]

Orbis terrarum: ***the "londe and the see ys all round"***

As an ostensible guide to a Jerusalem-centered pilgrimage, Mandeville's *Travels* has been compared to a *mappamundi.* However, the *Travels*' actual focus on lands and customs well beyond a Holy Land pilgrimage opens up a de-centered world-view. True, the marvels of the text can be "mapped" on or "geo-rectified" with the Hereford Map, as has been usefully done in the Mapping Mandeville project,[69] but even if its anecdotal geography of *mirabilia* adheres to that of the *mappamundi,* the *Travels* is not its ideological equivalent. While the *Travels* precedes the introduction of Ptolemy's *Geography* into Europe by approximately 50 years, it indirectly draws upon one of Ptolemy's other works, the *Almagest*, which was already in circulation in European book culture. Seymour points out that there are "two traditions" present in Mandeville's geography: that of the *mappamundi* and that of the Ptolemaic round-Earth principle. For the round-Earth material, Mandeville draws upon Sacrobosco's *De Sphaera* (c. 1220), "a comprehensive explanation of the Ptolemaic system ... within a traditional Christian framework,"[70] which, in turn, drew upon on the astronomical theories of the *Almagest*, works that Mandeville refers to as including a spherical Earth, " after opynion of olde wyse philosofres and astronomers."[71] Perhaps because of its geographical content, Mandeville's *Travels* shows a similar plasticity to that of maps and geographic texts of the Ptolemaic revival, subject to active augmentation and revision. Indeed, Higgins observes that in many respects "the work's closest formal analogue is a map."[72]

Because there are many versions of the *Travels* in English, Kohanski urges caution about any sweeping claim about "the book." She contends that conflating variants misses versioned nuances and "fosters an acceptance of generalizations about the *Book* as a whole" that "may not be supported by the bulk of the texts."[73] Rather than imposing a fiction of single authorship over a controlled, closed work, it is essential to acknowledge the *Travels* as "continuously metamorphosing over the course of its transmission."[74] Of the versions of the *Travels* that circulated in England during the era of the Ptolemaic revival in Latin, French, and English, some of the English versions contain Mandeville's chapter on theoretical geography which emphasizes the theory and practicality of a round Earth, but some omit the chapter altogether, significantly modifying the weight of the work's geographical impact.

The key chapter on theoretical geography in the *Travels* addresses the sphericity of the Earth and how the reckoning of the Pole Star, or Antarctic star, affirms this principle:

> But ther is another sterre, which is y-clepid Anteryke, and that is evene agenst that other sterre. And by that sterre shipmen beth y-ladde toward the south, for that is noght y-seye toward the north. And therfore men

> may wel se that the lond and the see ys all round, for partyes of the firmament which sheweth in oo place and apereth noght in another place. And that may men preve thus: for yf a man myght fynde shypyng and men wolde go to see, men myght go all aboute the worlde above and bynethe.[75]

This emphasis on the experience of "seeing" as physical proof of a theoretical premise is key to Mandeville's eye-witness-based narrative. As part of his proof of the Earth's rotundity, Mandeville relates a popular anecdote of a traveler's unwitting circumambulation of the world derived from *Image du Monde*, the text Caxton chose to print as *The Myrrour of the World* (see Chapter 2). Mandeville explains:

> And therfore Y have y-thought many tymes of a tale that Y hurde when Y was yong, how a worthy man of oure contré wente a tyme to se the worlde. And he passyde Ynde and these iles byyonde Ynde, where beth mo than 5000. And he wende so longe by londe and by see seynge aboute the worlde, and he fonde an yle where he herde his owen speche … But Y say he hadde y-go so longe on londe and see, goynge aboute the worlde, that he was y-come into his owen marches.[76]

Mandeville's anecdote, in other words, verifies the possibility of the circumambulation of the spherical Earth. As in *Image du Monde*, Caxton's source text for *The Myrrour of the World*, Mandeville also provides an estimate of the Earth's circumference that his "worthy man" traversed: "For the erthe is gret and long, and hit holdeth in roundnesse aboute, above and bynethe, 20,400 and 25 myle"[77] He professes he has "lytel wyt," but his estimation is not too far off the mark: the Earth's circumference is now calculated at 24,901 miles.

Near the conclusion of the work, Mandeville reaffirms the round-Earth principle: "And therfore Y sey sykerly that a man myght go all the worlde aboute, above and bynethe, and com agen to his owen contré."[78] As he nears the limits of his travels, Mandeville stops short of the geographic Paradise to reprise his central tenet:

> And as moch as a man ryseth upward out of our contré toward Jerusalem, as moche shal he go donward toward the lond of Prester John fro Jerusalem, and that is for all the erthe is round.[79]

Twice more in the concluding pages of the text, the Earth's sphericity is emphasized such that no reader of any version that contained the theoretical chapter could fail to be convinced of the roundness of the Earth as a mathematical and experiential certainty.

Based on the presence of this chapter in modern editions, it would be tempting to argue that English readers during the Ptolemaic revival had ready access to a text which argues for the Earth's sphericity in Middle English. However, the particular textual history of the *Travels* in England during the fifteenth century is defined by what has come to be called the Defective Version, a transmission stream based on a French exemplar that omits the chapter on theoretical geography in its entirety. The Defective Version is so named because, earlier in the text, it omits a large passage, mid-sentence, about Egypt, a lacuna that has come to be called "the Egypt Gap," a distinguishing feature of manuscripts in this stemma. About half of the manuscripts in the Defective Version group and, significantly, the subsequent print editions by Pynson (1496) and de Worde omit the chapter with the excursus on the round Earth. One sub-group of the Defection Version, the "A-text" group, *does* contain this chapter, as do manuscripts that circulated in England from other recension groups, including the Egerton, Cotton, and Royal manuscripts. The presence of this chapter in almost half of the extant Middle English manuscripts makes any conclusion about the receptivity of English readers to the round-Earth principle or a possible religious censorship of the material on the part of English clerics difficult to assert.[80] In sum, the round-Earth chapter of Mandeville is unevenly distributed in the English manuscript tradition and absent in English early print editions.

Despite the uneven presentation of the round-Earth premise, all English editions affirm this principle in varying degrees. The Pynson edition of 1496 and the subsequent de Worde editions, for example, are based on a manuscript of the Defective Version that does not contain the chapter on theoretical geography. This is not to say, however, that the Defective Version manuscripts of the *Travels* do not affirm the Earth's sphericity: they do. The assertion of a round Earth is also mentioned in passages in other chapters. For example, in Pynson's edition, the discussion of Paradise as distinct from geographical east concludes:

> this wyldernes and myrke londe lasteth to the Earthly paradyse terrestre where Adam and eve were sette/ but they Paradise were there but a lytell whyle and that is towarde the est at begynnynge of the erthe/ but that is nat oure est that we call where the sonne rysethe in those countrees towarde paradyse. than it is mydnight in oure countre. for the roundnesse of the erthe for oure lord made the erthe all rounde in myddes of firmament.[81]

Thus, the premise of the Earth's sphericity is present but as a minor tone. A now-lost note by William Caxton copied onto the flyleaf of the Egerton manuscript version of the *Travels* indicates that Caxton planned an edition of the *Travels* based on the Egerton and so would have included the round-Earth chapter. A transcription of the original note, from 1803, reads, "Thys fayre Boke I have fro the abbey at Saint Albons in thys yeare of our Lord m. cccc.lxxx the sixt day of Aprylle. Willyam Caxton."[82] But by the time de Worde went to produce his edition, his former master's copy had either been returned to the abbey in Mandeville's hometown, or de Worde preferred to use Pynson's print edition as his exemplar.[83]

Aside from the inclusion or omission of the theoretical geography chapter, all of the manuscripts in English add a final interpolation in which Mandeville claims that he returned home by way of Rome to request the Pope's blessing of his text:

> after whan he/ and hys counseyle had examyned it all thorowe he sayde to me for certeyne that all was sothe. for he sayde he had a boke of laten that conteyned all that and moche more of the whych the Mapa mundi is made/ the whych boke I sawe. and therfor the holy fader the pope hath ratyfyed and confermed my boke in all poyntes.[84]

Despite the religious framing of this final passage in the *Travels* and the imprimatur of the Pope, the authenticating act is actually not the pope's spiritual authority, but his consultation of a *geography book and map*. Like the many marvels for which he claims eyewitness authority, Mandeville also claims to have seen this Latin book of maps and "moche more." The reception of the geographical assertions of the *Travels* in England is, like the byways and itineraries of our narrator, rhizomatic. Its textual history speaks to the necessity of tracing specific textual variants and reading communities rather than making broad-scale claims regarding the reception of Mandeville's geography as a single entity. For a significant number of English readers, the text reinforced medieval *mirabilia* and narrative *mappamundi*, suggesting that for those readers, Mandeville's *Travels* was read more for its marvels than for its contribution to emerging ideas on geography. Perhaps this focus on "mervailes and customes of contreis and diversiteis of folkys and diverse shap of men and of beistis"[85] accounts for the waning interest in the book once it is produced in print. De Worde's last edition appears c.1510, after which the *Travels* goes out of print until East's edition of 1568. While it is included in Hakluyt's first edition (1589) of *The Principall Navigations, Voiages and Discoveries of the English Nation*, it is dropped from Hakluyt's second edition. Mandeville's reputation as a verifiable traveler unraveled by the seventeenth century and his name became associated with exaggeration and false claims.[86] For readers interested in the new geography, other works such as imported copies of Peter Apian's *Introductio geographica* (1533), an illustrated synopsis of the *Geography* in Latin, and William Cuningham's *The Cosmographical Glasse* (1559), a précis of the *Geography* in English, would hold greater geographical interest. Despite the growing influence of maps based on the Flemish cosmographer Gerardus Mercator's projection and world map of 1569, Ptolemy's prestige in both humanistic and pragmatic English geographical works was vital through the mid-sixteenth century and, to a lesser degree, through the mid-seventeenth century.[87]

Mappings

Noting the innovative and "active recombination" of sources in the *Travels,* Higgins likens the *Travels* to "a series of maps."[88] This analogy holds true not only for the work's geographic matter but also because, much like a map or globe, the *Travels* is not rigidly prescriptive: "Y wole say no more of mervayles that beth ther, so that other men that wendeth theder may fynde many nywe thynges to say."[89] Like a map or globe, the traveler's gaze opens up possibilities of multiple routes, via land or sea.[90] It should come as no surprise, then, that the cartographic aesthetic of Mandeville's *Travels* should be reproduced on a globe, a geographic object that invites the traveler's omniscient gaze across "many waies."[91] In the late fifteenth century, a team of Nuremberg artists, crafts people, and civic promoters, created a unique, but unacknowledged edition of the *Travels* in the form of a globe under the diffuse authorship of Martin Behaim.

Martin Behaim's Earth-apple, 1492

In the fall of 2010, *New York Times* writer Patricia Cohen offered a witticism about the nature of the digital humanities as a new "alliance of geeks and poets."[92] She noted the exhilaration, anxiety, and productivity of collaborations between scholars working in the disciplines of science and the humanities as well as the potential for scholarly work to supersede the boundaries of what have come to be defined as traditional academic fields. The Behaim globe, produced in 1492, similarly represents a collaboration of geographers and artists. As the oldest extant terrestrial globe, the Behaim globe has a special place in European cartographic history and has been studied primarily on that basis—as a transitional cartographic object. However, the language of the Behaim globe calls for it to be studied as a *literary text* as well.[93] The Behaim globe is an icon of the fluid collaboration of geographic and poetic expression in the late fifteenth century.

Günther Görz, computer scientist and lead researcher for a digitalization project of the Behaim globe, considers the map image of the globe "primarily Ptolemaic" and the globe as "an early masterpiece of different scientific and technological achievements."[94] The globe resides in its native Nuremberg, now in the Germanisches Nationalmuseum (Figure 3.1). A collaborative effort of artists and crafts people, the globe was constructed around a ball of clay, later removed to leave a hollow core, and then built up with multiple layers of linen and parchment.[95] Martin Behaim, the "author" and geographer of the globe, created a two-dimensional map that formed the basis of the globe. The delightful miniatures on the globe of cities and people were hand drawn by Georg Glockendon and the globe was suspended, with an axial tilt, within a wooden frame (later replaced by a metal one) within which the globe could rotate. A member of a noble, mercantile family, Behaim (1459–1507) provided the geographical "content" based on his participation

Figure 3.1 The Behaim globe, 1492
Source: Courtesy of the Germanisches Nationalmuseum/Bridgerman Images.

in Portuguese explorations as well as his reading. The endeavor was financed by the Nuremberg City Council, displayed in one of the chambers of city hall, and overseen by one of its members, George Holzschuher. The civic context of the globe is a distinguishing feature, and Görz speculates that it may have been made with the intention of attracting investors for German-Portuguese seafaring expeditions.[96] The globe depicts a pre-Columbian world-view; hence, there is no American continent and a somewhat vague delineation of southern Africa, India, and the East. There are lines representing the equator, the polar circles, the two tropics, and the ecliptic along which 12 zodiacal disks appear. A color facsimile of the globe in four fold-out sections, or gores, was produced in two-dimensional form as part of E. G. Ravenstein's English edition of 1908, *Martin Behaim: His Life and His Globe* (Figure 3.2).[97] Only 500 copies of Ravenstein's edition, inclusive of the gores, were produced; hence it has become, itself, a rare geographic artifact, held by rare book libraries. Ravenstein's facsimile was made via tracing and photographing an 1847 facsimile of the Behaim globe, then collating it with the original in Nuremberg. Ravenstein's facsimile has a completely different color registry from the original, but allows for close study of the multiple pictograms and inscriptions that are central features of the globe.

Measuring 50 centimeters in diameter, there are over 2,000 places named on the Behaim globe.[98] It is also a delight of artistic merit with 111 miniatures, including 48 seated kings, Marco Polo, Christian saints, and national flags (10 from Portugal). The globe continues the medieval cartographic tradition of using ideographs, or pictograms, to depict places, including London and St. Patrick's Purgatory (discussed in Chapter 2) in the British Isles; Lisbon and Strasburg, on the Continent; Syria, Alexandria, and Jerusalem in the Middle East; the Ganges Mountains in India, and Cathay in Northern China. The legendary Christian ruler of the east, Prester John, is represented in two places, a bi-location that reflects growing uncertainty and conflicting information, akin to Caxton's historiographical practice of providing alternate information for the reader's active judgment (see Chapter 2). In addition, there are a smaller number of winsome sea creatures, mermaids and mermen, parrots and serpents, 11 ships and, in Africa, two sciapods, also known as the umbrella-footed peoples. Though not visually depicted, sirens, satyrs, and dog-faced men are mentioned in the written inscriptions. Görz and a team of computer scientists studied the Behaim globe as the basis of a hypertext and

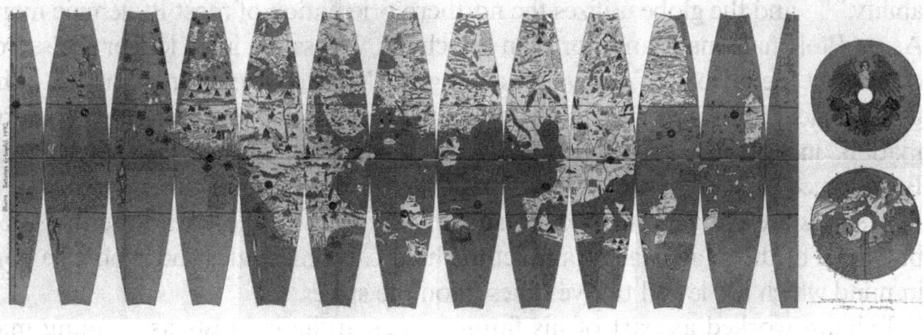

Figure 3.2 E.G. Ravenstein, *Martin Behaim: His Life and His Globe*, 1908
Source: University of Freiburg Library.

multimedia project that assessed "the value of Behaim's work in the history of cartography."[99] To do so, Görz's team made high-resolution "orthophotographs" along with 34 photographs shot in polarized light to create the image database.[100] Their work revealed the globe to be an edited document: over time, the original inscriptions have been overwritten, sometimes in varying hands, such that the globe has become a kind of palimpsest[101]—fragile, multivocal, and beautiful in its coolly-lit display room in the Germanisches Nationalmuseum.[102]

One of the distinguishing features of the globe is its articulation of the geographic aesthetic of the late fifteenth century: it synthesized geographic frameworks that include Ptolemy's *Geography*, medieval travel narratives, and Portuguese seafaring charts. Together, they create a fluid geographic hypothesis with evocative text-image interactions, at once a work of mathematics, literature, and painting. The citational language of the globe is in productive tension with the Ptolemaic fretwork of latitude and longitude, creating a complex multi-paradigm. In describing the globe held by the figure of Ptolemy in Raphael's famed painting, *The School of Athens* (1511), James Sykes describes the mix of geographic representation as "a great puzzle" and as "curiously erratic geography,"[103] terms that have been similarly applied to the geographic cross-currents of Behaim's globe and, as we have seen, to Caxton's edition of *The Myrrour of the World*. What the artifacts and texts of the late fifteenth century reveal to us is the way that major paradigm changes are rarely the break that they appear to mark in the annals of history but, rather, fluid, polyvalent periods that mark curiosity about and integration of new models of conceiving the world as much as they do a departure from the past. Görz notes that the globe, poised on the brink of Columbus' new world encounters, is "one of the very few existing cartographic works where different traditions of late medieval mapmaking are bound together."[104] Visually, the geography of the globe is informed by the 1482 Ulm edition of Ptolemy's *Geography* and Portuguese sea charts while textually it is informed by the work of Isidore of Seville, and, especially, the travel narratives of Polo and Mandeville. The land masses of Europe and northern Africa are derived from Ptolemy's projection (the calculation of latitude and longitude to convert a two-dimensional map to a three-dimensional sphere), embodying the principles of "contiguity and reachability,"[105] and the globe utilizes the northern orientation of most Ptolemaic maps. As on Ptolemaic maps and portolan sea charts, Jerusalem is no longer the sacred center, but merely one of many cities along the Mediterranean. In addition to this geographical framework, the globe presents ethnographic and commodities information, including a lengthy quotation on the Indian spice trade derived from Bartolomeo of Florence, in which Behaim outlines 12 "supply-chain" specifics of how spices are bought and sold, concluding: "Twelfthly, those who use the spices buy them of the retail dealers, and let the high Customs duties and profits be born in mind which are levied twelve times upon the spices."[106]

Behaim worked as part of his family's mercantile interests as a young man and subsequently traveled to Portugal where he became involved in the exploration-and-cartographic circle of the court of King John the Second.

Behaim subsequently settled and married a Portuguese woman. In one inscription on the globe, Behaim claims he participated in a Portuguese sea journey off the coast of Africa. In his early thirties, Behaim returned to Nuremberg where he was welcomed home as a world adventurer, self-styled as a cosmographer-humanist with a corresponding cultural cachet. On that basis, he was commissioned by the burghers of Nuremberg to create a globe, updating the Ptolemaic depiction of Africa with information from Portuguese exploration and mapping. In his 1908 edition of the text of the globe, Ravenstein, a member of the British Royal Geographical Society, cast doubt on Behaim's claim as a navigator as well as his "utter failure to portray the results of contemporary Portuguese discoveries upon his globe."[107] Agreeing with Oscar Peschel's assessment of Behaim, Ravenstein considers Behaim a "geographical dilettante" rather than a cosmographer or an innovator in astronomy.[108] Ravenstein's search of Portuguese exploration logs and accounts found scant mention of Martin Behaim. Görz, however, speculates that the Lisbon earthquake of 1531, known to have destroyed maps and documents, may also have destroyed records listing Behaim.[109] Like the accounts of Polo and Mandeville before him, Behaim's account of travel experience is open to speculation and doubt: unlike Mandeville, his identity is not in question,[110] but it is uncertain how much of his geographic knowledge was gained from actual expeditions or from the aggregate of avid reading and creative writing.

By the late nineteenth century, Behaim's reputation had been burnished in his native Nuremberg such that his purported navigational accomplishments were celebrated as a near equivalent to those of Columbus (a reputation Behaim no doubt would have enjoyed and promoted). Behaim had become a celebrity in Nuremberg where, in 1890, town officials erected a bronze statue of "Behaim the Navigator." At the public unveiling, a choral group sang a cantata in Behaim's honor and a procession was staged, with all fanfare, to his family home.[111] The globe was considered a national treasure such that Hitler forced the sale of the globe by the Behaim family and placed it with the Germanisches Nationalmuseum. Under the scrutiny of scientific rationalism in the early twentieth century, however, Behaim's reputation as a geographer eroded to a level comparable to that of one of his most significant literary sources: John Mandeville.

To date, the globe's written inscriptions have been considered charming but secondary to the interest in the globe as a cartographic artifact of the transition from medieval to early modern cartography. However, if we read the globe as a multi-dimensional literary text—intended to delight and entertain, as well as to inform and inspire—we can approach the globe as an assemblage of geographical treatises, cultural anecdotes, and travel romance. The Behaim globe was not intended as a navigational tool, but rather, as a geographical spectacle and showcase of Nuremberg's aspirations as a global city. Beyond its role as a celebrated object by the leading merchants, government officials, and townspeople of Nuremberg, the globe has had a resurrected role as entertainment in twentieth-century film: a replica of the globe appears in the Charlie Chaplin film, *The Great Dictator*, in Cruella de Ville's attic in *101*

Dalmatians, and in the Hogwarts Library in the *Harry Potter* films.[112] Its visual mingling of images and text across the surface of the globe continues to evoke a spatial mystique and otherworldliness.

Despite its visual materiality, the Behaim globe is as much a textual artifact as it is cartographic: short, prose passages are inscribed in black, red, gold, or silver ink, covering the surface of the globe, literally filling the seas (Figure 3.3). The inscriptions, in German, provide commentary on geographical features of the globe, but they also tell the poignant, wistful story of Martin Behaim himself while also creating a redaction or edition of Mandeville's and Polo's texts in a new material form. The globe theorizes an oceanic, trade-based Earth, based on Ptolemy's geographic principles, but inscribed across the surface of the globe, passages from the travels of Polo and Mandeville are replete with cultural contact and wonder. The inscriptions float upon the open ocean of late medieval geographic thought, creating a geo-poetic space. The globe is more than a

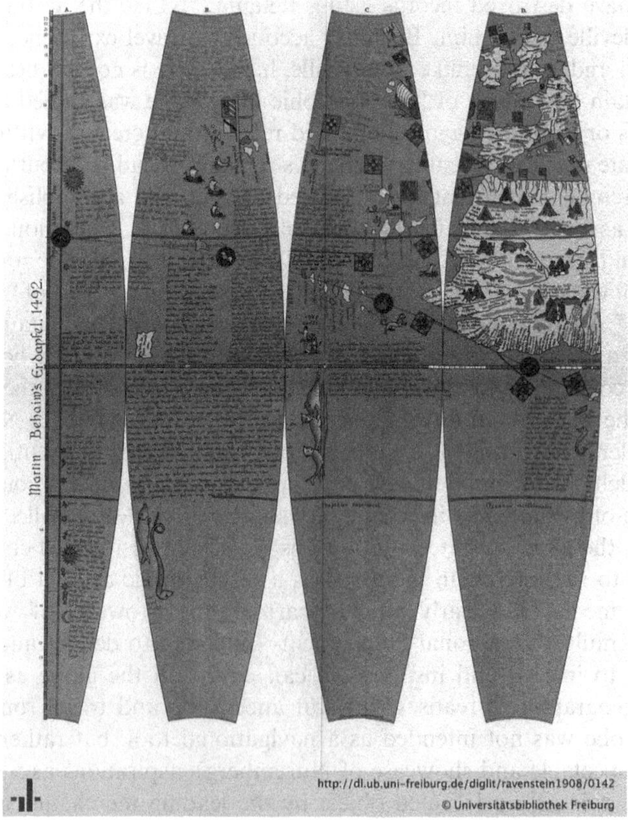

Figure 3.3 E.G. Ravenstein, *Martin Behaim: His Life and His Globe*, 1908, detail
Source: University of Freiburg Library.

jumbled matrix of transitional geographic information; it is a collaboratively created, speculative, fluid, literary, and artistic endeavor. In an homage to fifteenth-century multi-paradigmatic modeling, the Ptolemy Project at the University of California, Berkeley investigates the "heterogeneous mixture of models of computation"[113] to develop polyglot metadata. At opposite ends of the Cartesian divide, the heterogenous modeling of the two projects share a common ethos. The Behaim globe performs a geographic aesthetic of the Ptolemaic revival, one which embraced a flexible accommodation of multiple geographic frameworks as part of the assimilation of Ptolemaic geography as it merged with and gradually transformed medieval models. As a geo-literary edition, the globe synthesized Mandeville's medieval travel narrative with the cartographic, visual, and oceanic culture of the Ptolemaic revival.

The Behaim globe as a literary text

The globe functions as a literary text in a variety of ways: it is densely textual, has a prologue (actually two), a narrative persona, and a narrative arc; Behaim uses descriptive and affective language that is intertextual and rich in citation, self-consciously situated in the practice of reading; and, like the *Travels* and Caxton's *Myrrour*, the globe invokes the imagery of an apple as a leitmotif. Ultimately, the globe records the death of the author within a frame narrative which recounts a global story of peoples, commodities, and expeditions. The Behaim globe serves as an icon of the synthesis of medieval cosmology and Ptolemaic cartography, a linking of "poets and astronomers" that is one of the hallmarks of the late fifteenth century. The inclusion of the traditional marvels and quoted passages from Mandeville sustains the possibility of a world that harbors wonders and diversities, situating the globe not only as a remarkable geographic artifact but as an edition in the tradition of the many Mandeville redactions. While the globe may explicitly cite Polo more frequently, its narrative strategy is thoroughly in the tradition of Mandeville.

Prologues: inscriptions on the sea

The sea is the primary site of the inscriptions on the Behaim globe. On the surface of the seas, the inscriptions float free from their original texts and create an oceanic mosaic of prose. The inscriptions can be read in a narrative arc, beginning in the Atlantic Ocean, or in a fluid, fragmentary, or rhizomatic trajectory. If the reader/viewer begins in the Atlantic Ocean, there are two lengthy "prologues," side by side: the first tells the "back story" of the globe, and is in the narrative voice of Holzschuher, the city council member and financier who supervised the production of the globe:

> [T]his figure of an apple has been achieved and made by the skill, according to the directions and by the industry of the worshipful and

> honourable D. Martin Behaim, Knight, who is much experienced in the art of Cosmography and has circumnavigated one third of the world ... This (work of) art and apple has been achieved and made in the year 1492 after the birth of Christ, and has been left behind him by the said D. Martin Pehaim [*sic*] for the honour and enjoyment of the commonality of the city of Nürnberg in order that he may be thought kindly of for all time, when he shall have gone back to his wife, who lives 700 miles away, where he keeps house, in order to end his days in his island where is his home.[114]

This first prologue establishes Behaim's credentials as a cosmographer and explorer and the civic patronage of the project. It also creates a narrative frame that will introduce and conclude the global story represented on "this art and apple." The metaphor of the Earth as an apple draws directly from the imagery of the *Travels*, revealing the broader influence of the travel narrative beyond direct citation. Adjacent to the introductory prologue in the Atlantic Ocean, a second prologue follows, in which Behaim—who will now "narrate" the rest of the globe—presents his seafaring credentials as well as the literary and cosmographical sources for the globe, including Claudius Ptolemy:

> Be it known that on this Apple here present is laid out the whole world according to its length and breadth in accordance with the art geometry, namely, the one part as described by Ptolemy in his book called 'Cosmographia Ptolemaei,' and the remainder from what the Knight Marco Polo of Venice caused to be written down in 1250. The worthy Doctor and Knight Johann de Mandavilla likewise left a book in 1322 which brought to the light of day the countries of the East, unknown to Ptolemy, whence we receive spices, pearls and precious stones, and the Serene King John of Portugal has caused to be visited in his vessels that part to the south not yet known to Ptolemy in the year 1485, where I, according to whose indications this Apple has been made, was present.[115]

The prologue authenticates Behaim as a credible traveler/narrator via the assertion of eyewitness experience while also acknowledging his reading debt to books by Ptolemy, Polo, and Mandeville. The globe functions as an example of the ultranational, newly Europeanized context of Ptolemy's *Geography,* merging the prestige and textual tradition of the Latin *Geography* with Portuguese exploration accounts, German instrumentation and metalcraft, humanist scholarship, and the literary travel narratives of Italian and English authors. As the globe's "author," Behaim utilizes narrative strategies in his prologue that include the creation of an authorial persona, textual citation, descriptive language, and eyewitness authentication to prepare the reader/viewer for the reading and visual journey of the globe.

Author/narrator

On the globe, Behaim is sometimes referred to as the author to distinguish his contribution from that of the craftspersons and artists engaged in the project. Behaim, like Mandeville, constructs an author as the guide for the global journey, stating that he gained much of his knowledge in Portugal "in the year 1485, where I, according to whose indications this Apple has been made, was present."[116] In the prologue to Mandeville's *Travels*, a similar authorial authentication is provided:

> I shal telle somwhat that I have sey. John Maundevyle knyght, thow Y be nought worthi, that was bore in Engelond in the toun of Seynt Albones … Y be long tyme over the see, and have Y seye and go thorgh many londes … wher that dwelleth many diverse folk of maneris and diverse lawes and shappes. Of which londes and iles Y shal speke more plenerly.[117]

Like Mandeville, Behaim uses the narrative "I" to establish his eye-witness credibility. In inscriptions across the globe, Behaim recounts his travels, in one example stating, "the writer of this was sent further by the King of Portugal in the year 1485."[118] As a result, a clear sense of Behaim as both narrative persona and character in the drama of the globe is constituted, as in Chaucer's *Canterbury Tales*, within a frame story. The narrative "I" serves as an authentication bid, one of several narratives strategies in early modern geographic practice. The narrative persona of the globe calls for it to be read not only as a cartographic object but as a literary text both in its own right and as a cartographic redaction of Mandeville's *Travels*. There can be few cartographic objects with as distinct a literary voice as that of the Behaim globe.

Citation and reading

Behaim scrupulously provides a bibliography and cites his textual sources at the end of the globe's second prologue: "All this has been extracted with industry from the books of Ptolemy, Strabo, and Marco Polo, and put together, the sea as well as the land, each according to its shape and form."[119] In establishing the globe within in a geographical-textual tradition, Behaim takes particular care to note the specific chapters from Polo's *Travels*, probably a 1485 print edition based on Fra Pipino's Latin edition,[120] and a late fourteenth-century Latin edition of Mandeville's *Travels*.[121] Behaim integrates the *oikumene* of Ptolemy's Greco-Roman world with the travels of Polo and Mandeville, stating that "Ptolemy has described the world no further, but the remainder has been described by Marco Polo and Mandavilla."[122] One example of Behaim's citational practice can be seen in regard to Zanzibar, about which Behaim writes, "Marco Polo in the 3rd book and the 41st chapter describes all this."[123] He utilizes humanist, scholarly practices in a secular, fragmentary text. In so doing, Behaim provides a readerly provenance for his description of the world, a practice in keeping with David Rundle's definition of a vernacular humanist collaborator.[124] Behaim also references details from Mandeville's text, though typically without the same textual specificity of his Polo citations, suggesting that he had a copy of Polo's text close to hand while composing the globe but that his references to Mandeville were from memory and familiarity. Reinforcing the implied readership of the Behaim's citational practice, an illustration on the island of Zanzibar portrays a delightful image of two readers, open books in laps, with a third figure listening.

From Behaim's citational practice, we learn that he was well read in the literature of the East, in classical geography, and in medieval travel narrative, thus setting his globe in a readerly, bookish context. Significantly, he encourages the reader/viewer to embark on a similar textual excursus. For example, in a citation written over part of the Indian Ocean, Behaim remarks:

> Here are found sea monsters, such as Sirens and other fish. And if anyone desire to know more of these curious people, and peculiar fish in the sea, or animals upon the land, let him read the books of Pliny, Isidor (of Seville), Aristotle, Strabo, the 'Specula' of Vincent (of Beauvais) and many others.[125]

Note that Behaim does not encourage his reader/viewer to set forth on a *physical* journey to the lands he describes, but rather to turn to classical geographical texts and the authority of ancient geographical writers. In this regard, the globe is not intended as a navigational tool so much as an imaginative journey of further reading. Like Mandeville's *Travels*, the globe is a substitute for "the restless" who cannot travel.

Figurative language/prose poetry

In addition to the lengthier introductory passages, the globe is filled with brief inscriptions providing information about locations or additional background about the globe's author. For example, Behaim recounts a three-year voyage sponsored by King John in which Behaim claims he took part. The tale is no dry navigational account:

> Thus fitted out we sailed from the port of Lisbon in Portugal to the Island Madeira, where the sugar of Portugal grows, and through the Fortunate Islands and those of the savage Canarians. And we found Moorish Kings to whom we gave presents, for which they gave return presents. And we came to the kingdoms of Gambia and Golof, where grow the grains of Paradise, and which are distant 800 German miles from Portugal. And subsequently we arrived in King Furfur's country, which is 1,200 leagues or miles distant, and where grows the pepper called Portugale pepper. And far from there we

> came to a country where we found cinnamon growing. And when we had sailed about 2,300 miles or leagues from Portugal we turned back, and in the 19th month were once more with our king.[126]

What reader of the globe wouldn't want to know more about the savage Canarians or the place where the grains of Paradise grow? Behaim's geo-literary style is lively and detailed, conveying a sense of epic travel romance in his account of a global voyage that covered 4,600 miles.

Most of Behaim's inscriptions are marked by a potent compression of language; they shimmer with spare sensory detail that read like modern prose poems. About the Strait of Malacca in Indonesia, for example, Behaim writes, "If you sail from the kingdom of Loach to the southward, you reach this island Pentan, where there are large forests of delicious perfume. The sea around is less than ten paces deep. Thus writes Marco Polo in the 9th chapter of the third book. All the inhabitants go naked."[127] Behaim, I would like to suggest, "wrote" the globe in a genre both familiar to readers of medieval travel narratives and, in its fragmentary, open-ended citational practice, akin to the modern genre of prose poetry: compressing and amplifying language. In this case, geographical experience is redolent with the inference of spices, island topography, and naked others, a geo-textual experience that utilizes the language of literature to expand the toponyms of Ptolemy's tables into the realm of the imagination.

Audience and imagination

As declared in the first prologue, Behaim's globe is expressly intended for a public audience, made "for the honour and enjoyment of the commonality of the city of Nürnberg in order that he may be thought kindly of for all times."[128] The globe as public spectacle performed Behaim's explorations and, by implication, Nuremberg as a cosmopolitan, global city. It may have had a social practice, with the inscriptions meant to be read aloud, as is suggested in the pictogram on Zanzibar where two readers have a third companion, listening. Cosgrove has speculated on the imaginative impact of a "kinetic vision of a turning earth,"[129] and the movement of the Behaim globe on its axis would certainly have evoked a powerful performativity. Whether the globe served a broadly public function or as an elite object of art for the city leaders has been debated;[130] either way, part of the enjoyment of the globe for viewers—whether ordinary citizens or city leaders—was a cartographic speculation, as Alfred Hiatt denotes, that includes filling blank spaces of maps with discursive, speculative textuality.[131] Behaim's globe recounts his purported journey, but in the cartographic tradition of Mandeville's *Travels*, is suggestive of other journeys that might be made.

We can imagine the bodily, somatic experience of reading and viewing the globe, the act of touching its surface—a haptic experience of geography, art, and textuality. The viewer could slowly spin the globe and, in that act, conjure an imaginative experience of travel and exotic others, as well as conceive of the space of the entire Earth. As Cosgrove observes, "Earthbound humans are unable to embrace more than a tiny part of the planetary surface. But in their imagination they can grasp the whole of the earth."[132] At once a story of one traveler and a speculative bird's-eye view of the whole world, the Behaim globe created an aesthetic of cartographic pleasure, both informative and poetic. This spatial aesthetic defined fifteenth-century geographic practice, one in which literature, romance, astrology, astronomy, geographic narrative, travel accounts, book illustration, and cartography were conceived as integrated ways of knowing, consciously fashioned "after poyetes and astronomyrs."[133] An alliance of poets, scholars, and scientists collaborated in the imaginative representation of the "whole Earth;" together, they created the Behaim globe within the pre-disciplinary geographic culture of the Ptolemaic revival.

"Mandavilla likewise left a book": Mandeville's *Travels* and the Behaim globe

In all of the ways discussed above—prologues, authorship, citation, imagery, and the construction of a reader—the globe embodies the style and genre of Mandeville's travel narrative within a frame story. But what is the specific literary and geographic debt of the globe to Mandeville? Because of the many textual citations to Polo, the influence of Polo's work on Behaim's globe is more readily apparent, especially in its identification and inclusion of Japan: "This island Zipangu lies in the east of the world ... This is stated by Marco Polo of Venice in his 3rd book."[134] The authority of Marco Polo's *Travels* derived from its articulation of first-person verification which, in a similar manner, was the basis for humanists' interest in realism and verifiable accounts. Polo's work was incorporated into the "continuous struggle" of fifteenth-century humanists to "verify or augment" the information of the ancients with contemporary accounts.[135] Polo's description of Cipangu (Japan) was of particular interest to fifteenth-century humanist cartographers, for scant information was available. While Polo's book influenced the geography of the Catalan *mappamundi* (c. 1450–1460), Fra Mauro's *mappamundi* (c. 1450), and Henricus Martellus Germanicus' *mappamundi* (1489/1490), Behaim's globe was the first cartographic representation to specifically mention Polo by name. In addition, Larner suggests that the Behaim globe "was constructed from two maps based on different translations of Marco's Book."[136]

While Behaim cites Polo's text with greater frequency, Mandeville's "strategies of heterogeneity" guide the ethos of the globe."[137] Moseley claims that Mandeville should be seen as "a major source" for Behaim's globe,[138] a claim I wish to amplify. Beyond the seamless juxtaposition of marvels with technical detail, the Behaim globe also reproduces Mandeville's pastiche practice, cultural curiosity, and, ultimately, the articulation of the round Earth coupled with the imagery of the globe itself as an *Erdapfel*, or Earth-apple. An example of the globe's Mandevillian *heteroglossia*, mixing the *mirabilia* with the scientific, can be found in a single sentence which knits together the mythology of the dog-headed men with polar theory and navigational instrumentation:

> Here was discovered in the time of John de Mandavilla an island inhabited by men having dogs' heads and there the Sea-star, called Polus Arcticus by us, is not visible, and those navigating the sea must do so with the help of the astrolabe, for the compass does not point.[139]

As Paul Smethurst has argued, Polo's text conveys a "sense of realism and cartographic unity" in contrast to "Mandeville's indeterminacy, ironic presentation of wonders, digressive ramblings and textual borrowings," attributes that strike a familiar chord in postmodern literature.[140] In Mandeville's text, Smethurst argues, "there is no unified 'real,'" but instead "loosely arranged facts and fables shift unevenly between a Cartesian world and symbolic worlds of myths, legends, and the Scriptures."[141] He notes Mandeville's "self-consciously digressive" mode of textual citation and his use of pastiche, attributes that give rise to an assemblage, "not the product of an individual consciousness but a *text* that synthesizes and produces meaning through other texts."[142] In this assemblage of texts and genre, Behaim's globe not only follows the footsteps of Mandeville, but reads as an homage to Mandeville's textual pastiche, now in three-dimensional geographic and visual form.

Ravenstein suggests that Behaim's citations of Mandeville may have been culled from an Italian map rather than a specific edition of Mandeville. Since Ravenstein's study, however, Mandeville scholars such as Moseley, Seymour, and Josephine Bennett have demonstrated the extent of Mandeville's influence on late medieval and early modern geographic thought and cartography,[143] a conceptual framework that extends to the Behaim globe. Moseley argues that "there are many more elements—for example, the mythical sciapods—that could have come from Mandeville without acknowledgment."[144] He notes the globe's references to the Pole Star excurses from Mandeville's chapter on theoretical geography and suggests that "it is extremely unlikely that any map would even go to the trouble of explaining that Mandeville was 'doctor und ritter' and that he set out in 1322, let alone his rather complex astronomical idea."[145] In identifying Mandeville as a "ritter," or knight, Behaim implicitly situates himself within the tradition of knight-geographer. In all, Behaim draws on Mandeville's chapter on theoretical geography for his references to the Pole Star, the Antipodes, the dog-faced peoples, the roundness of the Earth, and the central image of the Earth-apple. The influence of Mandeville is also evident in the globe's description of the pepper forests,[146] the two competing locations for the mythical Prester John, and the depiction of the islands of *Masculina* and *Feminea*, each inhabited by their respective single sex, but who enjoy an annual visit.[147]

Behaim adopts an authorial persona similar to that established in Mandeville's *Travels*, in which the writer/knight imparts a very human portrait of himself as a world traveler, now home from his journeys and at ease to recall his adventures—augmented perhaps by travel lore and other texts. Behaim's narratorial voice hews much more to the tradition of the wine-imbibing, chatty Christian knight, John Mandeville, than the cooler, predominantly third-person narrative style of Polo. Bennett points out that Mandeville "does not pretend to confine himself to the experience of an actual traveler," as did Polo.[148] With Behaim, as with Mandeville, the literary persona exudes a sense of adventure and zest for life, a narrator not averse to taking certain liberties in the interest of the greater story.

Behaim similarly adopts Mandeville's relatively balanced sensibility when describing his encounters with ethnic others. Mandeville writes with scathing hostility about the Jews, but then shows surprising tolerance for other racial or religious groups, what Lochrie calls a "radical privileging of non-Western Christian faiths."[149] In discussing the Christian faith with the Sultan, Mandeville comments: "But the Sarasyns beth trywe, for they kepe truly the comaundementz of her *Ackaron*."[150] Like Mandeville, Behaim uses observational statements rather than religious judgment to narrate other cultures or religious practices, as in this description of Java Minor: "the inhabitants worship idols. Various spices, such as pepper, mace … [and] cinnamon abound, and those sold are afterwards distributed throughout the world, for which reason many merchants are generally to be found there."[151] Here, as in other citations, Behaim notes a religious or cultural practice without comment, then quickly turns to matters of trade. In contrast to Mandeville, Behaim's world-view is more thoroughly secular. While Behaim cites Jerome's Bible, mentions St. Brendan's voyage, and attributes the Earth to the creation of God, he does not narrate a pilgrimage to Jerusalem, but, rather, mercantile exploration: his journey is highly descriptive of potential goods such as sugar, cinnamon, pepper, and gold. Unlike Mandeville, Behaim makes no explicitly vicious commentary about Jews. Indeed, one of Behaim's early-life experiences foreshadowed his more secular perspective: as a young man he was charged and convicted of dancing at a Jewish wedding in 1483, an offense amplified in severity to the authorities by the fact that it occurred during Lent.[152] Clearly, strict religious observance was not a priority for Behaim and his openness to other cultures extends to his having married a Portuguese woman as well. The Garden of Eden, which Mandeville describes as near-geographically possible to reach, is not depicted on Behaim's globe; instead, the lands of the East hold the promise of wealth in this life. The Behaim globe adopts a Mandevillian ethos and reconstitutes it within the latitude and longitude of Ptolemy's mathematical projection. In addition, the globe's retention of Mandeville's world-view is evidence of the way in which the paradigm of the Eastern Other was still deeply embedded in European culture; the cultural narrative of the East remained persistent and potent despite the visual transfer to a new geographic worldview.

The Earth-apple

Behaim's debt to Mandeville exceeds the citational theoretical and mythological geography of the *Travels* and creates what I argue is *an edition* of

Mandeville's *Travels*, one that also adopts Mandeville's literary persona, pastiche genre, geographic and cultural heteroglossia, cultural curiosity, and, ultimately, his defining metaphor of the Earth as an apple. Introduced in Holzschuher's prologue as "this figure of an apple" and in Behaim's prologue as "this apple," the globe embodies the cultural anecdote from *Image du Monde* and Mandeville's *Travels* that explained the sphericity of the Earth. Ravenstein translates "*Erdapfel*" into "apple," perhaps to distinguish it from the linguistic association of an "Earth-apple" as a potato, as in the French *pomme de terre*. While the imagery of the Earth-apple is often remarked upon, it has not been clear how Behaim came to use such a term. In the Christian tradition, of course, the apple had long served as a symbol of knowledge, of the Fall, and of the worldliness of man.[153] The orb as a symbol of Roman imperialism was transformed into the iconography of Christianity as a cruciform orb, or *globus cruciger* (the cross-bearing orb), symbolizing Christ's dominion of the world. Also known as the *Reichsapfel* or Imperial Apple, the cruciform orb was part of medieval visual culture and often associated with imperial rule. A beautiful apple chalice made in Nuremberg shortly after the globe's construction reveals a similar nexus of craftsmanship and the Earth-apple motif.[154] Certainly, the representation of the Earth as an apple would have been replete with symbolism for Behaim, his collaborators, and his reader/viewers. As noted in Chapter 2, the association of the Earth with the shape of an apple is referenced in the thirteenth-century *Image du Monde*, the basis for Caxton's *Myrrour of the World* and a possible source for Mandeville's anecdote of the man who circumambulates the world.

More specifically, Mandeville's *Travels* is an overlooked source for the globe's metaphoric representation of the Earth as an apple. Behaim expressly cites Mandeville as his authority for the sphericity of the Earth: "All of this is because God has created the world, together with the water of a round shape, as described by Johannes de Mandavilla in the third part of his voyage."[155] In addition, the specific imagery of the imperial apple is suggested in the first chapter of Mandeville's account when he describes the famous sculpture of the Emperor Justinian, outside the church of Saint Sophia:

> And before that chirche is the ymage of Iustynyan the emperor couered with gold, and he sytt vpon an hors ycrowned. And he was wont to holden a round appelle of gold in his hond, but it is fallen out thereof ... And men wolden many tymes put the appulle into the ymages hond ayen, but it will not holde it. This appulle betokeneth the lordschip that he hadde ouer alle the world that is round.[156]

Not only does Behaim note Mandeville's description of the Earth's "round shape," but he adopts the imagery of the apple which Mandeville invokes to denote mastery of the world. Smethurst suggests that this anecdote of the emperor's fallen apple may be read as Mandeville's "vision of a world misshapen but redeemable."[157] The possibility of redemption on Behaim's globe, however, is expressly secular. For Behaim, the marvels and commodities of the world have become the seductive apple now available to the "commonality" of Nuremberg citizens who hold this new Earth-apple in their gaze.

In light of the influence of Mandeville's text on the globe's literary qualities and central imagery, the Behaim globe must be read as part of the textual transmission of Mandeville's *Travels*, a heretofore unidentified "isotope" in the network of redactions, abridgements, and adaptions of this medieval "best-seller." As Mary Campbell notes, the *Travels* as a work is "a continuous, if continually mutating, textual tradition;" the Behaim globe, I argue, is part of that mutating, isotopic tradition.[158] Thus, the Behaim globe is not only an artifact in the history of cartography, but a cartographic memoir in the tradition of Mandeville's *Travels* and an enduring instantiation of the poetic cartography of the Ptolemaic revival. While John Gilles has called attention to the poetic geography of Tudor England, there is a tendency to identify this aesthetic as part of a "revolution in ways of thinking about cartographic space" associated with the Renaissance.[159] A "reading" of the globe as a version of Mandeville's *Travels*, however, calls for an emphasis other than the globe's fidelity to a new geographic paradigm and reaffirms, instead, the "inseparability of imagination, moral reflection, and practical action apparent in representations of the globe."[160] Mandeville's *Travels* and the Behaim globe are literary and cartographic "siblings," mutually constitutive of the literary and spatial aesthetic of the Ptolemaic revival.

"Mirabiles elaciones maris": wonderful are the risings of the sea

In addition to the literary qualities of the globe, one additional and foundational element is attributable to Mandeville's influence: Behaim's emphasis on the sea. The Behaim globe provides one of the first truly hydrographic depictions of the Earth, a stunningly original assertion of the Earth as fundamentally oceanic. While Mandeville's text cannot be claimed as the sole source of this perspective, it supports the great argument of Behaim's project: that the world is oceanic. In describing islands of the Indian Ocean, Mandeville writes:

> [The sea] is so high that it semeth as though it henge at the clowdes and that it wolde coueren alle the world. And that is gret meruaylle that it myghte be so, saf only the wille of God that the eyr susteyneth it. And thefore seyth Dauid in the Psautere, *Mirabiles elaciones maris.*[161]

In some editions a translation of the phrase is added, "Wonderful are the risings of the sea." Visually and textually, Behaim's globe draws the reader/viewer's attention to the sea as the expansive "covering" or contiguous network of the Earth's surface. While Behaim credits God with the creation of the oceanic world, he credits the narration of that navigable world to "Johannes de Mandavilla in the third part of his voyage."[162] In his prologue to the globe, Behaim refers to the "Sea Ocean" of the West, which he claims has now been "navigated further that what is described by Ptolemy,"[163] and links Mandeville's *Travels* with the emerging oceanic principle of the early modern era.

The Behaim inscriptions float like ice floes, drifting away from the edges of terrestrial models, simultaneously textual, speculative, and imaginative; medieval and modern. In recalling the ethnographies of the medieval past and the exploration expeditions of the present, the inscriptions float off-site from the Ptolemaic landmass amid the space of the sea, a space Iain Chambers defines as "ambiguous, heterotopic, uncontrollable space."[164] A comparison with Matthew Paris' map (c. 1250), the Hereford Map (c. 1284), the Gough map (c. 1360), and even Fra Mauro's map of 1459 demonstrates the way in which in these earlier medieval maps the visual field is dominated by a focus on land mass. As Cosgrove notes, the Mauro map "is powerfully unitary, closing its single landmass tightly within the circular frame and reducing oceanic space to a narrow belt scattered with islands at the earth's edges."[165] Likewise, the Ptolemaic-based maps depict bulky landmasses that frame the world-view. In contrast, on the Behaim globe, the inscriptions on the seas represent more than a convenient use of the blank space: in both location and content they call attention to the increasing importance of the sea in the late fifteenth century. In his prologue, Holzschuher writes, "All of this has been extracted with industry from the books of Ptolemy, Strabo, and Marco Polo, and put together, the sea as well as the land."[166] Behaim builds his central hydrographic argument when writing of the north countries (Norway and Iceland), stating that they are "now known to us, and are visited annually by ships, wherefore let none doubt the simple arrangement of the world, and that every part may be reached in ships, as is here to be seen."[167] Behaim's thesis of a fluid, navigable Earth represents a conception of the world attuned to the lure of exploration and trade, "a highly complex representation of a network of territories, ports, goods, and people."[168] Though the globe includes citational references to the eastward, land-based journeys of Polo and Mandeville, the globe simultaneously announces the ascendancy of the sea and the coming era of the merchant ship, a development that would pressure the traditional land-based commercial routes in which Nuremberg had played a vital role. In the context of the medieval land-based pilgrimage or trade-route, the Behaim globe can be read both as elegy and harbinger.

Death of the author

The Behaim globe tells the tale of an adventurous explorer—whether real or exaggerated, a young man who traveled to Portugal, possibly sailed on voyages of exploration, and fashioned a life based on a new vision of a global world, one bound together by commodities and that presaged imperialism and colonialism. Like Mandeville, Behaim returned home to share the stories of his travels. In Behaim's case, it took the form of a multi-voiced, densely citational globe. Just northeast of the Azores, where Behaim had made an expatriate life for himself, a posthumous inscription is added—a death notice for Behaim. It states simply: "Martin Behaim died at Lisbon in the year of the Lord 1506 on the 29th July."[169] There is evidence that near the end of his life, Behaim died alone, in poverty.[170] The death notice on the globe brings the reader/viewer back to the initial framework established by Holzschuher's first prologue, reawakening the reader/viewer to the fictive world in which they have been traveling. Like Mandeville, Behaim is a "disappearing author,"[171] and the globe itself both *vita* and *memento mori*.

Despite his celebrity status as the returned adventurer and "author" of the globe, prior to his sea-based explorations, Behaim had not only been arrested for dancing with Jews but had also mismanaged his finances such that his brother Wolf wrote to their cousin Michael expressing relief that Martin had sought his livelihood overseas. Wolf commented, "Here at Lyons they say things about Martin which make me ashamed of him, and I wish very much we were rid of him altogether."[172] In Behaim's prologue to the globe, he expressed the wish that his work on the globe would ensure "that he may be thought kindly of for all time;"[173] in this regard, the globe becomes a form of cartographic redemption.

Fluid geographies

The transitional period of amalgamated geographic paradigms has led cartographic historians such as Robert J. Mayhew to observe that "maps of the period … tended to happily juxtapose unaltered medieval descriptions of the world with maps which made a nonsense of them."[174] But this implied cartographic naïveté belies the aesthetic of capaciousness, polyvalence, and interpretation that distinguishes the Behaim globe and late fifteenth-century cartography, an aesthetic integration of the literary and the geographic. In *Liquid Modernity*, Zygmunt Bauman links fluidity with modernity, casting the "long pre-modern centuries" as "locked in a stable and apparently invulnerable one-to-one correspondence"[175] of time and space. But fifteenth-century cartography disrupts this fiction of pre-modern temporal-spatial stability and, instead, models a "liquidity" of geographical thought, a disruptive intellectual process of reframing the representation of time and space that resists the determinism that will come to define the scientific era. Postcolonial and mobilities studies in literature and the "scale debate" in cultural geography[176] have queried the possibility of concurrent yet alternative geographic frameworks, or what Iain Chambers proposes as "fluid cartographies,"[177] a means to disorder the rationalism that resulted in colonialism and oppressive, instrumental hierarchies of global power. The theory of fluid cartographies, however, has a productive analog in fifteenth-century geography in the way that works such as Mandeville's *Travels* and the Behaim globe complexly structure both Ptolemaic and medieval spaces. In Thomas Malory's *Le Morte Darthur*, Arthur marches to Rome in clear, linear time but later steps aboard a rudderless boat that transports him to a dream-like experience outside of the bounds of measurable space and time (see Chapter 1). These alternative geographical frameworks embody a fluid geography—one that eschews fixity, encompasses emotion, allows for speculative wonder—fusing the stories of the cartographic with the spatial explorations of the literary. As contemporary critical studies seek to sail beyond the legacy of colonialism and exploitation, the multiple geographical frameworks of fifteenth-century literature and cartography offer an alternative model to that of our "naturalized" time-space continuum and, in so doing, offer a rich field of inquiry from across the Cartesian divide.

The current alliance "of poets and geeks" in the digital and spatial humanities has a clear antecedent in the production of the Behaim globe. As part of the collaborative production, Holzschuher kept a detailed account of the costs and procedure of the manufacture of the Behaim globe, including the following note:

> Item, paid for a loam mould over which the sphere was to have been made … also for linen for the first sphere; also for wine and beer, and other things, for the limner's dinner whilst painting the globe and occasionally also for [B]ehime.[178]

Clearly, the making of the globe was both intellectually productive and social. Craftspersons, artists, author, and financier gathered over wine and dinner to create a cartographic, literary, and artistic artifact, an iconographic synergy of the wonders of the world and the story of one traveler in that world. Collaborations fostered in the digital humanities today similarly call upon the cross-disciplinary skills of computer scientists, artists, and humanities scholars—collaborations which increasingly reveal that geographic thought is deeply contingent, culturally situated, poetic, and, as in Holzschuher's account, potentially convivial. As Gilles has asserted, the "difference between the poetic mapmaker and the cartographic poet is less important than their similarity;"[179] each draws upon the imagination to create new worlds.

Globes in England

If England was the putative birthplace of John Mandeville, popularizer of the round Earth theory in the late Middle Ages and inspiration for Behaim's Earth-apple, what role did globes play in the dissemination of geographic thought within England during the Ptolemaic revival? By the late sixteenth century, globes increasingly signaled prestige and imperial power in England, although, like Ptolemy's *Geography*, they had not been previously produced in England. As discussed in Chapter 2, the earliest woodblock depictions of a globe in England appear in Caxton's 1480 edition of *The Myrrour of the World*. The earliest production of globes in England, however, is officially dated far later, to 1592, with the famed Molyneux globes, a joint endeavor of Emery Molyneux and the

Dutch engraver Jodocus Hondius, living in London at the time. But the impact of globes in English geographical culture should not be equated solely with the production of the Molyneux globes, although they mark a significant cultural and political moment. The Molyneux globes, comprised of a celestial and terrestrial globe, were a visual spectacle—at over 2 feet in diameter, they were the largest globes made in Europe since antiquity.[180] Smaller "student" and navigational copies of the globes were also produced for more pragmatic uses. The completion of the globes met with nationalistic fanfare, championed "for the honor of his countrey, and good of the Schollers, Gentrye, and Marriners of the same."[181] Red and blue lines are drawn on the terrestrial globe marking the route of English navigators Francis Drake and Thomas Cavendish, and the globe is dedicated to Queen Elizabeth. Elizabeth had already been paired with cartographical images in the famed Armada portrait (artist unknown) of 1588 and the Ditchley Portrait (Marcus Gheeraerts the Younger) of 1592 in which she wears an earring in the shape of an armillary sphere and stands upon a map. Earlier, Elizabeth's coronation portrait of 1559 shows her holding an imperial orb. The imperial-scale Molyneux globes, according to Adam Max Cohen, "announced the arrival of England as a global empire while transforming the globe itself into an English symbol,"[182] and influenced the naming of the Globe Theatre by the Lord Chamberlain's Men, Shakespeare's company. In contrast to the scholarly focus of the world map in Ptolemy's *Geography*, the purpose of the Molyneux terrestrial globe was far more specific: to "immortalize English overseas exploration and to inspire Englishmen to continue such ventures."[183] While the production of globes in England may have lagged behind the Continent, the symbolic uptake of the Molyneux globes with English identity as an emerging global power was rapid.

The advent of the production of English globes did not announce the first use or ownership of globes in England. As Helen Wallis points out, prior to the Molyneux globes, English scholars and gentlemen purchased the globes of Gemma Frisius and Gerard Mercator, produced on the Continent, for their libraries,[184] and globes, as we have seen, appeared in English book illustration. By the mid-sixteenth century, maps and globes were important components in the libraries of English gentlemen as evidenced by account books and wills.[185] The 1556 inventory of William More's estate, for example, includes a perpetual calendar, world maps, and a globe. The globe figured as metaphor or motif in a variety of Tudor literary works presupposing "a familiarity with maps and globes amongst their various types of audience."[186] As globes gained "iconic status in northern Europe"[187] by the mid-sixteenth century, England participated in this European fascination via globe ownership and the depiction of globes in artwork, most famously in *The Ambassadors.* The 1533 painting of the foreign-born but English court painter Hans Holbein the Younger depicts a globe thought to be a replica of a globe from Nuremberg which used print technology to produce printed gores that were then affixed to the wooden sphere. Holbein's globe shows divisions of the New World based on European colonial treaties and reinforces the developing perception of the globe as symbolic of imperial aspirations and territorial possessions.[188]

Neither the absence of the production of globes or world maps in England during the Ptolemaic revival nor the absence of the round-Earth passage in a major strand of the English language version of Mandeville's *Travels* tell the story of the circulation of the round-Earth principle in England during the Ptolemaic revival. English readers had texts, illustrations, artwork, and imported globes, a result of the trade networks between England and northern Europe, that promulgated the new geography for an interested citizenry and constructed a "cultural history of imagining, seeing, and representing the globe."[189] The production of Molyneux globes and the construction of Shakespeare's Globe Theatre shortly thereafter were not a matter of a spontaneous geographical combustion in the late sixteenth century, but rather the result of a long imaginative relationship with the representation of the Earth as a globe through texts such as Mandeville's *Travels,* images in books such as Caxton's *Myrrour of the World*, and the importation of texts such as Ptolemy's *Geography* and European-made globes. Evidence of the cultural uptake of globes in English geographic culture, as will be discussed in Chapter 4, can even be found in a stage globe made by the English humanist and playwright John Rastell in 1518, made to entertain and inform his Tudor audience.

Notes

1 Denis Cosgrove, *Apollo's Eye: A Cartographic Genealogy of the Earth in the Western Imagination* (Baltimore, MD: Johns Hopkins University Press, 2001), 3.
2 Iain Macleod Higgins, *The Book of John Mandeville, with Related Texts* (Indianapolis, IN: Hackett, 2011), x.
3 Ibid., xii. Higgins proposes that the work is "an accomplished forgery" which obfuscates the actual author in order to allow for greater freedom of expression and a critique of the Christian world-view, ibid., ix–x.
4 Tamarah Kohanski and C. David Benson, eds. *The Book of John Mandeville* (Kalamazoo, MI: West Michigan Press for TEAMS, 2007), ln 56–57. Also available online via the Robbins Digital Library. Their edition is based on a Middle English Manuscript, British Library MS Royal 17 C. xxxviii.
5 Ibid., ln 68–69.
6 Ibid., ln 74–75.
7 Ibid., ln 1735–1736.
8 Ibid., ln 1716–1722.
9 Typically, this chapter on theoretical geography is Chapter 20 or 21. In the Cotton manuscript, it is Chapter 21.
10 Bernard Kline, *Maps and the Writing of Space in Early Modern England and Ireland* (New York: Palgrave, 2001), see especially 1–11.
11 It is perhaps no coincidence that the earliest use in English of the French word "equipollent" originates in the early fifteenth century.
12 Cosgrove, 35–37.
13 See Jeffrey Burton Russell, *Inventing the Flat Earth: Columbus and Modern Historians* (New York: Praeger, 1997).
14 C.W.R.D. Moseley, trans. *The Travels of Sir John Mandeville* (London: Penguin, 1983), 32.

15 M.C. Seymour, ed. *Mandeville's Travels* (Oxford: Clarendon Press, 1967), xiii.
16 "In Ethiope beth soch peple that han but oon foot, and they goo so faste that hit Is mervayl. And that foot is so large that hit maketh shade and covert to his body for the sonne." Kohanski and Benson, ln 1501–1503.
17 Seymour, *Mandeville's Travels*, xix.
18 Cosgrove, ix.
19 Higgins, *The Book of John Mandeville*, xxi.
20 The Defective Version is a Middle English manuscript recension of Mandeville's *Travels* distinguished by a missing passage on Egypt as well as on the Earth's sphericity.
21 David Wallace, ed. *Europe: A Literary History, 1348–1418*, vol. 1 (Oxford: Oxford University Press, 2016), v-vi.
22 Stephen Greenblatt, *Marvelous Possessions* (Chicago: University of Chicago Press, 1991), 38.
23 M.C. Seymour, *Sir John Mandeville* (Aldershot: Variorum/Ashgate, 1993), 11.
24 *The Book of John Mandeville: An Edition of the Pynson Text with Commentary on the Defective Version*, Medieval and Renaissance Texts and Studies, vol. 231, ed. Tamarah Kohanski (Tempe, AZ: Arizona Center for Medieval and Renaissance Studies, 2001), 51.
25 Karma Lochrie, *Nowhere in the Middle Ages* (Philadelphia, PA: University of Pennsylvania Press, 2016), 100.
26 Lochrie, 100.
27 For a discussion of the work's complex textual history, see Iain Macleod Higgins, *Writing East: The 'Travels' of Sir John Mandeville* (Philadelphia, PA: University of Pennsylvania Press, 1997); for Higgins definition of the work as a "multi-text," see 17–25. Tamarah Kohanski, C.W.R.D. Moseley, and M. C. Seymour have all written extensively about the rich textual history of the work.
28 Kohanski and Benson, 2.
29 See Higgins, *Writing East* and *The Book of John Mandeville.*
30 Kohanski and Benson title their edition *The Book of John Mandeville* "because that was the title under which the work was usually known in the Middle Ages … as opposed to the later *Mandeville's Travels*," n. 1 on p. 1. Higgins follows this practice, using the shorthand *The Book*. Other scholars, such as Karma Lochrie, refer to the work as *Mandeville's Travels*, while Seymour and Moseley use the modern editorial title of *The Travels of John Mandeville.* I use the *Travels* both for brevity and to denote its uncertain authorship.
31 Kohanski and Benson, ln 1–4.
32 Ibid., xvi.
33 For detailed analysis of the various manuscript stemma, see Higgins, *Writing East*.
34 John Larner, "Plucking Hairs from the Great Cham's Beard: Marco Polo, Jan de Langhe, and Sir John Mandeville," in *Marco Polo and the Encounter of East and West*, eds. Suzanne Conklin Akbari and Amilcare Iannucci (Toronto: University of Toronto Press, 2008), 133–155; 133–141.
35 Michael C. Seymour, "Mandeville in England: The Early Years," in *A Knight's Legacy: Mandeville and Mandevillian Lore in Early Modern England*, ed. Laden Niayesh (Manchester: Manchester University Press, 2011), 15–27; 15–16.
36 Ibid., 16.
37 Kohanski and Benson, ln 70–72.
38 Ibid., ln 1496–1499.
39 Ibid., ln 1519–1522.
40 Ibid., ln 299–301.

41 Sir John Mandeville, *Mandeville's Travels: The Cotton Version,* British Library. Manuscript. Cotton Titus C.16., ed. Paul Hamelius, 1:133. Online: Corpus of Middle English Prose and Verse.
42 Ibid., 1:210.
43 Kohanski and Benson, ln 2832–2835.
44 Moseley, *The Travels*, 36.
45 British Library MS Cotton Titus C. xvi, British Library MS Egerton 1982, and British Library MS Royal 17 C. xxxviii. The Metrical version has the round Earth information as well; Coventry City Record Office MS. For a discussion of these and other extant manuscripts in England, see Kohanski and Benson, "Introduction," in *The Book of John Mandeville.*
46 Seymour, "Mandeville in England," 18–19. Seymour itemizes 31 extant manuscripts of the Defective Version; in *The Book of John Mandeville,* Kohanski identifies approximately 40 extant manuscripts in Middle English of which 35 are part of the Defective Version, 12. Seymour's "sub-group A" manuscripts of the Defective Version contain the round Earth chapter. In sum, the majority of Defective Version manuscripts omit this chapter but about a quarter contain this chapter as do other Middle English, French, and Latin versions that circulated in England during this time.
47 Seymour, "Mandeville in England," 19.
48 See M.C. Seymour, "The Early English Editions of Mandeville's *Travels*," *The Library* 5:9 (1964), 202–207.
49 Seymour, "Mandeville in England," 19.
50 Ralph Hanna, III. "Mandeville," in *Middle English Prose: A Critical Guide to Major Authors and Genres*, ed. A.S.G. Edwards (New Brunswick, NJ: Rutgers University Press, 1984), 121–132.
51 See Kohanski, *Pynson.*
52 Kohansky and Benson, "Appendix: Sources."
53 Ibid., ln 742–744.
54 Paul Smethurst, "The Journey from Modern to Postmodern in the *Travels of Sir John Mandeville* and Marco Polo's *Divisament dou Monde*," in *Postmodern Medievalisms, Studies in Medievalism XIII*, eds. Richard Utz and Jesse G. Swan (Cambridge: D.S. Brewer, 2004), 159–179; 169.
55 C.W.R.D. Moseley, "Availability of Mandeville's *Travels* in England: 1356–1750," *The Library, Fifth Series* 2 (1975): 125–133, 125–126.
56 Seymour, "Mandeville in England," 23.
57 Moseley, "Availability," 132.
58 For an explication of the Christianized historical geography of the Hereford *mappamundi*, see Jocelyn Wogan-Browne, "Reading the World: the Hereford *mappa mundi*," *Parergon* 9:1 (June 1991), 117–135.
59 In referencing Jerusalem, Mandeville asserts that the "vertu of thynges is in the myddel" (ln 32). For a discussion of Mandeville's discussion of "middle" and "center, see Lochrie.
60 See David Woodward, "Maps and the Rationalization of Geographic Space," in *Circa 1492: Art in the Age of Exploration*, ed. Jay A. Levenson (New Haven, CT: Yale University Press, 1991), 81–87.
61 Claudius Ptolemy, *Geography of Claudius Ptolemy,* trans. Edward Luther Stevenson (New York: Cosimo Classics, 2011), 26.
62 Lochrie, 7.
63 Ibid., 10.
64 Ibid., 90–91.
65 Jamie K. Taylor, "Toward Premodern Globalism: Ocean Exemplarity in Chaucer's Men of Law's Tale," *PMLA* 135:2 (March 2020): 254–272; 264.

66 Zur Shalev, "Main Themes in the Study of Ptolemy's *Geography* in the Renaissance," in *Ptolemy's Geography in the Renaissance*, eds. Zur Shalev and Charles Burnett (London and Turin: The Warburg Institute – Nino Aragno Editore, 2011), 1–14; 11.

67 R.J. Mayhew, "Material Hermeneutics, Textuality and the History of Geography: Print Spaces in British Geography, *c.*1500–1900," *Journal of Historical Geography* 33:3 (July 2007): 466–488; 473.

68 John B. Friedman suggests that world maps are inherently didactic cultural objects which reveal cultural conflicts in "Cultural Conflicts in Medieval World Maps," in *Implicit Understandings: Observing, Reporting and Reflecting on the Encounters Between Europeans and Other Peoples in the Early Modern Era*, ed. Stuart B. Schwartz (Cambridge: Cambridge University Press, 1994), 64–95.

69 See the Mapping Mandeville Project at http://historiacartarum.org/john-mandeville-and-the-hereford-map-2/

70 Seymour, *Sir John Mandeville*, 18.

71 Kohanski and Benson, ln 1771–1772.

72 Higgins, *The Book of John Mandeville*, xi.

73 Kohanski, *Pynson,* xli.

74 Kohanski and Benson, 5.

75 Ibid., ln 1689–1695.

76 Ibid., ln 1733–1740.

77 Ibid., ln 1755–1756.

78 Ibid., ln 1709–1710.

79 Ibid., ln 1716–1722.

80 For the reception of Mandeville's *Travels* and the wide interest in the book as a trove of information rather than strictly religious inspiration, see Rosemary Tzanaki, *Mandeville's Medieval Audiences: A Study on the Reception of the Book of Sir John Mandeville, 1371–1550* (Aldershot: Ashgate, 2003).

81 Kohanski, *Pynson*, 88–89.

82 Seymour, "Early English Editions," 203. The note about Caxton's lost note is on f. 2 of British Library, Egerton MS 1982.

83 Whether de Worde's choice of copy text for the *Travels* was simply pragmatic or a theologically conservative choice is open to speculation. De Worde chose not to reprint Caxton's *Myrrour* with its detailed information on the Earth's sphericity. Here again, this could be simply pragmatic, as Caxton had reissued a second edition in 1491, or it may reflect a more cautious position regarding geographical science. The *Myrrour* was not reprinted until Andrewe Lawrence's 1527 edition.

84 Kohanski, *Pynson*, 92.

85 Kohanski and Benson, 3–4.

86 Charles W.R.D. Moseley, "'Whet-stone Leasings of Old *Maundeuile*': Reading the *Travels* in Early Modern England," in *A Knight's Legacy: Mandeville and Mandevillian Lore in Early Modern England*, ed. Ladan Niayesh (Manchester: Manchester University Press, 2011), 28–50.

87 Lesley B. Cormack, "Ptolemy at Work: The Role of the *Geography* in Geography and Mathematics Teaching in Early Modern England," in *Ptolemy's Geography in the Renaissance*, eds. Zur Shalev and Charles Burnett (London and Turin: The Warburg Institute – Nino Aragno Editore, 2011), 152–156. Mercator's maps better suited ocean navigation and were therefore of greater practical use.

88 Higgins, *The Book of John Mandeville*, xi.

89 Kohanski and Benson, ln 2830–2832.

90 For a discussion of the central action of the traveler's gaze, see Francis Tobienne, Jr., *Mandeville's Travails: Merging Travel, Theory, and Commentary* (Newark, NJ: University of Delaware Press, 2016).

91 Kohanski and Benson, ln 74.
92 Patricia Cohen, "Digital Keys for Unlocking the Humanities' Riches," *The New York Times*, November 16, 2010. Available at: www.nytimes.com/2010/11/17/arts/17digital.html.
93 This essay was first developed as part of The Language of Maps colloquium hosted by the Bodleian Library at Oxford University in 2013, convened by Keith Lilley (Professor of Geography, Queens University, Belfast) and Nick Millea (Map Librarian, Bodleian Library). The colloquium focused on the language of the Gough Map and brought together cartographic historians, art historians, classics scholars, and literary scholars to consider the shared language of maps and texts.
94 Günther Görz and Martin Scholz, "Semantic Annotation for Medieval Cartography: The Example of the Behaim Globe of 1492," in *22nd General Conference of the International Council of Museums*, eds. H. Rengmann et al. (Shanghai: ICOM/CIDOC, 2010), 1–19.
95 Günther Görz and Norbert Holst, "The Digital Behaim Globe (1492)," in *Museum Interactive Multimedia 1997*, eds. David Bearman and Jennifer Trant (Pittsburgh, PA: Archives & Museum Informatics, 1997), 155–172; 159.
96 Görz and Scholz, "Semantic Annotation," 2.
97 E.G. Ravenstein, *Martin Behaim: His Life and His Globe* (London: George Philip & Sons, 1908).
98 Görz and Holst, 158.
99 Ibid., 159.
100 Ibid., 163.
101 See Jim Siebold, "Behaim Globe," *Cartographic Images Network*. Available at: www.cartographic-images.net.
102 You must go there. Short of that, a 3D exploration of the globe is available at www.youtube.com/watch?v=Tz8UDg6ctUc. For a discussion of the 3D modeling process, see F. Menna et al., "High Resolution 3D Modeling of the Behaim Globe," *International Archives of the Photogrammetry, Remote Sensing and Spatial Information Sciences*, vol. XXXIX-B5 (2012).
103 James Sykes, "The Terrestrial Globe in Raphael's 'The School of Athens,'" *Globe Studies* 55/56 (2009): 53–72; 59, 70.
104 Görz, "Semantic Annotation," 1.
105 Alfred Hiatt, *Terra Incognita: Mapping the Antipodes Before 1600* (Chicago: University of Chicago Press, 2007), 158.
106 Ravenstein, 69 and 89.
107 Ibid., 58.
108 Ibid., 58.
109 Personal conversation, July 2013.
110 For the question of Mandeville's identity, see, for example, M.J. Bennett, "Mandeville's *Travels* and the Anglo-French Moment," *Medium Aevum* 75:2 (2006): 273–292.
111 Ravenstein, 52.
112 The English firm, Graves and Thomas creates reproductions of the globe for individuals and films.
113 The Ptolemy Project, University of California, Berkeley. Lead Researcher: Edward Lee. Available at: https://ptolemy.berkeley.edu/.
114 Ravenstein, 71.
115 Ibid., 71.
116 Ravenstein, 71.
117 Kohanski and Benson, ln 57–67.
118 Ravenstein, 74.
119 Ibid., 70.

120 Ibid., 63.
121 C.W.R.D. Moseley, "Behaim's Globe and 'Mandeville's Travels,'" *Imago Mundi* 33 (1981): 89–91; 90.
122 Ravenstein, 84.
123 Ibid., 104.
124 David Rundle, "Humanism Across Europe: The Structure of Contact," in *Humanism in Fifteenth-Century Europe*, ed. David Rundle (Oxford: Society for the Study of Medieval Languages and Literature, 2012), 307–336.
125 Ravenstein, 86.
126 Ibid., 72.
127 Ibid., 87.
128 Ravenstein, 71.
129 Cosgrove, 42.
130 Ravenstein notes that the globe was in the upper reception rooms of the Town Hall, 59.
131 Alfred Hiatt, "Blank Spaces on the Earth," *The Yale Journal of Criticism* 15:2 (2002): 223–250.
132 Cosgrove, ix
133 *The Compost of Ptholomeus*, printed by Robert Wyer, 1530?, STC 20480, sig. O (2)r.
134 Ravenstein, 89.
135 John Larner, *Marco Polo and the Discovery of the World* (New Haven, CT: Yale University Press, 1999), 136–137; 141.
136 Ibid" 149.
137 Smethurst, 171.
138 Moseley, "Behaim's Globe," 90.
139 Ravenstein, 88.
140 Smethurst, 169.
141 Ibid., 162.
142 Ibid., 168, 171.
143 In addition to Moseley and Seymour, see also Josephine Bennett, *The Rediscovery of John Mandeville* (New York: MLA Monograph Series:19; reprinted. Krause, 1971).
144 Moseley, "Behaim's Globe," 90.
145 Ibid., 90.
146 "In þat lond groweth the peper in a Forest þat men clepen COMBAR & it groweth nowhere ell in all the world but in þat Forest … And summe men seyn þat whan þei wil gadre the peper þei maken fuyr & brennen aboute to make the serpentes and the cokedrilles to flee." Cotton, 1:111–112.
147 Ibid., 1:102–103.
148 Bennett, 49–50.
149 Lochrie, 90.
150 Kohanski and Benson, ln 1329–1330.
151 Ravenstein, 87.
152 Ibid., 9.
153 The biblical passage does not specifically refer to an apple, but to a "forbidden fruit." See Genesis 3:6–7.
154 The apple chalice or goblet is found at the Germanisches Nationalmuseum, c. 1515. See also *Gothic and Renaissance Art in Nuremberg, 1300–1550*, exhibition catalog (New York: Metropolitan Museum of Art, 1986), 411.
155 Ravenstein, 88.
156 I use here Seymour's edition of *Mandeville's Travels*, 6. Note that the final phrase of this passage re-affirming that the world is round is not in the English

Defective Version nor in the Pynson printed edition. Behaim, however, likely consulted the Pepino Latin edition.

157 Smethurst, 172.

158 Mary Campbell, "Introduction," in *A Knight's Legacy: Mandeville and Mandevillian Lore in Early Modern England*, ed. Ladan Niayesh (Manchester: Manchester University Press, 2011), 3.

159 Virginia Mason Vaughan, "Preface: The Mental Maps of English Renaissance Drama," in *Playing the Globe: Genre and Geography in English Renaissance Drama*, eds. John Gilles and Virginia Mason Vaughan (Madison, NJ: Fairleigh DickinsonUniversity Press, 1998), 11. See also John Gillies, *Shakespeare and the Geography of Difference* (Cambridge: Cambridge University Press, 1994).

160 See Cosgrove, xi.

161 Seymour, *Mandeville's Travels,* 145. This passage does not appear in British Library MS Royal 17 C. xxxviii, which is the basis for Kohanski and Benson's edition.

162 Ravenstein, 88.

163 Ibid., 71.

164 Iain Chambers, "Maritime Criticism and Theoretical Shipwrecks," *PMLA* 125:3 (May 2010): 678–684; 679.

165 Cosgrove, 98–99.

166 Ravenstein, 71.

167 Ibid., 71.

168 Jerry Brotton, *Trading Territories: Mapping the Early Modern World* (London: Reaktion Books, 1997), 70.

169 Ravenstein, 75. Records indicate that the actual year of Behaim's death was 1507.

170 Ibid., 51.

171 For Mandeville's positionality as an author, see Smethurst, 173. See also Michel Foucault, "What Is an Author?" in *Modern Criticism and Theory*, ed. David Lodge (Harlow: Longman, 1969) and Roland Barthes, "The Death of the Author," in *Image Music Text*, trans. Stephen Heath (New York: Hill and Wang, 1977) ,142–147.

172 Ravenstein, 11.

173 Ibid., 71.

174 Mayhew, 473. See also R.J. Mayhew, "Geography, Print Culture and the Renaissance: 'The Road Less Travelled By,'" *The History of European Ideas*, 27:4 (2001): 349–369.

175 Zygmunt Bauman, *Liquid Modernity* (Cambridge: Polity Press, 2000), 8–9.

176 Andrew E.G. Jonas, "Pro Scale: Further Reflections on the 'Scale Debate' in Human Geography," *Transactions of the Institute of British Geographers*, New Series, 31:3 (September 2006): 99–406.

177 Chambers, 681.

178 Ravenstein, 112.

179 Gillies, 182.

180 See Helen M. Wallis, "Further Light on the Molyneux Globes," *The Geographical Journal* 121 (1955): 304–311.

181 M.S. Harley 5208, 50–51. See also Helen M. Wallis, "The First English Globe: A Recent Discovery," *The Geographical Journal* 117:3 (1951): 275–290; 275.

182 Adam Max Cohen, "Englishing the Globe: Molyneux's Globes and Shakespeare's Theatrical Career," *The Sixteenth Century Journal* 37:4 (Winter, 2006): 963–984; 968.

183 Ibid., 966.

184 Wallis, "The First English Globe," 275.

185 See, for example, John Evans, "Extracts from the Private Account Book of Sir William More of Loseley, in Surrey, in the Time of Queen Mary and Queen Elizabeth," *Archaeologia* 36 (1855): 284–310.
186 Victor Morgan, "The Literary Image of Globes and Maps in Early Modern England," in *English Map-Making, 1500–1650*, ed. Sarah Tyacke (London: The British Library, 1983), 46–56; 46.
187 Cohen, 966.
188 For a reading of the objects in Holbein's painting, see Lisa Jardine and Jerry Brotton, *Global Interests: Renaissance Art Between East and West* (London: Reaktion, 2000).
189 Cosgrove, 3.

4 The painted world

John Rastell's stage globe and geographic pleasure in early Tudor England

Geography's moral qualities, especially in relation to representations of the globe, are an overlooked aspect of late medieval and early Tudor geographic practices. Anathema perhaps to nineteenth- and twentieth-century rationalist geographer-scientists, pre- and early modern geography sought to intellectually edify but also to morally uplift and to entertain. Indeed, the contemplation of geographical representation as part of a moral practice was accompanied by a sense of wonder and pleasure.[1] The study of geography, in short, gave rise to delight (then, as now). In addition to the moral association of astronomy as the contemplation of the celestial realm, the conjoining of geography and pleasure arose, in part, from geography's literary heritage and thus its long association with poetic sensibilities. In Gavin Douglas' *Palice of Honour* (1553), he describes poetry as engendering "pleasaunce and half-wonder"—an equally apropos descriptor of late medieval and early modern responses to the beautifully colored maps of Ptolemy's *Geography*, as well as to the pan-European landscape of Malory's *Le Morte Darthur*, the geographical abundance of Caxton's *Description of Britain*, and the exotic marvels of Mandeville's *Travels* or Behaim's globe. Part armchair imperialism, such responses also elicited a more personal and recognizable pleasure continued today in contemplating, say, travel plans and the deep emotional responses that maps and globes engender. An affective, emotive response to maps and literature framed late medieval and early modern carto-bibliographic practices in which the moral, the imaginative, the imperial, and the cartographic were not incompatible experiences. This aesthetic stands in contrast to the later development of geography and the humanities "as isolated and autonomous disciplines."[2] Despite such disciplinary divides, Western culture's continued love of maps, now extended to the digital via Google Earth, still contains the trace of an aesthetics of wonder.

The mutual influence between geographic production and early modern drama has been well established. John Gillies has drawn attention to the geographic context and metaphors of Shakespeare's plays and early modern theatrical space,[3] identifying a "poetic geography" in Tudor drama and the mutually constitutive cultural production of new dramatic forms and new world exploration. Rhonda Lemke Sanford draws attention to the influence of early modern maps on literature while Tom Conley pays particular

DOI: 10.4324/9781003096245-5

attention to early modern French printing and the way in which "cartographic writing" opened up new discursive spaces and influenced early modern literature in England.[4] As these scholars have clearly demonstrated, the measured spaces of the new cartography developed in tandem with literary practices such as print, drama, and literature. Peter Barber categorized Henry VIII's gradual strategic integration of the use of maps for administrative and governmental purposes, as evidence of a newly emerging "map consciousness"[5] a term that has come under critical scrutiny for its focus on the instrumental use of maps. The assessment of English cartographic "belatedness"[6] presumes as a corollary that cartographic interest in England did not "blossom" until late in the sixteenth century with the publication of Hakluyt's *Principal Navigations* (1589), the construction of Shakespeare's Globe Theatre (1598), and the print publication of Edward Wright's *Chart of the World* (1599), all under the auspices of Elizabeth's territorially-conscious reign. The impact of Ptolemaic-influenced cartography, however, can be found far earlier in a strand of early Tudor literary-geographical culture that was in active dialogue with the geographical developments of the Continent, introducing the new geography to London audiences as performative and moral.

William Caxton's acumen in assessing and shaping English book culture resulted in his translating Latinate scientific matter, via a French intermediary, into the popular, accessible *The Myrrour of the World* for a literate but lay readership. This geographically interested audience of readers has been all but invisible in the narrative of English geographic thought, one that points to the gap in English cartographic representation that stretches from the Hereford Map (c. 1300) to Wright's (1599) world map. A lively geographic interest can be found, however, not just in the production of cartographic forms that have come to define the discipline of post-Enlightenment geography. Despite a sometimes-colloquial nature, a variety of late medieval and early modern geographic expressions in England demonstrate curiosity about the new geography while still linked to wonder and literary geography. The new geographic learning found expression, as we have seen, in Caxton's geographical editions and continued to find cultural purchase in book importation, book illustration, cosmological ceilings, paintings, and in humorous "interludes"—an evening's entertainment of vernacular drama.[7] In other words, geography had not yet settled into a solely scientific domain; infused with poetics and morality, geography functioned within a polysystem of poetics, art, translation, and drama, mutually constituting a spatial-temporal relationship of literature, geography, and early print culture.

In tracing the development and dissemination of geographical thought and imagination in early modern England, a key figure in this transitional period was the printer and man of affairs, John Rastell. Rastell translated literary texts, oversaw a print business, and circulated in the world of international business and law in early sixteenth-century London. A key figure in the More circle and in Henry VIII's court, Rastell is credited with the early composition of the English dramatic form known as the interlude. Rastell was also active in cosmology and geography but his work in this area is treated more as a curious side-line than as an influential element in the transmission of emerging geographic thought in England. As with his predecessor William Caxton, Rastell's place in the dissemination of continental geographic thought in England has gone largely unrecognized. An under-explored figure in the transmission of geographic thought in England during the era of the Ptolemaic revival, Rastell's cosmological work reveals an English geo-poetic response to the new geography and to global navigation.

John Rastell: geography as spectacle and entertainment

Around 1518, Henrican insider and brother-in-law of Thomas More, John Rastell wrote the cosmological play *A New Interlude and a Mery, Of the Nature of the Four Elements.* The interlude includes a pedantic geography lesson and presents a figure, or globe, on stage. Rastell's humorous, moral interlude has been thoroughly mined as a record of geographic knowledge in the early sixteenth century in England, as an assessment of English knowledge of Columbus's exploration of "new founden lands," and as one of the earliest printed references to "America" in English. The interlude has also received critical attention for its role in the early development of Tudor drama—specifically as a hybrid form of medieval morality play and early modern drama. A vexed question about the work centers around its self-referential title as an interlude, typically performed in private venues for an elite audience, in tension with the play's suggestion of an intended popular audience. In the main, the play has suffered from what J.A.B. Somerset has called "critical hostility,"[8] and, indeed, disparagements have been leveled at the play both from geographic historians, citing geographic lacunas, and scholars of early modern drama, citing the work's heavy-handed moralizing. Despite these bi-focal critical appraisals, the *Four Elements* has rarely been considered as a coherent form of poetic geography, both cartographic *and* literary, nor as a key work in the transmission of English poetical geographic thought. The role of the *Four Elements* in terms of early Tudor geographic culture bears further inquiry: I do not refer here to Rastell's personal geographic knowledge as an attestation of geographic knowledge at the time, but rather, to the way in which his work functioned as an agent of the dissemination of geographic thought in early 1520s London via the medium of a literary art form. What did it mean—in the tradition of Caxton—to produce a work that "translated" texts from the domain of Latin or Greek scholarship for courtly entertainment and dissemination in popular print? How was the English language conceived as a vehicle for this cultural transmission, especially on the eve of the Reformation? And what can the fusion of literary forms, popular entertainment, and scholarly geographic information reveal about emerging early modern geographical thought in England among a wide spectrum of readers?

Rastell fashioned "an energetic career"[9] of "wide-ranging enthusiasms:"[10] he was a well-known Londoner, an active member of the "More circle" by dint of marriage, employed by the court of Henry VIII, a lawyer, and publisher. Married to Elizabeth, Thomas More's sister, Rastell's interconnections with English humanists was further reinforced by the marriage of his daughter to playwright and member of the More circle, John Hewyood. A lawyer by profession, Rastell also owned a print house and was the creative force behind a variety of royal pageants and plays. Like William Caxton, his predecessor in the print trade, Rastell was part of an emerging intellectual and mercantile class that was also closely associated with the commercial, humanistic, and diplomatic activities of the court. Later in his career, Rastell included maps as part of his print shop business as well. Though a member of a failed exploratory expedition that left him temporarily stranded in Ireland, Rastell's interest and nationalistic pride in the explorations of Newfoundland by the Cabots in 1497 are clearly evident in the *Four Elements.* Rastell later served as a Member of Parliament and, in addition, combined his skills as an artist and engineer in the production of "cosmological ceilings" for diplomatic events such as the famed Field of Cloth of Gold[11] and in the design of "stages and awnings for royal entertainments."[12] Because of a later lawsuit, we also know that Rastell constructed a stage at his house and rented a considerably-sized theatrical wardrobe collection to other producers of theatrical entertainments.[13]

As part of his diplomatic stagecraft, Rastell collaborated on the ornate ceremonies held at the "Field of the Cloth of Gold," a temporary phantasm of diplomatic accord constructed near Calais in June of 1520 to mark the meeting of Henry VIII and Francis I of France. Temporary pavilions and a fabric "palace" were created from unimaginable yards of the highly prized "cloth of gold" such that the festive meeting grounds have ever since born this commemorative title. Rastell is credited with the creation of a temporary circular theater whose domed ceiling "depicted in extraordinary fashion a vast *mappa mundi* displaying the four elements."[14] In addition, "propositions" from Aristotle's *Meteorologica* accompanied the *mappa mundi* as well as Latin mottos and aphorisms on friendship.[15] Temporary structures were erected for the Field of the Cloth of Gold by both the French and the English in what Richardson terms the "politics of display."[16] The impact of the English *mappamundi* and cosmological ceilings was not limited, however, to the elite audience for whom it was intended. Though the diplomatic rendezvous was ostensibly between Henry VIII, Francis I, and their royal retinues, Richardson estimates that during the two weeks of the event, 12,000 people—from the royals down to the cooks and scullery maids—were at the Field of the Cloth of Gold in innumerable capacities. Richardson thus argues the Field of the Cloth of Gold had a wide-ranging cultural and financial impact not only on the two young kings, but also on the thousands employed to create what was the Renaissance equivalent of the Olympics Games Opening Ceremony.[17] Geographic exposition was a key visual element in what can only be considered *the* event of the 1520s, and Rastell was at the heart of this event. With his interest in geography and cosmology, Rastell went on to produce a pageant in 1522 on the "Elements, the Planettes and Starres in their places," no doubt influenced by his 1518 *Four Elements* as well as a number of public geographical theatrical spectacles. Rastell's works thus exemplify one of the earliest connections between theatrical and geographic space usually associated with later English Renaissance drama.[18]

In 1527, Rastell was involved in yet another Anglo-French diplomatic pageant, "The Father of Heaven," in which he collaborated with Nicholas Kratzer, the king's astronomer, and the painter Hans Holbein (the Younger) to create a theatrical narrative and astrological roof for an inner banquet chamber.[19] Such painted cosmological ceilings suggest the continued close association between the celebration of an astronomy of wonder and the new geography in elite as well as more public forms. The work merited approbation by Edward Hall in his chronicle when he noted that the ceiling was "conninglie made by the kynges Astronimer, for on the grounde of the rofe, was made the hole earth environed with the sea, like a very Mappe or Carte."[20] The painted ceiling included "the fiue circles or girdlelles and the two poles apered on the earth and water compassing the same ... it was a connyng thing and a pleasant sight to beholde."[21] Like the Behaim globe, Hall's description emphasizes the conception of the Earth as "oceanic," a conceptual shift that reimagined the Earth as essentially fluid.

Rastell's engagement with geographical representation was such that John Bale, in his catalog of British writers, refers to Rastell as a "cosmographus,"[22] a Latin term suggestive of Rastell's work as a cosmographer, writer, and printer. Across these various endeavors, Rastell's energies seem prodigious, and the range of his intellectual activity bespeaks a man vitally engaged with the Renaissance ideal of engagement in art, politics, engineering, natural philosophy (i.e. science), and cosmography.[23] The amalgam of geographic knowledge during the Ptolemaic revival and the era of Columbus's explorations has been described as a time of "chaotic empiricism" for its juxtaposition of competing models and conflicting information.[24] In cosmographical projects such as Rastell's, we can also detect a form of "cartographic latency,"[25] an expression of geographic knowledge deeply connected to art and literature, yet simultaneously inclusive of eye-witness geographic information. Robert J. Mayhew argues that the transitional geography of the late medieval and early modern period geography "remained in the thrall of broader currents in the history of ideas."[26] To modern eyes, the connection of geography to astronomy and astrology seems, at best, quaintly folkloric, but as Shalev argues, there is "abundant evidence of the use of Ptolemy's *Geography* for astrological purposes."[27] Thus, the highest expression of early Renaissance cartography and geographical scholarship is also linked to a culture of cosmology and astrology. More than merely "in thrall" to other ostensibly disparate cultural forms, the emerging discipline of geography co-formed a geo-poetic aesthetic particular to the late fifteenth and early sixteenth centuries, an aesthetic intertwined with the development of early print. The full-on entry of English cosmographers into Renaissance world map-making is typically marked by Emery Molyneux's (1592) globes, Wright's (1599) *Chart of the World*, and John Speed's *Prospect of the Most Famous Parts of the World*, printed in 1627. But early in the sixteenth century, Rastell created a theater prop globe which, though ephemeral, embodied the hybrid form of the geographical culture of the era, the purview of both "poets and astronomers."[28]

A New Interlude and a Mery, Of the Nature of the Four Elements

As wryly observed by the character of the Messenger, Rastell's play, in stanzaic verse, was written "Without great eloquence, in ryme rudely" (line 129) and, indeed, Rastell's drama is composed in "the simpler tail-rhyme stanza, staple of popular romance and of popular drama."[29] Aside from the opening seven-line stanza, the majority of the play presents the action in six-line stanzas. Published sometime between 1518 and 1520, the *Four Elements* marks an interest in drama that Richard Axton characterizes as "unique in this period."[30] For example, the widespread and private reading of a printed play text was a new phenomenon in contrast to public spectatorship at the annual performances of medieval dramas whose playscripts were not widely distributed.

Barber argues that Tudor court pageantry, of which Rastell was an active director, continued "to employ maps in the medieval manner" and should not be seen as geographically innovative.[31] Barber cites an emerging interest among members of the court in maps as a tool of propaganda and governance and acknowledges Rastell's interest in the "new geography" as extending beyond "medieval theories exemplified in the court pageants of the early 1520's."[32] Rastell's productions integrating theater and geography straddled the typical markers of the medieval and the early modern, confounding designations such as morality play versus secular theater, sacred cosmology versus mathematical geography. Rastell's interlude reveals an aesthetic of poetic geography unique to the era of the Ptolemaic revival, drawing on literary, cosmological, and Ptolemaic geographies as well as the functional geography emerging in the court of Henry VIII. Barber draws a firm distinction between strategic mapping activity and the cosmology as entertainment in popular court pageants and interludes. To Rastell, however, theatrical performance seemed a fitting venue for new geographic expression in which moral cosmology was integrated with secular and mathematical geographic knowledge, a "provocative syncretism"[33] that defines late fifteenth-century and early sixteenth-century geographic practice.

In terms of genre, the designation of Rastell's *Four Elements* as a "play" or an "interlude" is contested,[34] with certain critical repercussions in terms of the audience and circulation of the play. Those issues aside, Rastell's humorous interlude presents a geography lesson which includes the presentation of a "figure" on stage. In all likelihood, this figure is one of the earliest globes produced in England. Typical of a morality play, Rastell's *Four Elements* has a didactic rather than plot-driven focus. In essence, the eight characters—the Messenger, Natura Naturata, Humanyte, Studyous Desire, Sensuall Appetyte, the Taverner, Experiens, and Yngnoraunce—enact the common morality play convention of temptation and restoration. Studyous Desire begins his studies with Natura Naturata, falls in with Experiens and Sensuall Appetyte, but is presumably restored to his good senses by the play's end (the last part of the play text is missing but the direction of the action toward this moral end seems clear). The play largely consists, as Axton observes, of "a geography lecture with practical demonstrations" of the geographic principles.[35]

As the play begins, the Messenger introduces the theme and offers an explanation of why the anticipated scholarly content is set within a "mery" context:

> But because some folke be lytyll disposyd
> To sadnes, but more to myrth and sport,
> This phylosophycall work is mixed
> With mery conseytis, to gyve men comfort
> And occasyon to cause them to resort
> To here this matter, wherto yf they take hede
> Some lernynge to them thereof may procede.
> (lines 134–140)

In this way, the Messenger imparts a principle of constructivist learning; that is, that people learn best by mixing learning with "mery conseytis"—a radical concept in English educational theory in the early sixteenth century.[36] Nature Naturata is then introduced as the embodiment of the classical conception of nature (the character's name will appear, subsequently, as Nature), followed by entrances by Humanyte and Studyous Desire, each eager to learn more of nature and the Earth. Before long, Sensuall Appetyte enters the scene and woos Studyous Desire and Humanyte to accompany him to a tavern where they meet Experiens. Experiens provides a demonstration of early sixteenth-century geographic knowledge with the aid of a "fygure," a map or globe, by which the audience imaginatively circumnavigates the globe. The plot continues with a wrestling for the mind and soul of Humanyte, followed by a merry song in the tavern, and the indication that the play is moving toward a moral resolution. Though a bit staid, the interlude provides an enjoyable geography lesson: its exotic geographical prop and lively tavern scene surely made for a pleasant evening's entertainment.

Several geographical references in the play have been of particular interest to cartographic historians: a reference to "new landis" (line 855), a reference to America, and an enigmatic line stating that of "the south parte on the other syde … we knoew nothynge at all" (lines 868–870). As Romuald Ian Lakowski notes, Rastell's geographic knowledge, in particular regarding America, has been "extensively analyzed"[37] during the mid-twentieth century, arising from progressivist geography's interest in determining the development of geographical "accuracy" via specific references to Columbus. The voyages of the Cabots to Newfoundland in 1497 are synthesized into Rastell's geographical world-view as is that of Columbus, though the latter is not specifically mentioned in his play text other than a generalized statement that

> Westward be found new lands
> That we never harde tell of before this
> By wrytynge nor other meanys
> Yet many nowe have ben there.
> (lines 747–750)

Other than a single reference by Thomas More's son, John, Rastell's *Four Elements* is the only More circle writer to make specific reference to "newe landys" and the coining of the term "America" by Amerigo Vespucci:[38]

> But this newe landys founde lately
> Ben callyd America by cause only
> Americus dyd furst them fynde.
> (lines 838–840)

While More clearly integrated his geographic knowledge into the imaginative realm of *Utopia*, he did not take up geographical discoveries explicitly. In the early twentieth century, George B. Parkes dismissed Rastell's geographic knowledge as extremely limited and ventured that Rastell "may never have read any geographical works at all."[39] Parkes' assessment that Rastell's information was "probably based on hearsay"[40] stems from three related premises: (1) a presumed lack of English geographic culture; (2) no specific mention of Columbus; and (3) the apparently frivolous context of the interlude. But Parkes' disdain misses the mark, pointing to a misreading of the geo-poetic multisystem that marked the dissemination of English geographical thought in the age of the Ptolemaic revival: specifically, a reliance on an international book trade for the importation of maps and scholarly geographic books, the printing of vernacular narrative geographies, and the embedding of geographical thought in poetic and artistic expression.

In the mid-twentieth century, Johnstone Parr refuted Parkes' dim view of Rastell's geographic knowledge arguing instead that Rastell "was by no means ignorant of contemporary geography,"[41] and placing Rastell's knowledge in the context of maps and geographical works currently in print and circulation in England and the Continent,[42] including seven geographical and cosmological works based on Ptolemy and very likely multiple globes. No diplomat or merchant involved in continental trade could have been ignorant of such texts and globes. Parr argues that the "Columbus criteria" used to assess Rastell would founder the reputation of many a reputable contemporary geographer. In Rastell's case, his nationalistic desire to emphasize the exploration of the Cabots over that of Columbus may have been a factor in his silence regarding Columbus. Parr finds Rastell's description of the location and native cultures of America as "amazingly accurate"[43] and assesses Rastell's geographic knowledge as what "one could expect from an intelligent and well-read man of affairs in the first two decades of the sixteenth century."[44] Elizabeth Nugent notes that only one scientific work, Bartholomaeus Anglicus' *De Proprietatibus Rerum*, translated by John Trevisa, had been translated into English and printed by this time.[45] Nugent's observation, however, overlooks Caxton's (1482) edition of the *Polychronicon*, which contained a world geography or "natural science" section, as well as Caxton's edition of *The Myrrour of the World* (see Chapter 2). It is a fair judgment that the printing of scientific works in England was small in comparison to the Continent and, in particular, Germany. However, a significant number of Latin editions of important texts on natural science were printed in Rome, Venice, Strasbourg, and Augsburg, including works by Strabo, Ptolemy, Euclid, Proclus, Aristotle, Pliny, Sacrobosco, Albertus Magnus, and Vincent de Beauvais, and Rastell would have had easy access to such Latin editions through the London book trade. As a lawyer and intellectual, his knowledge of Latin is unquestioned, and two Latin geographical texts by Gregorius Reisch and Martin Waldseemüller are key sources for the geographical information found in the *Four Elements.*[46] Rastell drew upon Reich's *Margarita Philosophica*, of which 16 editions had been printed on the Continent between 1503 and 1515, and Waldseemüller's *Cosmographiae Introductio*, first printed with a map and globe in 1507, with five subsequent editions by 1518—about the time Rastell is thought to have been working on the *Four Elements.* Nugent also identifies several other key sources for the *Four Elements*, including *Textus de Sphaera*, Caxton's *Myrrour of the World*, and *De Proprietatibus Rerum*, adding that there "can be no doubt" that Rastell was well aware of the discoveries of Columbus, the Cabots, and Vespucci.[47] In light of these influences, Parkes' estimation of Rastell's geographic knowledge must be set aside; instead, the play can be reappraised as a creative "translation" of up-to-date, scholarly geographical knowledge in the form of popular entertainment. Rastell's light-hearted play artfully disguised his scholarly geographic sources—the result of which may have pleased Rastell as proof that his translation of formal, Latin culture into popular culture was highly successful, enough so that he put later scholars off-track in identifying the geographic scholarship that underlaid his "mery interlude."

Rastell's play invited the uninitiated into an encounter with scientific principles that were significantly changing European and English world-views. A cultural figure akin to late twentieth-century science popularizers Carl Sagan and Neal DeGrasse Tyson, Rastell found traction for his scientific popularism in early modern English theater. Just as Sagan grasped the power of the medium of television for his popular series *Cosmos*, Rastell likewise saw the potential for secular drama and print publication to reach non-scholarly audiences previously sequestered from Latinate scientific discussion. C.R. Baskerville captures the forward-looking nature of Rastell's interlude when he states, "Though a strong didactic purpose in the drama would seem to associate Rastell with the old spirit of the moralities ... Rastell's affiliations are with the new spirit of humanism."[48] Rastell's humanistic play refashioned the newest geographical treatises from the Continent into a hybrid dramatic form intended for the edification and enjoyment of a public audience, in performance and in print.

"and compass them everywhere orbicularly:" presenting the round Earth

As the play commences, the Messenger explains that although "poyntys of phylosophy naturall," (line 9) that is, natural science, "be not so well declaryd/As a great clerke coude do, not so substancyall" (lines 10–11) nonetheless the play will take up an exposition of the "four elementis" (line 124). Tellingly, this introduction of the geographical subject touches upon the perceived limitation of the English language as a substantial vehicle for scholarly discourse. Rastell's scholarly moment cedes the stage to his love of theatricality with the dramatic entrance of what is referred to in the play as a "fygure." The Messenger dramatically calls out, "Beholde I prey you, see where they are! / The players begyn to appere in presence" (lines 143–144) at which point Natura Naturata, Humanyte, and Studyous Desire enter the stage "*portanus figuram*," (stage direction, subsequent to line 147), that is, carrying a stage device of some kind. Rastell provides no further detail on the "figure," but the players clearly use it as the basis of the geography lesson that forms the core of the play, leading scholars to conclude that the figure was either a map or, more likely, a globe. Nature amplifies the dramatic entry of the figure, declaring:

> The great worlde beholde, lo, devydyd wondersly
> In to two regyons, wherof on I call
> The etheriall region with the hevyns hye,
> Conteynynge the planettys, sterris and speris all;
> The lower region callyd the elementall.
>
> (lines 162–166)

Indicative of Rastell's sense of the play's function as a vehicle of translation from scholarly Latinate discourse to popular entertainment, the stage directions are provided in Latin, but the action is in English.

In the stage directions listing the characters, there is an enigmatic line stating that "Also yf ye lyst ye may brynge in a dysgysynge" (lines xi–xii). Axton speculates that this "disguise" may refer to a masque and therefore indicated masked dancers in the play. Alternatively, he suggests it may have been a reference to a "mechanical spectacle" of some kind.[49] With Rastell's engineering skills, some form of stage wizardry was certainly possible. Another possibility is that the "dysgysynge" might have been a representation of the celestial sphere to serve as a contrast to the "fygure" of the Earthly globe. This possibility is suggested by Nature's situating the Earth in the Ptolemaic and medieval Earth-centered cosmos:

The yerth as a poynt or center is sytuate
In the myddys of the worlde, with the water joyned,
With the ayre and fyre rounde and hole invyronyd.
(lines 229–231)

Nature suggests several "proofs" that the oceanic Earth is surrounded by air and fire that "compasse them everywhere orbycularly, / As the whyte aboute the yolke of an egg doth lye" (lines 249–250). Rastall's penchant for using domestic and everyday metaphors to explain scientific concepts adopts a strategy in the tradition of Lucretius's *On the Nature of Things.* For example, Nature turns to the figure, stating: "wherefore this fygure beholde / For a more manifest demonstracyon" (lines 283–284). In so doing, Nature suggests that the "figure" will inspire the audience to "more scyens to adquire" (line 288). Humanyte, playing the fall guy to Nature's geography lecture, innocently declares:

For I can no maner wyse perceyve nor see,
Nor prove by reason why the yerth sholde be
In the myddys of the fyrrmament hengyng so small
And the yerth with the water be rounde with all.
(lines 336–339)

Evidently, the tenet of the new geography that most stretched "common-sense" credulity was the concept of a watery, oceanic Earth as a coherent sphere, a principle the playwright takes great pains to address. This oceanic principle, as we have seen emphasized on the Behaim globe (see Chapter 3), is a key construct of Ptolemaic-era geographic thought even while maintaining the medieval and Ptolemaic cosmology of an Earth-centered universe.

The geography lesson of the play is interrupted by a comical interlude in which Humanyte is tempted away from such serious matters, joining Sensuall Appetyte for an excursion to a tavern. Madcap drinking humor ensues, but before too long Studyous Desire seeks out Experiens who has traveled "In Affryk, Europe, and Ynde" (line 671), representing the classical and medieval conception of the tripartite habitable Earth in medieval *mappaemundi* and "T-O" maps. Experiens broadens this traditional geographic space, however, to claim that "Both est and west I have ben farre" (line 672), suggesting, as we have seen in Caxton's illustration and Mandeville's Earth-apple, the concept of a round, circumnavigable Earth. The exchange between Experiens and Studyous Desire reinforces the spatial practice of contemporary early sixteenth-century rutters, or navigational itineraries, that provided distance in miles from port to port, and, as we have seen in Malory's romance and Caxton's geographical editions, the increasing attention to the measured distance of English miles (see Chapters 1 and 2). Studyous Desire wonders rhetorically about a pilgrimage to Jerusalem, then asks specifically, "How many myle is it I you pray / From hens theder to goo?" (lines 682–683). In his answer, Experiens acknowledges the importance of "cosmogryphy," but also advocates for the importance of cartography:

> Ye shall never lerne then more surely
> Than by that fygure yonder
> For who that fygure dyd fyrst devyse,
> It semeth well he was wyse.
>
> (lines 688–691)

Jerusalem as a destination is no longer merely symbolic, but spatially quantified. As the "deviser" of the figure, Rastell's self-referential wit reveals itself when Studyous Desire rhetorically asks, "Who thynke you brought here this fygyre?" and Experiens comically answers "I wot not" (lines 696–697). Rastell steers the audience back to the play's geographic matter, allegorically attributing the figure to the hand of "lorde Nature" (line 698).

The detailed geography lesson begins in earnest at this point, with Studyous Desire asking the location of "this realme ye know is callid Englande, / Sometyme Brettayne, I understonde," (lines 704–705) and requesting Experiens to "point with your hande / In what place it shulde lye" (lines 705–706). The play's dramatic action underscores the school-room quality of the scene and confirms that a globe or map was being used to convey the geography lesson. Experiens points out places on the globe, using locational phrases such as "Here than is the narrowe seey" (line 715) or "This contrey is callyd Italye" (line 721). After locating England first—a starting point that echoes Caxton's reorganization of the *Description of Britain* to begin in England—Experiens loosely narrates the southward journey from England to Rome, one followed both by pilgrims and King Arthur in the well-known Roman War account alike (see Chapter 1). Experiens identifies Calais and Boulogne in northern France, Flanders, central France, makes a passing reference to Spain and Portugal, then locates Italy, exclaiming, "Beholde where Rome in the myddes doeth ly" (line 731). Nugent makes the argument that this confirms Rastell's reliance on Waldeseemüller's 1507 map on which the centrality of Rome is marked by a double eagle and the papal keys.[50] The dramatic exclamation emphasizes the iconic geo-cultural itinerary from England to the locus of western cultural authority—Rome.

Experiens then calls attention to Germany, Denmark, Norway, and Iceland before turning to identify "the great Occyan" and the discovery "within this twenty yere / Westwarde be founde new lands" (lines 736–737). Like Martin Behaim on his globe, Experiens highlights the potential for commodities that could be extracted from these new lands, then "turns" to the globe, reinforcing the turning of the spherical globe, stating:

> Than wyl I torne agayne to my matter
> Of Cosmogryfy where I was err.
>
> (lines 826–827)

Moving eastward, he indicates the Mediterranean Sea, "Soudans contrey" (line 832), and the site of lingering European unease: "The great Turke on the north syde doth ly/ A man of merveylous strengthe" (lines 833–834). Like any good teacher, Experiens summarizes the geography lesson just presented by affirming the knowledge his audience was already familiar with—the tripartite world—and then expanding upon that traditional conception with new information:

> This sayde north part is callud Europa,
> And this south parte callyd Affrica,
> This eest parte is callyd Ynde,
> But this newe landys founde lately
> Ben callyd America by cause only.
> (lines 835–839)

Experiens rounds out the geography lesson by pointing out Jerusalem, India Minor and Major, which he identifies as the "lande of Prester Johan" (line 846), drawing attention eastward to the lands of the "Cane of Catowe" (the Kahn of Cathay) (line 852), made famous by the *Travels* of Marco Polo. The globe is then apparently turned to its original position as Experiens continues:

> And this is called the great eest see,
> Which goth all alonge this wey
> Towardes the newe landis agayne;
> (lines 853–855)

After describing "all this parte of the yerth" (line 865), Experiens has narratively circumnavigated thc globe and brought his audience back to England, adding somewhat cryptically:

> But the south parte on the other syde
> Ys as large as this full, and as wyde,
> Whiche we knowe nothynge at all.
> (lines 869–870)

Parkes interpreted Experiens' statement as a disavowal of knowledge of the southern antipodes, lands south of the equator, and deemed it an indication of Rastell's lack of geographic sophistication. Parr, however, clarifies that Experiens is not distinguishing northern hemisphere from southern, but rather, is referring to the "other side" *of a globe* and that the region in question would have been the South Pacific Ocean, a region that geographic writers and cartographers, indeed, knew little about in the early sixteenth century. In Parr's estimation, this monologue confirms that the "fygure" is a globe and not a map. If so, Rastell's stage globe is the first known globe produced in England and the context for that globe is not a university or scholar's study, but an evening's entertainment.

Experiens concludes the geography lesson with a nod to the classical and medieval trope of *mirabilia*, seemingly channeling Mandeville and Behaim when he claims: "Lo, is not this a thynge wonderfull" (line 877). Experiens might have continued to pontificate on geographic wonders, but the lagging didactic scene is fortuitously interrupted by Sensuall Appetyte, Humanyte, and the Taverner. Rastell reinforces his philosophy of learning through entertainment by cleverly including in the tavern scene a danced "round," at which the Taverner exclaims,

> For by God, it is a pretty gyrle!
> It is a worlde to se her whyrle
> Daunsynge in a rounde.
> (lines 957–960)

The danced round and, indeed, the whole tavern scene, replicate the motion of the global Earth. Experiens continues to buttress the play's round-Earth thesis:

> Why, what doute have ye therin founde?
> Thynke ye the yerth shulde not be rounde,
> (lines 1049–1050)

To which Humanyte assents:

> One wey it is rounde, I must consent,
> For this man provyd it evydent.
> (lines 1052–1053)

Experiens takes his new pupil through another brief recapitulation of the previous tour of the globe and concludes: "Thys provyth of necessyte / That the yerth must nedis rounde be" (lines 1082–1083). He also demonstrates the way in which the North Star appears to "ryse" as one sails north and "falleth" as one sails southward (lines 1091–1092) as yet another round-Earth proof. To hammer his point home yet again, he introduces "myne instrumentis" (line 1122)—a candle flame and telescope—replicating a woodcut illustration of Reisch's *Margarita Philosophica* (1512 and 1517 editions) in which two perspectives—that of a sailor in a ship's crow's nest and one on deck—reveal the curvature of the Earth.[51]

It seems clear from the play's internal evidence that the "fygure" was certainly, as Parr suggests, a globe rather than a flat map. Rastell had the geographic knowledge, engineering, and technical theatrical skills to construct England's first globe, far earlier than the 1592 production of the Molyneux globes. Though ephemeral, of particular significance is that Rastell's globe was created within the form and context of the arts. Rastell's globe thus unsettles the narrative of English "belatedness" in geographic production during the Ptolemaic revival. His theatrical globe, a prop no doubt ultimately dismantled or tattered after the last performance, is attested to in the linguistic evidence of his playscript. Both the text and the "figure" of the *Four Elements* affirm that geographic cultural production occurred in England in the context of vernacular, artistic expression.

After the song *Time to Pas*, of note in and of itself as the "only song with music to survive in an early Tudor interlude,"[52] Nature returns to the stage and intervenes in the play's descent into tavern revelry to reclaim the dignity of the scholarly matter. He gently chides Humanyte's dalliance, providing a clear indication that by play's end, Humanyte will return to the studious fold. Though the final lines of the play are lost, the trajectory of its geographically framed moral outcome is clear.

While the dichotomy between the moral aspect of Nature, in contrast to Ygnoraunce and Sensuall Appetyte, is clearly drawn, the figure of Experiens represents a new role representative of the increased secularity of learning and the importance of eye-witness experience and accounts. In addition, the play presents science, including geography, as a newly emerging and increasingly legitimate source of knowledge. The moral characters of the play embody the various forms of "authorized" knowledge: Nature functions as the embodiment of divinely ordered creation. The Messenger similarly expresses the traditional Christian world-view when he states, "And so by lytyll and lytyll ascendynge / To know Goddys creaturys and marvelous werkinge" (lines 97–98), a position shortly reaffirmed by Nature in attributing the generative nature of the Earth to "The perfection and furst cause of every thynge" (line 151), ordained and created by "The hye, mighty, most excellent of all" (line 148). Nature, in this recapitulation of a medieval world-view, models the narrative practice of a universal history in which geographic writing expresses the marvelous world of divine creation and generativity. Studyous Desire, emblematic of classical geography and humanism, uses theoretical astronomy as "proof" that the world is round. He cites the "sone, mone, and sterris in theyr movynge" (line 358) and twice affirms the Ptolemaic geo-centric universe, asserting that the Earth "hangyth" in "the myddys of the fyrmament" (lines 345–346). These two sources of authority—the Christian world-view and the revival of classical scholarship by humanists—were well known to Rastell's early modern audience. But to this dyad, Rastell adds "Experiens," the new lynch-pin of geographic knowledge with roots in the scientific work of Francis Bacon and experiential confirmation in the voyages of Columbus, Vespucci, and Cabot.

When Experiens first appears on stage he is "ryght welcom to this contrey" (line 661) by Studyous Desire, who comments:

Syr, I understonde that ye have be
In many a straunge countree.
(lines 665–666)

To which Experiens replies:

Ryght farr, syr, I have rydden and gone,
And seen straunge thynges many one.
(lines 669–670)

Thus, the Christian, classical, and experiential geo-epistemological modes are blended, in much the same way as in Caxton's *Myrrour of the World*, with respect accorded to each mode as complementary sources of knowledge of the Earth.

On the Nature of Things

A classical influence on Rastell's world-view includes, of course, the Earth-centered Ptolemaic-based cosmology, but the influence of humanistic engagement with classical cosmological writing is also evident, and specifically that of *De rerum natura* [*On the Nature of Things*] by the Roman poet and philosopher Lucretius, written in the first century BCE. Influenced by Epicurean philosophy, Lucretius set forth his philosophical and cosmological insights in a poetic treatise esteemed by Cicero and Ovid, early modern, and contemporary scholars alike. W.R. Johnson admires the "austere beauty" of Lucretius's "poetic model of the physical world," a work Stephen Greenblatt describes as "a near-perfect integration of intellectual distinction and aesthetic mastery."[53] Of particular interest in regard to Rastell is Lucretius' integration of cosmological thought into poetic form and his use of a non-specialized language within this poetic form to speak to the broadest possible audience, scholars and laypersons alike. Rastell similarly embraces both of these approaches, an influence which can be tied, at least in part, to the dissemination of classical texts such as *De rerum natura* through European humanists. The Italian intellectual and clergyman Poggio Bracciolini rediscovered the sole surviving manuscript of Lucretius' poem in 1417 and set in motion the copying, printing, and influence of the text throughout Europe in the fifteenth and sixteenth centuries.

That Lucretius' poetic inquiry into the nature of the universe, religion, and happiness was influential in the More circle is indisputable. After his recovery of the manuscript, Bracciolini saw to it that a copy, no longer extant, was made of Lucretius' lengthy poem (7,400 lines of dactylic hexameter). Two of the core tenets of Lucretius' poem—that there is no divine, prime mover of the universe nor any afterlife—were, unsurprisingly, considered heretical such that, in 1516, the Florentine Synod banned the use of the text in schools.[54] But by then the text had been reproduced in manuscripts as well as print editions by humanist presses in Bologna, Paris, Venice, and Florence such that the church-based prohibitions on the dangers of reading such a "pagan" text "could not quell interest in Lucretius." The spread of Lucretius' radical, Epicurean ideas, antithetical to the Christian beliefs in a divine being and a judged afterlife, certainly made its way through the book trade and international contact to English humanists and the More circle, "the center of intellectual and cultural life at court."[55] At the invitation of the Cardinal of Winchester (Henry Beaufort), Bracciolini spent almost five years in England (1419–1423),[56] just after his discovery of Lucretius' manuscript, and his English scholarly contacts were certainly aware of this controversial, yet elegantly written, Latin poem. Erasmus wrote a fictional dialogue entitled *The Epicurean* that attempted to integrate the ideals of Epicureanism into Christianity, and his English colleague Thomas More skewers Epicurean ideals by situating Utopia in "no-where." Greenblatt suggests More's creative inspiration for *Utopia* may have been sparked by Amerigo Vespucci's comment about the peoples of the new found lands: "Since their life is so entirely given over to pleasure, I should style it Epicurean."[57] Rastell's poem is indicative of the purchase of *On the Nature of Things* during the Ptolemaic era and the effort to negotiate Epicurean physics, as expressed in Lucretius' poem, with the precepts of Christian divinity and belief in an afterlife.

Rastell's interest in the concepts of *De rerum natura* is evident in three key features of the interlude: (1) his use of the character "Natura Naturata" to introduce the cosmological precepts of the play; (2) his association of the four elements with atomic, indestructible matter (a key thesis of Lucretius' work); and (3) his use of a poetic form as a vehicle for imparting physics and cosmology to a broad, non-specialist audience. The phrase "*natura naturata*" posits nature as a world created and finished whereas the term "*natura naturans*" implies the continuously creative, generative power of nature. Rastell's character's name implies, at first blush, a conventional Christian view of nature as a finished, divine creation, but Rastell's text suggests that he was attempting to synthesize the Epicurean sense of atoms as the basic component of universal matter with the Christian precept of a divinely ordained cosmos. The idea of atoms as the basic building blocks of nature had been proposed in the fifth century BCE by Leucippus of Abdera and, over the centuries, was part of a debate about the make-up of the universe as composed of atoms or of four elements—earth, air, fire, and water. While Rastell's play explicitly names the four elements as the "matter" of the play, he clearly embraces the Lucretian concept that each element is a form of indivisible matter. In the opening of the play, the Messenger plays with the sense of a "high" intellectual subject matter being a worthy topic for a "low," or unlearned, audience but extends the pun to include clerks focusing on "high" subjects, i.e. celestial study, at the expense of "low" matters such as the four elements and the study of the Earth.

> Nor a matter more lowe can not be arguyd,
> For though the elementis Goddys creaturis be,
> Yet they be most grose and lowyst in degree.
> (lines 110–112)

The Messenger urges a synthetic approach, arguing that "it semyth nothynge convenient" for a man to study "hye thyngys excellent" and not to similarly attend "these four elementis here below" (lines 120–124). Nature takes up this emphasis on physical matter and distinguishes the "etheriall region" from the "lower region callyd the elementall" (lines 164–166). This lower region, Nature explains, is composed of "The fyre, the ayre, the water, and yerth also" (line 168). While this seems a traditional refutation of the morally suspect theory of atomic matter, Nature goes on to state that:

These elementis of them selfe so single be
Unto dyvers formys can not be devydyd
Yet they commys togyder dayly ye see,
Wherof dyvers kyndes of thyngys be ingenderyd,
Whiche thyngys eftsonys whan they be corruptyd
Yche element I reduce to his furst estate,
So that nothynge can be utterly adnychelate.
(lines 176–182)

Here, Nature argues the basic premise of Epicurean matter theory—that is, that "things" are composed of combinations of a basic building block, a unit of matter that in its elemental form cannot be "annihilated." Rastell's use of the word "adnychelate" is an intriguing combination of the words "annihilate" and "chelate," a term that has since come to be associated with the bonding of chemical substances rather than the sundering of parts, but that here explains the Lucretian hypothesis of indestructible, atomic matter. One of the earliest uses of "adnychelate" in this sense appears in Caxton's *Vitae Patrum* and, subsequently, in More's *Life of Pico della Mirandola*, written in 1504.[58] Rastell's passage reinforces the concept of matter as invincible to the forces of annihilation.

After setting forth the basic principle of "generation" and "corruption," Nature suggests the Lucretian principle that

For though the forme and facyon of any thynge
That is corporeall body be destroyed,
Yet every matter remaynyth in his beynge,
Wherof it was furst made and formyd.
(lines 183–186)

Without venturing into the question of the human soul, as Lucretius does, Nature affirms the immutable character of matter within the construct of the four elements:

So the elementis can never be destroyed.
For essencyally ther is now at this tyde
As muche fyre, ayre, water, yerth, as was
Ever before this tyme, nether more nor las.
(lines 200–204)

Such principles of the immutability of matter are found in Book I of Lucretius's *De rerum natura* in his description of "fixed seeds:"

Confess then, naught from nothing can become,
Since all must have their seeds, wherefrom to grow,
Wherefrom to reach the gentle fields of air.

Hence too it comes that Nature all dissolves
Into their primal bodies again, and naught
Perishes ever to annihilation.[59]

Rastell's *Mery Interlude of the Four Elements* indicates an indebtedness to the philosophy of Epicureanism as expressed in Lucretius' poem, a work in circulation among European and English humanists despite the Church's attempted suppression of its heretical principles. In addition to his use of the Latinate term for his character Natura Naturata and his emphasis on regenerative matter, Rastell's emphasis on the word "thynge" cannot help but bring to mind, in addition to Trevisa's translation of Bartholomaeus Anglicus' *On the Property of Things*, the philosophical questions raised in Lucretius' work with its English title, *On the Nature of Things*. When Nature introduces himself in the play, he states:

Wherefore I am the verey naturate nature,
The immediate mynyster for the preservacyon
Of every thynge in his kynde to endure
And cause of generacyon and corrupcyon
Of that thynge that is brought to distruccyon
Another thynge styll I brynge forth agayne.
(lines 155–160)

While scholarship on Rastell's play has largely concentrated on the play's reference to new lands or its role as a link between morality plays and Elizabethan drama, Rastell's interlude is also indicative of an engagement with Epicurean philosophy of matter and natural philosophy in relation to late medieval cosmology and emerging geographic principles in theatrical form.

Rastell also embraces Lucretius' interest in reaching a broad, non-elite audience. Lucretius' work was written with the intention of reaching a wide audience that he perceived as burdened by fear of the gods or a hellish afterlife and Rastell adopts this Lucretian model for his cosmological drama, if not with the same elegance of style. E.J. Kennedy argues that Lucretius was well aware "that there was resistance to be encountered which logic and reason unassisted by force could not overcome."[60] Such a "force" to overcome resistance in Lucretius' case was the beauty of his verse, his domestic examples, and his appeal to broad audience. Rastell similarly attempts to use the structure of entertainment, poetry, and pleasure to overcome any resistance to his presentation of an oceanic, global world.

Rastell's interest in a public cultural form as the vehicle for the discussion of natural philosophy is attested within the play and the form of the work itself. In introducing the "lytll interlude," the Messenger notes that "though the matter be not so well declaryd / As a great clerke coude do (lines 9–11), such a "matter substancyall" (line 18) needs to be brought to a public, non-academic audience:

And that it is a commyn good act to brynge
People from vyce and to use good lyvynge
Lykewyse for a commyn welth occupied is he
That bryngyth them to knowledge that yngnorant be.
(lines 88–91)

The Messenger reiterates this point when he declares:

And though some men thynke this matter to hye
And not mete for an audience unlernyd,
Me thynke for man nothynge more necessary
Than this to know, though it be not usyd.
(lines 106–109)

In this regard, Rastell shares Lucretius' interest in philosophy's ability to lighten the burden of humanity with knowledge of the universe, and in this case the working of the "hye" cosmos, and its ability to promote learning as a public good.

Rastell's assertion that his play is intended for "an audience unlernyd" is curious for a purported court entertainment and calls into question the nature of Rastell's audience and the form of the drama itself. By definition, interludes are an English dramatic form distinct, as Anne Coldiron argues, from the more public nature of French farce. Coldiron asserts that Tudor drama was "dominated by the Humanist interludes and household productions" and that the "consistently noble, court-based audiences" of Tudor interludes "form a very different set of theatrical contexts … than do the open public staging of the rowdy French farces."[61] But the social, didactic nature of Rastell's *Four Elements* seems to place it in a context other than the usual private, courtly setting for an interlude, despite the self-designation of the play as an interlude in its prefatory matter: "A new interlude and a mery, of the nature of the four elementis, declarynge many proper poyntys of philosophy naturall, and of dyvers strange landys."[62] Suzanne R. Westfall has argued, "[v]irtually all Tudor interludes address themes of vital interest to the aristocracy,"[63] and yet Rastell's *Four Elements* indicates an audience of the "unlearned" both in its declaration of intent and in the way that the play frequently reiterates the rotundity of the Earth, an assertion presumably unnecessary for an educated audience. David Bevington suggests that "in trying to bridge the gap" between his intellectual circle and a wider, public audience, in both performance and in print, Rastell "makes his distinctive mark."[64] This hybrid nature of the *Four Elements*—an English interlude meant for a general audience—echoes Rastell's own class subjectivity. While he moved in an elite, well-educated and courtly circle, he was from a more modest social background and negotiated a variety of social classes: the highly educated world of the More circle and the Inns of Court, the elite audiences of Henry's court, actors and stagehands, as well as the working-class collaborators who ran his print shop and fabricated his dramatic productions.[65] In her analysis of the lawsuit Rastell brought against Henry Walton regarding Walton's excessive leasing out of Rastell's theatrical costumes left in his care, Janette Dillon cites a line from the lawsuit—"stage plays in the somer and interludes in the winter"—as evidence that "plays" were outdoor, public dramatic forms whereas "interludes" were private, indoors, and thus for noble audiences.

In the prefatory matter of the play, Rastell suggests the play was intended for multiple audiences and various production settings. He advises that:

> [Y]f the hole matter be played, wyl conteyne the space of an hour and a halfe; but yf ye lyst ye may leve out muche of the sad mater, as the messengers parte, and some of Naturys parte and some of Experyens parte.
>
> (lines v–vii)

If this is done, he estimates that the play "wyll not be paste thre quarters of an hour of length" (line viii). Presumably, in a courtly, interlude setting, the audience may have been well acquainted with the precepts set forth by the Messenger, Nature, and Experiens and thus those parts could be shortened so as to fit into an evening that included other forms of entertainment and dining. Based on the proceedings of the lawsuit, we know Rastell left a collection of theatrical costumes in Walton's care during a six-month overseas trip, and that Walton subsequently leased the costumes to a variety of acting companies. Rastell found the costumes tattered and in disarray upon his return and initiated the lawsuit.[66] Dillon argues that "the sheer scale of the costume hire suggests the existence of numerous performances of both stage-plays and interludes beyond the court" and that Rastell's interlude had both courtly and civic audiences.[67] Axton reasons that the medium of print was an additional factor in the dissemination of dramatic works to a broader audience than private court gathering, suggesting that as a result of "publication, repeated playing and reprinting," some plays may have achieved popular status.[68] If so, the potential for Rastell's *Four Elements* to have reached a wider audience than the court of Henry VIII and the More circle establishes the play as an important part of the transmission of geographical thought in England other than ownership of the expensive, luxury editions of Ptolemy's *Geography.* Rastell's legacy, I argue, should include his contribution to the dissemination of Ptolemaic geographic learning in England outside of Latinate learning as well as the construction of the first English-made globe. In addition to his service for the court, Rastell's mission to reach an "audyence unlearnyd" (line 107) was both progressive and attuned to the new potential of print publication.

"Our tonge maternall:" the limitations of English in the translation of classical scholarship

The transmission of scholarly Latinate texts to this expanding lay readership required, perforce, translation into English—a topic of keen interest to Rastell. In fact, the *Four Elements* offers a possible explanation as to why English printers were disinclined to print geographical texts and maps, and why their output did not match the breadth of humanistic works, in Latin and in translation, published on the Continent during this time. Henry Burrows Lathrop speculated that early print readers of English from the time of Caxton until early in the reign of Henry VIII were "medieval" in their tastes and "practically untouched" by the Renaissance.[69] Lathrop identifies only 11 printed books that were translated into English from Latin classics from 1477 to 1517.[70] Stuart Gillespie has consequently tackled this questions, asking why the translation of classical works was a slower process in English than in Italian or French.[71] He suggests the lack of royal patronage for such efforts in England but also the lack of a prominent scholarly publishing house on a par with that found on the continent, for example, Aldus Manutius in Venice, Henri Estienne in Paris, or Christophe Plantin in Antwerp. Gillespie corroborates that book importation played a significant role in meeting the needs of English book buyers, noting that "the tide washed in new literary translations continuously, as well as all manner of practical, technical, political, polemical and in particular doctrinal translated material, to contribute to what was by 1600 an extensive translating culture."[72] In geographic works as in translations from the classics generally, English readers either read the works in their original Latin, relied on imported translations, or found their way to the handful of English print houses, with Caxton, de Worde, and Rastell figuring predominantly in this group. Crucially, Gillespie identifies that English was often seen at this time "as an inadequate language, lacking both authority and copiousness."[73] Indeed, early modern translators "often apologize that their work falls short of Latinate copiousness, eloquence or diction"[74] while simultaneously recognizing, as Rastell does in the *Four Elements*, that translations into English will augment and improve "the language's expressive power."[75]

Within the first stanzas of the *Four Elements*, Rastell identifies this self-consciousness regarding the unsuitability of the English language as a vehicle for the transmission of scholarly thought. Speaking "on behalf" of Rastell, the character of the Messenger addresses this conundrum early in the play:

> Whiche in his [the author's] mynde hath oft tymes ponderyd
> What nombre of bokys in our tonge maternall
> Of toyes and tryfellys be made and impryntyd,
> And few of them of matter substancyall.
> For though many make bokys yet unneth ye shall
> In our englyshe tonge fynde any warkys
> Of connynge that is regardyd by clerkys.
>
> (lines 15–21)

Thus, Rastell creates a pun regarding the lack of books in English on scientific matters and on matter itself. Rastell's Messenger also observes that English clerics, academics, and scholars were skeptical of the value of English translations. It wasn't merely that such texts were *unavailable* in English,

Rastell suggests, but that English clerics—one might think of More here—were wary of the intellectual *legitimacy* of scholarship translated into English. Such wariness stemmed from a deep cultural suspicion of English as a language that "corrupted" holy scripture and the conviction that the process of translation itself was fraught with the potential of introducing heretical errors. More is said to have famously castigated the Tyndale Bible, claiming that looking for errors in the text of Tyndale's translation was akin to looking for water in the sea. Within this ambivalent cultural attitude toward translation, English scholars would have deemed Latin geographies and cosmographies equally vulnerable to the ravages of the translation process. In this context, Rastell's call for translation of classical works into English in the *Four Elements* would have been a radical stance that hewed uncomfortably close to Protestantism and religious dissent amid a still staunchly Catholic England.

In the *Four Elements,* Rastell embarks on an impassioned argument that English "is now suffycent" to be the linguistic bearer of the highest intellectual works and makes the case that Greek and Latin were each once a "mother tongue" that gave rise to works of philosophy. Though England was officially Catholic at this time, the tenets of Protestantism, which stressed translation into the vernacular, can be discerned in Rastell's argument:

> The grekys, the romayns, with many other mo,
> In their moder tonge wrot warkys excellent;
> Than yf clerkys in this realme wolde take payn so,
> Consyderyng that our tonge is now suffycyent
> To expoun any hard sentence evydent,
> They might, yf they wolde, in our englyshe tonge
> Wryte workys of gravyte somtyme amonge.
> (lines 22–28)

Rastell deepens the argument when he praises the "wyttes" of Englishmen but notes that many—both of noble and "mean" estates—read and write only English. Rastell comes near to the looming controversy concerning the translation of the Bible into English, arguing that:

> Than yf connynge laten bokys were translate
> Into englyshe, wel correct and approbate,
> All subtell sciens in englyshe myght be lernyd
> As well as other people in their owne tonges dyd.
> (lines 32–35)

In a nation in the early stirrings of the Reformation—and which engaged in an active book trade with countries in which the Bible was already translated into the vernacular—the corollary debate of whether *any* learned Latin text could be appropriately translated into English would have struck a cultural nerve. Rastell stepped into the fray and has been credited as "the earliest outspoken champion of the vernacular."[76] If, as Rastell's Messenger argues, Ptolemy, Pliny, or any classical Greek or Latin writer could be beneficially translated into English, the corollary question of the possibility and appropriateness of an English vernacular Bible could not be far behind. England's stiff resistance to biblical translation may shed some light on the vexed question as to why English printers did not actively produce editions of Ptolemy's *Geography* or other Latin geographies. It was not that English readers had provincial, "medieval" tastes, but rather that Latinate geographical works were in the wash of a deeply rooted skepticism of translation and the capacity of the English language. Thus, the conservative resistance to translation of the Bible into English was matched by the relatively slow translation rate of Latin and Greek texts into English.[77] If Latin and Greek were well understood by English readers, there would there be little need to translate such works into "rude" English. Lamentably, such was not the case, as the Messenger makes clear:

> For dyvers prengnaunt wyttes be in this lande,
> As well of noble men as of meane estate,
> Whiche nothynge but englyshe can understande.
> (lines 29–31)

This state of affairs, Rastell suggests, results in dissolute reading habits among the English:

> But now so it is that in our englyshe tonge
> Many one there is, that can but rede and wryte,
> For his pleasure wyll oft presume amonge
> New bokys to compyle and balates to indyte:
> Some of love or other matter not worth a myte;
> Some to opteyn favour wyll flatter and glose,
> Some wryte curyous termys nothyn to purpose.
> (lines 36–42)

As a social reformer and public intellectual, Rastell takes his countrymen to task for their resistance to vernacular English as a suitable language for works of original scholarship and for the translation of the classics. In the *Four Elements*, Rastell attempts, at the level of adaptation and entertainment, to breach the gap between lightweight entertainment in English and scientific, scholarly works in Latin.

Rastell's oceanic Earth

Rastell's enthusiasm for the potential of English as a vehicle for scholarly learning and in the value of an educated citizenry can be linked to his interest in the new globalism. Not only did Rastell use the *Four Elements* as a vehicle to assert, tirelessly, the rotundity of the Earth in a literal geographical way, but also to advance a cosmopolitan, global perspective for English citizens. In constructing his globe and crafting an entertaining drama, Rastell's world-construct is that of a fluid, oceanic Earth, a literary representation of geographic space with the concomitant implications of travel, trade in commodities, global exchange, as was the thesis of the Behaim globe. To this, Rastell adds the importance of transnational scholarship and a learned citizenry. Rastell's work embodies the cosmographical outlook of the Ptolemaic-revival era: he locates the English nation within this expanding geographic space and as part of an emerging oceanic trade. Just as Behaim emphasized an emerging oceanic world order defined by shipping, Rastell similarly proposes that even a humble English tavern is networked to a commodities-based globe.[78]

Though Sensuall Appetyte waylays Humanyte to a local tavern, even here the sense of a global world is emphasized—but now through the food and drink Rastell chooses to introduce into the scene. When Humanyte expresses skepticism on how "the yerth with the water be rounde with all" (line 339), it provides Studyous Desire with the opportunity to provide the astronomical and mathematical proof and sets the stage for Sensuall Appetyte's maxim to "let the wyde wordle wynde!" (line 418). As the scene gains momentum, the Taverner proclaims, "Ye shall have spayneshe wyne and gascoyn" (line 561), and he enumerates an international catalog of wines, including "sak" (white wine imported from Spain and the Canaries),[79] and "Greke ipocrase" (line 565), a spicy Greek wine. The "wide world" of the tavern humorously echoes the geographic globalism introduced by Studyous Desire. This sense of global fluidity and the movement of commodities is further amplified when Sensuall Appetyte asks the Taverner for "a dyshe of quales / Smal byrdes, swalowes, or wagtayles" (lines 589–590) because they will be "light of dygestyon" (line 591). When the Taverner queries this, Sensuall Appetyte responds that it is "because those byrdes fle to and fro / And be contynuall movynge" (lines 594–595). Like the rotation of the physical Earth, Rastell creates a sense of gustatory global movement based on international trade that, like de Worde's illustrations that included birds in flight and Behaim's rotating globe, announces a world in "contynuall" motion.

Amid the pandemonium of the tavern scene, Experiens expounds on the central oceanic theme of the play, using the stage prop globe to remark that "Ynglande lyenge here" (line 708), and that it is "Compassyd about every where / With the occian see rownde." (lines 710–711). This emphasis on the Earth as a sphere of ocean-surrounded land masses is the starting point for his land-based itinerary from England to Rome, followed by an intensive discussion of the raw commodities of the New World, including copper, "pyche and tarre and sope ashys" from the great forests (line 802) as well as fish in "great plente" (line 805). In this way, Rastell's globe and play recall the great thesis of the Behaim globe—that is, that the world is fundamentally fluid, oceanic, and commodity-driven.

Rastell's focus on commodities also follows the geographical-mercantile interests of his predecessor, William Caxton, who similarly sought to reach a new audience via print editions. According to Axton, the *Four Elements* "offers the plainest kind of popular instruction in science, in the style of Caxton's *Myrrour of the World*,"[80] one of Rastell's sources.[81] In addition to his interest in print, in reaching a popular audience, in mercantile geography, and in introducing Londoners to images of a globe, Rastell follows his predecessor's work in textual spatialization practices as well. Rastell provides a kind of table of contents that is clearly influenced by Caxton's practice of providing rubrics as part of the forematter of the book, even drawing upon the specific rubrics provided in the *Myrrour*. In the verse preamble of the play, Rastell lists the "diverse matters whiche be in this interlude conteynyd" (line xiii), including items such as "Of certeyn conclusions proveynge that the see lyeth rounde upon the yerth" (line xx) and, in a nearly direct quote from Caxton's *Myrrour*: "Of the cause of the lyghtnynge of blazyng sterrys and flamys fleynge in the ayre" (line xxxi).[82] In his rubrication and popularization of geographic knowledge, Rastell's playscript demonstrates the continued influence of Caxton's *Myrrour* on English geographic culture, an influence that has not been recognized. Rastell integrates the spatializing, narrative practice of a table of contents into the new literary form of a printed dramatic text. The tradition of popular geographic print works initiated by Caxton and continued by Wynkyn de Word in his editions of Mandeville's *Travels, Polychronicon, Kalender of Shepherds, De proprietatibus rerum*, and *Description of Britain*, formed part of Rastell's English geo-literary inheritance.

Rastell's concomitant passions for print technology and geographical literacy are perhaps best summed up in his printer's "device," one he used for most of his printing career beginning around 1513 (Figure 4.1). The device is cosmographical, illustrating the connection Rastell saw between his work as a printer, humanist, playwright, and geographical popularizer. Axton describes the device as "Christ in majesty over the sun, moon and planets; below, the four elements, a semi-circle of earth, surrounded by water, air and fire."[83] Included in the device are a merman and mermaid, a reference to Rastell's print shop "at the sign of the mermaid," near St. Paul's in London. Rastell's printer's device captures the spirit of English geographical thought during the Ptolemaic revival as influential printers such as Caxton and Rastell wrestled to resolve linguistic and ideological skepticism regarding the translation of Latinate texts into English and their estimation of an interested readership.

Rastell's light-hearted interlude and printer's device reflect his lively interest in cosmology and cartography, and that of his London circle, during the early Tudor period. Not only a printer of texts, Rastell also traded in maps. In 1536, roughly a year after the death of his friend Thomas More, Rastell was also imprisoned, the result of a financial rather than religious dispute with Henry VIII. Rastell had, in fact, worked on Henry's behalf in matters of religion, traveling to France to advocate for the legitimacy of the desired divorce.[84] Two years after Rastell's death in prison, an inventory of his estate

Figure 4.1 John Rastell's printer's device
Source: *The Commendations of Marriage*, 1517, STC 12798.7. Reproduced with permission from Durham University Library and Collections and the Lord Crewe's Trustees.

lists 110 "mappis of Europa," valued at a penny apiece,[85] evidence of Rastell's brisk trade in maps and indicative, at the very least, of his estimation that Londoners' interest in maps merited such a substantial inventory. Rastell's map inventory demonstrates, as well, the transformation of maps via print production from luxury item to penny-apiece purchase for a wider public. Rastell's stock of moderately priced maps matched his practice of printing play texts so as to make the works more widely available for a broad clientele. At his death, Rastell's print shop had "hundreds of unsold copies of various plays,"[86] but amid all these maps and play texts, no copy of Rastell's own cosmographical play, the *Four Elements*, was recorded. In reflecting on this state of affairs, Axton offers an elegiac possibility: "One would like to think that he had sold every copy of his own *Four Elements*."[87] Potentially, a good many Londoners had a text of Rastell's cosmographical play to hand, one of his inexpensive maps, and, with those, a basic grasp of Ptolemaic geography. In the printed pages of the playscript, readers would encounter Rastell's infectious enthusiasm for cartographic representation and the value of a globe: "Lo," Experiens remarks, "is not this a thynge wonderfull" (line 877).

Geography as pleasure

Ironically, two of today's icons of Tudor geographic thought—Thomas More's *Utopia* (1516) and Hans Holbein's painting *The Ambassadors* (1533)—had a scant presence in early Tudor England. More's *Utopia*, edited by Desiderius Erasmus and published in Louvain by Thierry Martens, utilized cartographic themes and voyages of exploration as the basis of his satire and humor. But More's work was published in Latin, on the Continent (reprinted in Paris and Vienna), and not translated into English or printed in England until Ralph Robinson's 1551 translation. Like Ptolemy's *Geography*, More's work, of course, circulated among English Latinate scholars via the London book trade, but its wider cultural impact in England is somewhat muted by its coterie readership and More's subsequent fall from grace with the King[88]—the antithesis of Rastell's public engagement via cosmological spectacles, inexpensive maps, and the vernacular geographic interlude. In a limited circulation similar to More's *Utopia, The Ambassadors* by Hans Holbein (the Younger) was painted in England, but shortly thereafter taken by its patron, the French ambassador Jean de Dinteville, to his chateau in Polisy, France. Both *Utopia* and *The Ambassadors* were appreciated by a relatively small, elite circle. Despite their limited circulation, both *Utopia* and *The Ambassadors* do, however, reflect the increasing impact of Ptolemaic cartography in early Tudor England: More's fantasy of no-where is replete with cartographic references that undergird the satire and make the puns, for a learned reader, all the more humorous, and Holbein's painting uses cartographic imagery to convey a powerful, almost menacing, diplomatic and cartographic imperialism as symbolized by the two globes.

More's title, *Utopia*, plays off of the Ptolemaic-influenced fascination, one might say obsession, with identifying ancient place names. In joining two Greek words (*ou-topos*), More's title inverts the cultural cartographic practice of locating places on the map, creating, instead, a no-where. A map, drawn by Ambrosius Holbein, the brother of Hans Holbein, accompanied the Louvain edition of *Utopia* and, like Rastell's globe, it is identified as a "figura." The map of "no-where" has received considerable attention for its visual spatialization of utopian principles, suggestion of New World geographic knowledge, and even for its possible function as a *memento mori.*[89] But while riffing, we may say, on cartographic trends, More's *Utopia* was primarily philosophical and political, almost anti-Ptolemaic, and certainly anti-vernacular. Published near the same time that Rastell was working on *the Four Elements*, More's philosophical work reflects his literate and scholarly audience and stands as an example of his clever use of a geo-poetic framework to explore philosophical, moral, and civic questions.

Interestingly, once *Utopia* was translated into English in 1551, the edition no longer included Ambrosius Holbein's map. More's co-mingling of literature and cartography served a similar purpose to Rastell's interlude, albeit for a different audience, linking geography with entertainment, in More's case, provoking inquiry into philosophic questions as to the nature of man and the role of government. The work's reception as both serious yet pleasurable is attested to by the contemporary humanist and friend of More, Richard Pace, in his work *The Benefit of a Liberal Education*:

> Ptolemy and Strabo, who described the regions of the worlds in their works, are the reason I shall not pass over geography in silence … For granted, *Utopia* doesn't have the spices the Portuguese buy, still, the way to it is far less dangerous, and it is full of its own unknown delights. But whoever takes the science of geography to heart either has to travel all over the world (which is extremely unpleasant, difficult, and expensive) or he has to read through Strabo, which is about as long and broad as the earth and is a world in itself—and in Greek too, since the translation is extremely corrupt. But that's what you have to do, unless this seems shorter: to study the sketches of the globe called colloquially maps of the world. But you ought to know that they also proceed from the learning of those two men I've already mentioned.[90]

Pace's good-natured resistance to slogging through Strabo's lengthy geography in Greek is resolved via the "unknown delights" of reading More's text or, more practically, perusing a map. Pace's wry commentary reinforces the carto-bibliographic aesthetic of the Ptolemaic era: geography was inextricably part of the liberal arts and humanistic study, and—at its best—engendered a sense of wonder and pleasure.

In her study of More, Karma Lochrie suggests More's text is less engaged with exploration and cosmopolitanism than has been presumed. She argues:

> In an age defined by the very discovery of cosmopolitan possibility through voyages of discovery and travel, it is curious that More's *Utopia* represents an anachronism, a world unto itself that is relatively insulated from the outside world—in short, an ideal of provincialism.[91]

Lochrie suggests that More's work may, in fact, be more traditionally medieval, despite its grounding in classical philosophy and humanism, than Mandeville's cosmopolitan, vernacular narrative of 200 years earlier. More's conservative approach to translation and his creation of "no-where" suggest his interest lay less in physical "new-founden lands" and more in the territory of the soul.

Holbein's painting demonstrates his long-standing interest in cartography and court culture, as evidenced by the earlier cosmological ceilings and pageants he had created with Rastell for the Field of Cloth of Gold and other state events. Holbein's lush and enigmatic painting depicts two wealthy, worldly courtiers either side of a still life of scientific objects, and the iconic *memento mori* created by the anamorphic skull which spreads across the foreground of the painting, only coming into focus when viewed obliquely. Two of the many scientific objects depicted with the gentlemen are globes: one celestial, one terrestrial. The terrestrial globe replicates a globe created from woodblock print gores, perhaps the now-lost globe of 1523 by Johannes Schöner; the celestial globe is identified with more certainty as modeled after a celestial globe by Schöner. The scientific instruments displayed, including the globes, are meant to convey innovation and scientific achievement, often associated with Tudor prowess by dint of Holbein's court association, but the instruments are all German-made and, specifically, many are linked to Nuremberg, with its famed craftmanship in globes and scientific instrumentation.[92] With its German-born painter, depiction of German scientific instruments, French patron, and quick relocation to France, the work stands as a testament to the geo-artistic culture in England at the time, but, as with More's *Utopia*, less to its impact on the transmission of Ptolemaic thought in early Tudor England.

In his 1531 work, *The Boke Named the Gouernour*, Thomas Elyot advocates a liberal arts education for young gentlemen, championing the humanistic tradition in England as essential to a gentleman's education. While not as progressive as Rastell's populist aspirations, Elyot's inclusion of the study of geography and cartography is indebted to the geo-poetic aesthetic of Rastell; both Elyot and Rastell, along with Holbein, were part of the More circle. Elyot advocates for the study of geography as part of the liberal arts but also argues for the value of graphical, cartographic images of the world rather than reliance on the traditional narrative forms of history and cosmography. In the first book of Elyot's treatise, he sets forth the proper components of a young gentleman's education, inclusive of the traditional liberal arts. But in his discussion of cosmography, Elyot advocates for the value of "figures and charts," that is, for visual maps. Elyot's handbook for the education of young men captures the new interest in geographic visuality while simultaneously situating the study of geography as related to the arts. Sandwiched between the discussion of the role of music in a young man's education and that of painting, Elyot's discussion of cartography clearly categorizes it as a form of artistic expression. While still valuing the narrative English geographical tradition, such as Caxton's *Polychronicon* or the *Description of Britain*, Elyot pushes for something more:

> Experience we haue therof in lernynge of
> geometry/astonomie/& cosmogrophie/ cal-
> led in engliss [t]he discrptio*n* of the worlde.
> In which studies I dare affirme/ a ma*n* shal
> more p*ro*fite in one wike by figures & chartis/
> well and perfectly made/ than he shall by

> the only reding or heryng the rules of that
> scie*n*ce/ by the space of halfe a yere at the left.[93]

To create these "visual figures & chartis," the study of history and geography is tied to that of painting. In Chapter XI, geography, history, and poetry are presented as an integrated, and pleasurable, course of study for a future gentleman:

> It shall be
> therfore and also for refreshing the witte/ a
> co*n*uenient lesson to beholde the olde tables
> of Ptholomee/ where in all the worlde is
> paynted/ hauynge firste some introduction
> in to the sphere/ wherof nowe of late be
> made very good treatises/ and more playne
> and easie to lerne than was wonte to be.
> All be it there is none so good lernynge/ as
> the demo*n*stration of cosmographie/ by ma-
> teriall figures and instrumentes/ hauynge
> a good instructour. And surely this lesson.
> is bothe pleasant and necessary.[94]

Like Rastell's interlude, Elyot emphasizes the study of geography as both pragmatic and pleasant. He advocates reading Ptolemy's "tables" and demonstrating the principles via maps and geographical instruments. As a younger member of the More circle, Elyot may well have attended a performance of or read Rastell's interlude. Elyot suggests the power of the imaginative bird's-eye view—traditionally represented in medieval *somnium* such as Macrobius' *Dream of Scipio* in which the narrator/dreamer rises above the Earth to gain the needed spiritual perspective—is now realizable with the aid of a sphere based on Ptolemy's "tables." The cosmology remains rooted in tradition, but the "materiall figures and instrumentes" offer a previously only dreamed-of view. As indicated by the Behaim and Rastell globes, Elyot replaces Scipio's spiritual dream with the experience of looking at a map or globe:

> For what
> pleasure is it/in one houre to beholde thole
> realmes/ cities/ sees/ ryeuers/ and mountay-
> nes/ that vneth in an olde mannes life can
> nat be iournaide and pursued: what incre-
> dible delite is take*n* in beholding the diuersi-
> ties of people/beastis/foulis/fisshes/trees/
> frutes/ and herbs: to knowe the sondry
> maners & conditions of peopple/ and the va-

> rietie of their natures/ and that in a warme
> studie or perler/ without perill of the see/ or
> aunger of longe and paynfull iournayes;
> I can nat tell/ what more pleasure shulde
> happen to a gentil witte/ than to beholde in
> his owne house euery thynge that with in
> all the worlde is contained.[95]

In Elyot's praise of the pleasure of geography, we hear echoes of Pace's humorous enjoinder to English readers to enjoy such speculative geography from the comfort of their armchair rather than by arduous travel. The compression of time and global perspective that such cartographic views afford provides information, aesthetic pleasure, and the gathering, potent gaze of empire.

Peter Barber argues that by 1527, a "seachange" in the use of cartography in England had taken place, suggesting that "maps were finally becoming familiar to the king and his ministers."[96] A broader cultural familiarity with maps within a geo-poetic aesthetic is evident from a collection of cultural spatial practices, including state cosmological spectacles, performances of the *Four Elements* with its stage globe, Rastell's inventory of over 100 printed maps, and the implication in Elyot's text that Ptolemaic maps were now commonly in the drawing rooms of young gentlemen. The 1527 peace accord between England and France which resulted in the Treaty of Westminster included a cosmological ceiling, as mentioned earlier in this chapter, done in collaboration with Rastell; it also included two festival maps painted by Holbein, works both of artistry and cartography.[97] Within the increasingly pragmatic understanding of the military and strategic use of maps, the aesthetic connection between geography, poetics, and religion suggested that there remained "many things to be maruailed at."[98]

"prynt well in thyne heart:" print and cartography

As a coda to this exploration of Rastell and the English geo-poetic aesthetic of Ptolemaic thought in early Tudor England, we can note the way in which cartography and print were intimately connected in the early print period, roughly 1475 to 1525. With the printing of geographical and cartographical texts, the previously memorial character of medieval *mappamundi* was no longer dominant. The Hereford Map,[99] for example, which served as a form of *rota* or memory wheel from which to recall Christian, universal history, was now gradually replaced by a cartography of print-based learning. In Rastell's *Four Elements*, Studyous Desire reflects on how he will utilize the "doctryne" of natural philosophy that Nature has expounded upon, pledging that he will "dayly put hym in remembraunce" (line 313) of these matters. Shortly thereafter, he challenges Humanyte to "call to your memory / The connynge poyntys that Nature hath declaryd" (lines 325–326). But in addition to the traditional educational strategy of memorization, Nature promotes a new metaphor for learning, utilizing a traditional aphorism but one that also cleverly references Rastell's immersion in English print culture,

> What I have shewid, man, prynt well in thyne hert
> And marke well this fygure that here shall remayne
> Whereby thou maist perceyve many thynges more playn
> (lines 319–321)

In this moment, Rastell offers a new metaphor for geographic education and transmission—that of print. While his cosmological interlude has been deemed a frivolous, retrograde medieval dramatic work, upon closer study, it is replete with emerging humanist and cartographic references, a barometer of the reach of Ptolemaic cartographic principles in early sixteenth-century England. Further, Rastell's "figure" rightly bears the designation as the first three-dimensional Ptolemaic globe produced in England. Following in the footsteps of Caxton, Rastell provides new images of a global Earth for English readers and playgoers, years in advance of the Molyneux globes. It is in keeping with England's particular carto-bibliographic tradition that this early English globe was part of a dramatic production meant to educate and to entertain.

Notes

1 For a discussion of moralized geography, see Tom Conley, "Early Modern Literature and Cartography: An Overview," in *The History of Renaissance Cartography*, vol. 3: *Cartography in the European Renaissance*, Part 1, ed. David Woodward (Chicago: University of Chicago Press, 2007), 401–411; 409.

2 Zur Shalev, "Main Themes in the Study of Ptolemy's *Geography* in the Renaissance," in *Ptolemy's Geography in the Renaissance*, eds. Zur Shalev and Charles Burnett (London and Turin: The Warburg Institute – Nino Aragno Editore, 2011), 1–14; 1.

3 See John Gillies, *Shakespeare and the Geography of Difference* (Cambridge, Cambridge University Press, 1994).

4 Rhonda Lemke Sanford, *Maps and Memory in Early Modern England: A Sense of Place* (New York: Palgrave Macmillan, 2002); Tom Conley, *The Self-Made Map: Cartographic Writing in Early Modern France* (Minneapolis, MN: University of Minnesota Press, 1996) and *An Errant Eye: Poetry and Topography in Early Modern France* (Minneapolis, MN: University of Minnesota Press, 2010).

5 Peter Barber, "England I: Pageantry, Defense, and Government: Maps at Court to 1550," in *Monarchs, Ministers, and Maps: The Emergence of Cartography as a Tool of Government in Early Modern Europe*, ed. David Buisseret (Chicago: University of Chicago Press, 1992), 26–56; 26.

6 R.J. Mayhew, "Geography, Print Culture and the Renaissance: The Road Less Travelled By," *The History of European Ideas* 27 (2001): 349–369; 355.

7 For a discussion of the role of early Tudor vernacular drama, see William Tydeman, ed. *Four Tudor Comedies* (London: Penguin, 1984), 27.

8 J.A.B. Somerset, ed. *Four Tudor Interludes* (London: The Athlone Press, 1975), 1.

9 Richard Axton, ed. *Three Rastell Plays: Four Elements, Calisto and Melebea, Gentleness and Nobility* (Cambridge: D. S. Brewer, 1979), 1. All quotations from the play are from this text.
10 Barber, 30.
11 See Glenn Richardson, *The Field of Cloth of Gold* (New Haven, CT: Yale University Press, 2013). For a brief description of the cartographic and cosmological elements of the decorations, see Barber, 30. For a contemporary account of the ornate festivities of the Field of the Cloth of Gold, see *Hall's Chronicles*, Henry Ellis, ed., original title: *The Union of the Two Noble and Illustre Families of Lancastre & York* (London: 1809, reprinted New York: AMS Press, 1965), 603–622.
12 Axton, 3.
13 Janette Dillon, "John Rastell's Stage," *Medieval English Theater* 18 (1996): 17–18.
14 Sydney Anglo, *Spectacle, Pageantry, and Early Tudor Policy*, 2nd edition (Oxford: Clarendon Press, 1997), 160.
15 Anglo, 161 and 163.
16 Richardson, 207.
17 Ibid., 202.
18 Virginia Mason Vaughan, "Preface: The Mental Maps of English Renaissance Drama," in *Playing the Globe: Genre and Geography in English Renaissance Drama*, eds. John Gillies and Virginia Mason Vaughan (Madison, NJ: Fairleigh Dickinson University Press, 1998).
19 Axton, 8.
20 Hall's *Chronicles*, 723.
21 Ibid., 723.
22 Axton, 10.
23 Ibid., 1.
24 Robert J. Mayhew, "Cosmographers, Explorers, Cartographers, Chorographers: Defining, Inscribing and Practicing Early Modern Geography, c. 1450–1850," in *The Wiley-Blackwell Companion to Human Geography*, ed. John A. Agnew and James S. Duncan (Chichester: Wiley-Blackwell, 2011), 23–29; 28.
25 Conley, "Early Modern Literature and Cartography," 409.
26 Mayhew, "Cosmographers," 26.
27 Shalev, 10.
28 *The Compost of Ptholomeus*, Robert Wyer, printer, 1530?, STC 20480, sig.O(2)r.
29 Axton, 13.
30 Ibid., 3. The "four elements" refer to earth, air, fire, and water, the components of matter.
31 Barber, 30.
32 Ibid., 30.
33 Iain Macleod Higgins uses this term to refer to Mandeville's compilation of sources in *The Book of John Mandeville, with Related Texts* (Indianapolis, IN: Hackett, 2011), xxi.
34 See Dillon.
35 Axton, 12.
36 Phillipa C. Maddern provides fuller insight into the term "mery" as indicating not only mirth in late medieval England, but also health, pleasure, and spiritual well-being. Further, she locates heaven as the "typical locale" for this sense of spiritual wholeness, an observation that would situate Rastell's "Mery" in the context of spiritual and cosmic well-being. See Phillipa C. Maddern, "It Is Full Merry in Heaven: The Pleasurable Connotations of 'Merriment' in Late Medieval England," in *Pleasure in the Middle Ages*, eds. Naama Cohen-Hanegbi and Piroska Nagy (Turnhout, Belgium: Brepols, 2018), 21–38; 33.
37 Romuald Ian Lakowski, "Geography and the More Circle: John Rastell, Thomas More and the 'New World'," *Renaissance Forum* 4:1 (1999), n.p. Par. 5.

38 Thomas More's son John More makes mention of new-found lands in his "Preface" to *The Legacy of Prester John*. See Elizabeth B. Blackburn, "The Legacy of 'Prester John' by Damião à Goes and John More," [a facsimile edition] *Moreana* 14 (1967): 37–98; 45–46.
39 George B. Parkes "The Geography of the Interlude of the Four Elements," *PQ* XVII (1938): 262.
40 Ibid., 261.
41 Johnstone Parr, "John Rastell's Geographical Knowledge of America," *Philological Quarterly* 27 (1948): 229–240.
42 Parr cites the following as in circulation in England at the time: Peter Martyr's *Decades of the Oceans*, circulating in manuscript and then in print with accounts of New World voyages, as well as print editions of Martin Waldseemüller's *Cosmographiae Introductio* (1507), Marcus Benventanus' *Geographiae Claudii Ptolemaei* (1508), the anonymous *Globus Mundi* (1509), Bernard Sylvanus' *Claudii Ptholemaei Liber Geographiae* (1511), Johann Stobnicza's *Introductio in Claudii Ptolomai Cosmographiam* (1512), the Strasburg edition of Ptolemy's *Geographiam* (1513), and Johann Schöner's *Luculentissima quaedam terra totius descriptio* (1515). Parr notes that "there were maps of the world in or attached to each of these cosmographies" and notes the circulation of the Contarini map of 1506 and Gregor Reisch's map in the 1515 edition of the *Margarita Philosophica*. Globes and gores were also in circulation, including the Lenox Globe (1503–1507), the Jagellonicus Globe (1510), the Hauslab Globe (1513?), the Green Globe (1515), the Schöner Globes (1515, 1520), the Acton Globe (1515?), the Liechtenstein globe gores (1507–1509), the Nordenskiold globe gores (c. 1518), the Boulengier globe gores (c. 1518) and the globe gores attributed to Leonardo da Vinci (1514–1519). See especially Parr, 230.
43 Ibid., 237.
44 Ibid., 239–240.
45 Elizabeth M. Nugent, "Sources of Rastell's *The Nature of the Four Elements*," *PMLA* 57:1 (1942): 74–88. *De Proprietatibus Rerum* was printed by Wynkyn de Worde in 1495.
46 Ibid., 75.
47 Ibid., 84.
48 C.R. Baskerville, "John Rastell's Dramatic Activities," *Modern Philology* 13 (1915–1916): 557–560; 560.
49 Axton, 136; note F, xii.
50 Nugent, 83.
51 Axton, 136; note F. 1098–1118.
52 Ibid., 137, note F. 1319–1324.
53 W.R. Johnson, *Lucretius in the Modern World* (London: Gerald Duckworth & Co., 2000), ix; Stephen Greenblatt, *The Swerve: How the World Became Modern* (New York: Norton, 2011), 151. For the impact of the rediscovery of the ancient "atomists" (Democritus, Epicurus, and Lucretius) and the revival of ancient materialism, see Catherine Wilson, *Epicureanism at the Origins of Modernity* (Oxford: Oxford University Press, 2008).
54 Greenblatt, 226.
55 Anne E.B. Coldiron, *English Printing, Verse Translation, and the Battle of the Sexes, 1476–1557* (Farnham: Ashgate,2009), 175.
56 David Rundle, "The Scribe Thomas Candour and the Making of Poggio Bracciolini's English Reputation." *English Manuscript Studies 1100–1700* xii (2005): 1–25.
57 Greenblatt, 227. Quoted in Thomas More, *Utopia*, ed. George M. Logan and Robert M. Adams (Cambridge: Cambridge University Press, rev. edition, 2002), 68.
58 Entry: "Annihilate." in *Oxford English Dictionary*, online.
59 Lucretius, *On the Nature of Things*, Book I, in *The Internet Classic Archive*, MIT, available at: http://classics.mit.edu/Carus/nature_things.1.i.html

60 E.J. Kennedy, "Lucretian Texture: Style, Metre and Rhetoric in the *De Rerum Natura*," in *The Cambridge Companion to Lucretius*, eds. Philip Hardie and Stuart Gillespie (Cambridge: Cambridge University Press, 2007), 76–91; 76.
61 Coldiron, 189–190.
62 Axton, 30.
63 Suzanne R. Westfall, *Patrons and Performance: Early Tudor Household Revels* (Oxford: Clarendon, 1990), 198.
64 David Bevington, *Mankind to Marlow* (Cambridge, MA: Harvard University Press, 1962), 47.
65 Dillon, 28–30. For details of the lawsuit, see William Ingram, *The Business of Playing: The Beginnings of Adult Professional Theater in London* (Ithaca, NY: Cornell University Press, 1992).
66 Ingram, 71.
67 Dillon, 35.
68 Axton, 30.
69 Henry Burrows Lathrop, *Translations from the Classics into English from Caxton to Chapman, 1477–1620* (New York: Octagon Books, 1967), 15.
70 Ibid., 27.
71 Stuart Gillespie, *English Translation and Classical Reception: Towards a New Literary History* (Chichester: Wiley-Blackwell, 2011). See also R.R. Bolgar, *The Classical Heritage and its Beneficiaries* (Cambridge: Cambridge University Press, 1954), and Stephen Medcalf, "Classical Authors," in *The Oxford History of Literary Translations in English*, vol. 1: *To 1550*, ed. Roger Ellis (Oxford: Oxford University Press, 2008), 364–389.
72 Gillespie, 7.
73 Ibid., 37
74 Ibid., 37.
75 Ibid., 38.
76 Baskerville, 560. Caxton makes a similar bid for the value of translating the story of King Arthur into English in his Preface to *Le Morte Darthur*. Karen Cherewatuk is rightly credited with making this connection.
77 Lathrop.
78 For a discussion of the production versus the *reproduction* of geographic "reality," see Ivo Kamps, "Introduction," in *Maps and Memory in Early Modern England: A Sense of Place*, ed. Rhonda Lemke Sanford (New York: Palgrave, 2002).
79 Axton 130 (note F. 561–566).
80 Axton, 11.
81 See Nugent for a discussion of Rastell's sources.
82 Oliver H. Prior, ed. *Caxton's Mirrour of the World* (Berlin: A. Asher & Co. for the Early English Text Society, Extra Series CX, 1913). For a discussion of Caxton's use of "rubrics" at the front matter of the book, see Chapter 2.
83 Axton, 8.
84 Ibid., 9.
85 See R.J. Roberts, "John Rastell's Inventory of 1538," *The Library*, 6th Series (1979): 34–42. A one penny map would cost approximately £2.58 in today's currency. See MeasuringWorth.com.
86 Axton, 10.
87 Ibid., 10.
88 More was executed in 1535.
89 See Peter Ackroyd, *The Life of Thomas More* (New York: Nan A Talese, 1998); Ronald Ian Lakowski, "Geography and the More Circle: John Rastell, Thomas More and the 'New World,'" *Renaissance Forum* 4:1 1999 (no pagination). For a unique perspective on the map's skull-like features, see M. Bishop, "Ambrosius Holbein's *Memento Mori* Map for Sir Thomas More's *Utopia*: The Meanings of a

Masterpiece of Early Sixteenth Century Graphic Art," *British Dental Journal*, 199:2 (2005): 107–112. For a consideration of the impact of the new geography in More's work, see Peter Hallberg, "Thomas More's Cosmopolitan Civil Science: The 'New World' and 'Utopia' Reconsidered," *History of Political Thought* 33:4 (2012): 578–606.

90 Richard Pace, *De Fructu qui ex doctrina percipitur*, eds. and trans. Frank Manley and Richard S. Sylvester (New York: Frederick Ungar Publishing Co. for the Renaissance Society of America, Renaissance Texts Series 2, 1967), 108–109, 172–173.

91 Karma Lochrie, *Nowhere in the Middle Ages* (Philadelphia, PA: University of Pennsylvania Press, 2016), 13.

92 Elly Dekker and Kristen Lippincott, "The Scientific Instruments in Holbein's Ambassadors: A Re-Examination," *Journal of the Warburg and Courtauld Institutes* 62 (1999): 93–125. Dekker and Lippincott note that, surprisingly, Holbein's globe does not show the track of Magellan's circumnavigation of the Earth (completed by his men in 1522).

93 Thomas Elyot, *The Book Named the Gouernor* [1531]. A Scholar Press Facsimile (Menston: England, 1970), No. 246, fol. 26.

94 Ibid., fol. 37.

95 Ibid., fol. 37.

96 Barber, 30.

97 Ibid., 26.

98 Elyot, fol. 48.

99 See Naomi Reed Kline, *Maps of Medieval Thought; The Hereford Paradigm* (Woodbridge: Boydell Press, 2001).

5 After “poyetes and astronomiers”

The *Kalender of Shepherds* and Ptolemaic geography in popular print

An obscure headwater of innovation both for early modern literature and popular cartography can be discovered in the early print tradition known as the *Kalender of Shepherds*, at once a “miscellany,” an almanac, a treatise of health advice, and a prognostication; it is a book of prayers, light verse, and popular imagery. The *Kalender* forms what might be called a “multi-text,”[1] a loosely affiliated body of early print books which share titles, images, or content but that contain wide variation between editions.[2] Anne Coldiron refers to such print phenomena as “early modern literary polystreams”[3] to denote their composite make-up, often inflected by or originating with foreign work translated into English, a process of transformation from “foreign-accented, to naturalized, to authoritative.”[4] And indeed, the *Kalender* functions in just this way. Initially a French Catholic manuscript calendar with the requisite “red-letter days” to mark saints’ days and holy days, the *Kalender* became a naturalized English Protestant text that synthesized contemporary moral and emerging scientific thought. Guy Marchant’s (1493) print edition of *Le Compost et Calendrier de Bergiers*, first translated into English in 1503 as *The kalendayr of the shyppars*, initiated a century-long print and reading practice in England, influential to Edmund Spenser, but overlooked in its influence on popular geographic thought. At the close of the sixteenth century, the *métier* of the *Kalender* continued forward in two strands: in Henry Gosson’s edition of *The Compost of Ptolomeus* and in Edmund Spenser’s poem, *The Shepheardes Calender*. In Gosson’s *Compost*, a simple woodblock world map includes instructions on how to read the symbolic codes of the map and signals that the almanac had long been functioning as a form of geographic discourse. Alternatively, in Spenser’s *Shepheardes Calender*, the proverbial and calendrical almanac tradition is transformed into an innovative English eclogue.

The *Kalender* was printed multiple times by nearly all the major early English printers and translators, including Wynkyn de Worde, Richard Pynson, Julian Notary, Robert Wyer, Thomas East, Thomas Colwell, and Henry Gosson. Despite such a robust print life, the material record is slim, with relatively few copies of each edition extant, “perhaps because of the miscellaneous and pragmatic nature of its content.”[5] As Oskar Sommer notes, “A great many editions have entirely perished.”[6] The varied content and pragmatic advice invited everyday use, and the texts, inexpensively printed, were literally read to

DOI: 10.4324/9781003096245-6

tatters. In its Derridaen "drift"[7] through early modern English print, the work emerged as thoroughly "Englished:" specific references to the Pope and Catholic practices were eventually eliminated, a French "rutter" (a mileage chart of port cities) was converted from "leagues" to English "myles," and, in one edition, a basic world map was added, complete with instructions for decoding the visual codes for English readers. This increasingly spatialized version of the *Kalender* looked not to Rome for guidance, but to a new world of English oceanic navigation and to the private reading and interpretive practices of Protestant England.

Fine-press quartos (de Worde and Pynson) as well as inexpensive octavos form the print corpus of the *Kalender*, a disbursed print "event" that spanned the late fifteenth century and continued through the early sixteenth. Collectively, these texts comprise an unrecognized development, *in situ*, of populist cartographic writing in England. Within the multi-text known as the *Kalender*, which includes a derivative branch entitled *The Compost of Ptholomeus*, one of the earliest "everyday" English world maps is printed, with a direct address to the reader: "Looke where you see the figure of 2 that is Affrica, the figure of 3 is Europe, the figure of 4 is Asia, and the figure of 5 is America."[8] Spenser also drew upon this native poly-stream for his innovative eclogue, *The Shepheardes Calender*. Published in 1579, Spenser's text adopts his title and structure of the seasonal calendar from these popular texts to fashion a new literary form. Spenser also appropriates the figure of the wise rustic as narrator, a move that mirrors the disavowal of intellectual learning seen in the *Kalender* tradition. In Spenser's eclogue, as in the *Kalender*, this trope of "natural wisdom" is promulgated through the very practice of reading and learning ostensibly eschewed. This twin textual phenomena of the *Kalender's* print history—the inclusion of a cartographic image of the world into the popular almanac genre and the adoption of a calendrical narrative structure and title by Spenser—illuminates the pairing of literature and cartography and reveals the mutual influences and shared history of literature and cartography in early modern culture within the material and theoretical perspective of book history. This transitional textual history—crossing boundaries between the late medieval and early modern periods—suggests the ways in which cartography, popular science, humanism, and literature were closely intertwined in the early print era in England.

The print history of the *Kalender* and *The Compost of Ptholomeus* (hereafter referred to as *The Compost of Ptolemy*) fills a significant gap in the narrative of English cartographic history, providing the transmission evidence of non-elite cartographic thought. David Woodward, editor of the definitive *History of Cartography* series, acknowledges that the focus of early modern cartographic study has been on "elite … figures engaged in cartography, to the exclusion of the everyday artisan or map consumer."[9] Further, he observes that the map market for and map reading practices of middle-class and working-class readers have been neglected in comparison to studies of the role of maps in European courts during the Renaissance. In light of this,

Woodward calls for new research on the everyday uses of maps during the Renaissance.[10] Maureen Perkins has similarly advocated for a methodology that seeks "to access the cosmology of ordinary people by looking at the things they used to think with."[11] The gradually increasing symbolic appropriation of the second-century Greek astronomer and mathematician Claudius Ptolemy as part of the textual polystream of the *Kalender* and the *Compost* thus fills a gap in the study of English cultural geography. As the text of Robert Wyer's (1540) edition states:

> I wyll nat saye that suche thynges shalbe. but that the sygnes haue such propertys, as it is the wyll of GOD. After Poyetes and Astronomiers/ Aries is the fyrst sygne that sheweth the fortunes of men and women, as sayth Ptholomeus.[12]

This kind of integration of Ptolemy as both sage, astrologer, and scientist makes explicit the connection between geographic thought and literary practices in the sixteenth century: a common body of knowledge whose source arises from both poets and astronomers.

Reading Ptolemy and the new geography in England

The trajectory of populist geographic thought in early modern England has received scant attention, perhaps due to the nature of its being embedded in quasi-literary texts with little reputational capital in post-Enlightenment histories of the "age of discovery." In such narratives, the *Kalender* seems more of an embarrassment to English intellectual history in the era of the Continental Ptolemaic revival.[13] Further, as Perkins has observed, "there have been few studies of European almanacs,"[14] which, humble as they are, were a primary vehicle for the dissemination of pseudo-Ptolemaic thought and popularized humanist texts. As a result, the influence of Ptolemaic cartography is obscure in England where no known edition of his influential second-century *Geography* was produced in manuscript or print in the fifteenth or sixteenth centuries. But the re-titling of the *Kalender* for the derivative branch entitled *The Compost of Ptolemy* refers to the second-century Greek mathematician and cosmographer, strongly suggesting the cultural influence of his works, including the *Geography, Almagest*, and *Tetrabiblos* (also known by its Latin title the *Quadripartitum*). Ptolemy's "tables" of locational coordinates, first translated from Greek into Latin in the early fifteenth century, formed the basis of early modern editions of the *Geography*, a text which exerted a powerful influence on European geographical thought in the mid-fifteenth and early-sixteenth centuries when maps drawn on Ptolemaic projections were first produced in manuscript and woodblock print editions.[15] By the late fifteenth century, the production of the *Geography* on the Continent coincided with the development both of print technology and navigational advances in a potent confluence of intellectual and material change. But the influence of the Ptolemaic revival, though less apparent in English print, is not insubstantial. It is my argument that this influence is evidenced via popular literature and book history as essential contexts for the study of the reception and transmission of Ptolemaic geography in England.

While no edition of Ptolemy's rediscovered work was produced by English scribes or presses from the fifteenth through the seventeenth centuries, the popular *Kalender* provides a textual and cultural witness of the increasing awareness of the prestige and influence of Ptolemaic-based geographic thought for English readers. While well-to-do, educated English readers could access Ptolemy's work in imported Latin editions produced on the Continent, complete with world and regional maps, non-elite literate readers could find, albeit in a syncretic, popularized and, some might say, bastardized form, material from Ptolemy's *Tetrabiblos* in the *Kalender of Shepherds*. These editions attest to the growing recognition of the prestige of Ptolemaic thought, and, eventually, provide a simple world map that constructed a globalized geography for everyday English readers.

The history of sixteenth-century scientific thought in England has been well documented by David Waters[16] and Henry Stanley Bennett[17] among others: scientific writers provided English-language texts for a scholarly and educated audience through works such as Robert Recorde's (1556) *Castle of Knowledge* and William Cuningham's (1559) *The Cosmographical Glasse*. These texts presented current scientific thought as well as instruction in astronomical and navigational instrumentation and yet, for all their erudition, these texts do not provide an image of a world map. Christopher Saxton's (1579) *An Atlas of the Counties of England and Wales* provided finely detailed county maps and, in 1583, he produced prints of a wall map of England. As David Waters points out, "By the 1580's … English hydrographers—Dee, Borough, and Norman to name only a few—were producing regional and oceanic manuscript charts of refinement and with characteristic features that distinguished them as English products."[18] But these compilations do not contain an image of a world map, nor were these texts produced for "laye people" as were the *Kalenders*.[19] The first early modern world map published in England is Sebastian Cabot's (1549) map, a collaboration between the Anglo-Spanish cosmographer and Clement Adams. Cabot, Spanish by birth, spent part of his childhood in England and sailed as part of his father John Cabot's famous English-financed explorations. Sebastian Cabot was truly a transnational cosmographer and returned to English service in 1548. While a previous edition of Cabot's world map had been published in Antwerp in 1544, Robert K. Batchelor hails the London publication as "a sign that the informational world of London had begun to change."[20] For global images prior to Cabot's (1549) map, English readers of financial means and with access to book importers relied on continental productions of the *Geography* or Waldseemüller's (1508) world map, as John Rastell seems to have done (see Chapter 4). After 1549, English readers had access to Cabot's world map, Abraham Ortelius' (1564) world map, Flemish cartographer Gerardus Mercator's (1569) double cordiform world map, or Ortelius' (1570) atlas, *Theatrum Orbis Terrarum*. None of these maps, however, would have been accessible or affordable to a majority of English readers.

Waters argues that Sir Walter Raleigh's explorations in the 1570s, in which new "charts" were made, mark the rise of the "oceanic expansion of England"[21] and thus an increased global and cartographic awareness, particularly for those in contact with the London book trade. Though preceded by Cabot's Bristol-based ventures, Raleigh's expeditions and subsequent "charts" signal a significant moment of material territorialization and geographic imagination in England.[22] Anticipating this oceanic expansion, however, was the publication of "rutters"—navigational tables which listed distances between port cities.[23] The first rutter was introduced to English readers with the publication by Robert Copland *The Rutter of the Sea* in 1528, a translation of an earlier French rutter by Pierre Garcie. Reprinted in 1535 and again in 1540, the translated versions added coastal English ports while retaining the Continental measurements of leagues and kennings, an example of a text in the dynamic process of being "Englished." By the end of the century, Thomas Blunderville's (1594) *Exercises* provides an intellectual English text on navigational tools, globes, and maps, such that, by 1600, science in England "had found its sea legs."[24] Here again, however, Blunderville's text does not provide an image of a world map.

The full-on entry of English cosmographers into Renaissance world map-making is marked by Emery Molyneux's (1592) globes, Richard Wright's (1599) *Chart of the World*, and John Speed's *Prospect of the Most Famous Parts of the World*, printed in 1627. The famed and much-anticipated Molyneux globes were produced in collaboration with the Dutch engraver, Jodocus Hondius and are thus considered an Anglo-Continental, geo-cultural hybrid.[25] While Adam Max Cohen notes that the "rising symbolic influence of the terrestrial globe in England during the 1580s is clear,"[26] we can also trace this symbolic influence far earlier—to Caxton's images of globes in his (1481) edition of *The Myrrour of the World* (see Chapter 2). Speed's *Prospect*, published in London, was based on Mercator's cartographic projection used for his (1569) world map, a projection which gradually superseded Ptolemy's projection. With the engraving work done in Amsterdam, there is speculation that Speed's world map is essentially an Anglicized version of world maps already in circulation in Amsterdam rather than an independent or innovative English production. All of these English cartographic endeavors demonstrate the international nature—both collaborative and competitive—of early modern cartography and speak to the circulation and influence of geographic texts in cosmopolitan London and in English literary culture more generally.

The absence of English-produced world maps until the mid- to late sixteenth century, however, did not forestall engagement with new models of geographic thought via importation and via English literary and popular texts, well before Wright's and Speed's cartographic publications. English editions of William Caxton's *Myrrour*, along with the many editions of the *Kalender* and *Compost*, were vital agents in the circulation and articulation of a narrative-based and populist geography, one influenced by natural philosophy and that sought to map the heavens, the body, the book, and, eventually, the Earth as part of an integration of the geographical authority of the classical past and the increasingly visual and spatial representations of the present. With their description of astrological influences, schematics of the body, a rutter, and, eventually, a world map, these texts represent a continuity regarding spatial relations that continued well into the early modern period—the perceived connection of the body, soul, the Earth, and the cosmos.

Of neglected significance in the history of sixteenth-century English cartography is a simple woodblock map of the world, a double cordiform shape that mirrors Mercator's general form, printed for Henry Gosson in 1600 or 1638 (the dating is uncertain) in an inexpensive octavo edition of the *Compost*. The publication of this basic world map, although likely published in limited copies, marks the emergence of an image of the world for an expanding print readership in England, one that arises out of a native geographic tradition and nearly contemporaneous with the esteemed English cartographic productions of global images by Speed and Wright. The possible publication date as early as 1600 of Gosson's edition of the *Compost* would place the Gosson edition map as among the earliest English-produced world maps. The Gosson map is significant (regardless of its dating at 1600 or 1638) for its introduction of map-reading instruction for the general public and its retrospective clarification of the *Kalender* and *Compost* as a proto-cartographic textual tradition. Bridging medieval and early modern spatio-temporal thought, editions of the *Kalender* and the *Compost* provide a native textual account of the translation, transformation, and reception of Ptolemaic-based geographic culture in England—from the medieval Catholic conception of time, as instantiated in Books of Hours with their cyclical and annual demarcation of time, to an increasingly mathematical model of distance and space. A socially constructed geography, operating tangentially to the then-emerging discipline of mathematically-based cartography, relied on narrative as well as visual allusions to Claudius Ptolemy as a key signifier of cartographic knowledge. Ptolemy's name and work attained a cultural cachet similar to twentieth-century terms such as relativity or chaos theory, terms that, while they may have a tenuous connection with specific scientific principles, have enormous resonance as cultural constructs. Part of the increasing cultural influence of Ptolemaic geography in England can be attributed to these popular, pseudo-literary and pseudo-scientific texts over a century of print publication. Although often disassociated from England's cartographic or literary histories, this strand of popular almanacs reflects an increasing awareness by English readers of the prestige and influence of Ptolemaic-based geographical thought and the interest in new representations of the world.

The *Kalender of Shepherds* and *The Compost of Ptolemy*

The long-standing popularity of the almanac tradition in England is reflected in the fact that by the 1660s approximately 400,000 copies were sold annually with, as Bernard Capp notes, "roughly one family in three" purchasing some kind of almanac each year.[27] As evidence of its particular popularity, the *Kalender* "remained in print in England in a variety of versions until 1631, going through some 19 editions."[28] Abigail Shinn notes that it was "unquestionably a publication which was sold on a large scale"[29] although the number of extant editions does not accurately reflect this scale. Versions of the text proliferated, such that the composite multi-text body of the various editions collectively comprise, in terms of reach and volume, nothing short of the most influential publishing "phenomenon of the age."[30] They are a treasure trove of early modern reading practices and a rich transcript of literary, visual, religious, and geographic culture.[31] Francis R. Johnson argues that "interest in all phases of science was widespread in Elizabethan England, extending to all literate classes."[32] These popular texts not only provide an insight into changing conceptions of religion, the cosmos, and natural phenomenon, but, in addition, the unique attributes of each edition provide a transcript of changing conceptions of time-space constructs on the part of non-elite readers. In Richard Pynson's edition of 1506, he notes in his Introduction that:

> [He] hath made it into playne englysshe to the entente that euery man may vnderstonde it/ that thys boke is very profytable bothe for clerkes and laye people to cause them to haue greate vnderstondyng/ and in espessyall in that we be bounde to lerne and knowe on peyne of auerlastinge deth.[33]

Printers of the text sought a broad readership and urged the value of cosmological knowledge as well as literacy in general, a not entirely disinterested position for those in the developing print trade, to be sure. De Worde's (1528) edition also urges the value of astrological knowledge for "every man:"

> Fyrst one ought to knowe what the fygure is/ the dysposycyon of the worlde/ the no*m*bre & order of the elementes/ & the mouynges of the skyes apperteyneth to be known of euery man of fre condycyon and noble wytte. For it is a fayre thynge/ delectable profytable/ and honeste/ and therewith it is necessary to haue dyuers other knowleges/ speciyally the astology of Shepeherdes/ which sheweth how þe worlde is round as a ball.[34]

Perkins has argued that English almanacs such as the *Kalender* "were influential cultural messengers, perpetuating beliefs in astrology, prophecy, and mysterious occurrences in the natural world."[35] But these texts should not be viewed solely as vehicles for mystification: they also perform the integration of scientific and geographic thought into popular culture and provide an optimal record of this process. If read as a poly-text with over a century of print permutations rather than as a monolithic text, the various editions, with their self-consciously constructed reader, reveal a great deal about the mechanism of dissemination of popular geographic thought. Beyond the impact of the text as a whole, each edition needs to be read as a unique textual event that instantiates specific conceptions of time and space, one that provides a compelling moment in the trajectory of the development of reading culture and geographic thought in early modern England.

Martha Driver has described the *Kalender* as a "miscellany of shepherd's lore, tables for finding the dates of moveable feasts, descriptions of the torments of purgatory, and explications of the Ten Commandments, Lord's Prayer, and Apostles' Creed, along with medical and other practical information."[36] The "tables" for identifying feast days would have had a generalized spatial and enumerative affinity with the locational tables that form the heart of Ptolemy's *Geography*. Ptolemy's *tabulae* did not, however, appear either directly or derivatively in editions of the *Kalender*. Initially a Catholic perpetual calendar,[37] a *computes* or calendar of Catholic feast days, the texts expanded into an amalgam of proverbial morality, health advice, and sententious or comic verse as well as "judicial" and "natural" astrology. They served to provide the reader with both "profyte and pleasure,"[38] both moral advice and entertainment. Judicial astrology, antithetical to orthodox Catholic doctrine, interpreted planetary and celestial phenomenon as the basis of prognostication and interpretation. Natural astrology, which was acceptable to the Church and which drew from an array of classical cosmologists, including Ptolemy, "applied information about planetary motion and lunar phase to physical phenomena such as the growth of crops or the administration of medicine."[39] Drawing upon the tradition of natural astrology, the *Kalender* is an amalgam of the cyclical feast days and seasonal topoi.

The connection to seasonal phenomenon is articulated in the *Kalender* with particular and traditional attributes for each month of the year:

> Than cometh September/ that wynes be made/and the fruytes of trees be gathered. And than there withall he dothe fresshly begynne to garnysshe his house/ & make prouysyon of nedefull thynges/ for to lyue in wynter/ whiche draweth very nere/ & than man is in his moost stedfast and courtous estate/ prosperous in wysdome/ purposynge to gather and kepe.[40]

Each month of the year is also associated with a corresponding six-year unit of time such that the text proposes that a typical human life span can be estimated at 72 years (6 x 12) and is represented as a metaphorical year, a metaphor Spenser adopts in his eclogue. The month of July, for example, is associated with mid-life. "At July, he is 42 and he begineth a lytell to declyne and feleth hym nat so prosperens as he was,"[41] a familiar sentiment, then as now. In accordance with this presumed life span, the text offers advice on how "to doe things possible and requisite for to liue long and in health."[42] In addition to advice on health and longevity, some editions also provide a description of "The Ten Christian Nations," the first of which is the "Laytyns," then "Grekis," and followed by other early Christian nations of the East and of Africa. The text, in other words, displays a movement from temporalization to a process of territorialization, replete, as we shall see, with a growing geographic consciousness over the life of the text as it develops in the sixteenth century. With multiple components: feast days, natural and judicial astronomy, astrology, religious doctrine, "trees" of vice, descriptions of the "peynes" of hell, moral verse, anatomy, palmistry, and health advice—editions of the *Kalender* are truly multivocal texts—informative, moralistic, winsome, prescriptive, and poetic. This wide range of subject matter is also visually replete with a diverse woodblock illustration scheme. It was a book to be dipped into and read again and again both for moral "profyte" and as well as "pleasure." For this reason, despite their presumed print runs and their popularity, relatively few intact copies are extant.

Both the *Kalender* and the *Compost* editions exhibit spatializing practices attuned to a wider cultural shift in Europe: images and the verbal text "map" the zodiac as well as the veins and the body (Figure 5.1). Some editions have an extensive palmistry section with images that depict the topography of an array of human palms. Many editions also adopt Caxton's innovation of rubrication, thereby mapping the text itself with a table of contents. These spatial practices were grounded in medieval antecedents which negotiated the self in relation to the cosmos[43] but which also express a form of nascent mapping practice in English reading culture.

Though its origin was a French Catholic calendar of holy days produced for aristocratic and popular audiences, over its 150-year textual drift, the *Kalender* gradually morphs into an English Protestant handbook of cosmological lore, self-help, lay literacy, and, ultimately, geography. As Capp has noted, while the *Kalenders* were "technically, a table of the astronomical and astrological events of the coming year: the movements and conjunctions of the planet and stars in the zodiac, and details of eclipses,"[44] these perpetual calendars came to be purveyors of "astrological medicine" and, further, a source of quasi-sacramental mystery, "supplying a need apparently ignored by the English Church after the Reformation."[45] The transformation to a Reformation text can be observed, for example, in one copy of an edition printed by Wynkyn de Word in 1528 in which references to the Pope are blacked out by hand,[46] a readerly "revision" that would be instantiated by later print editions in which the text's early association with Catholicism is obscured.

The Kalender *in English print: a complex textual history*

The first English translation of the work, published in 1503 by Antoine Verard in Paris,[47] was undoubtedly a curious specimen of translation work: a Scottish-French amalgam or "pseudo-Scottish dialect"[48] entitled *The kalendayr of the shyppars.* In 1506, Richard Pynson produced an independent translation, and in 1508 a translation by Robert Copland was printed by Wynkyn de Worde, Caxton's former assistant. The quality of the first

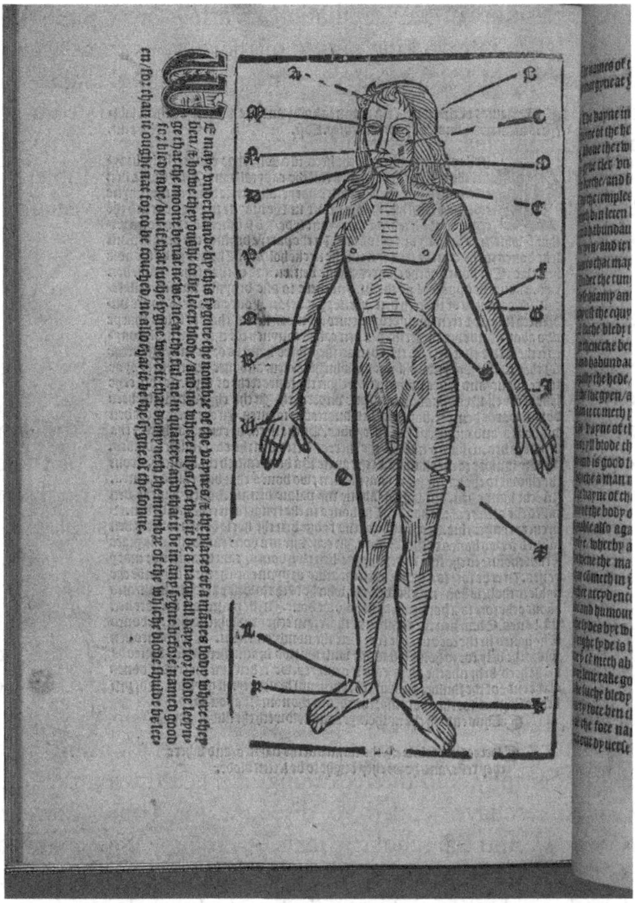

Figure 5.1 Here Begynneth the Kalender of Shepardes, Julian Notary, c. 1518 (STC 22410)
Source: RB 69474, The Huntington Library, California.

Scottish-French hybrid translation exasperated Julian Notary such that in his (1518) edition he sharply criticized the readability of the work and offered, instead, his own, more comprehensible English edition:

> Here before tyme this Boke was prynted in Parys into corrupt Englysshe/ & nat by no englysshe man/ wherfore the Bokes that were brought into Englande/ no man coude vnderstande them parfytely and no meruayle/ for it is vnlykely for a man of that countree/ for to make it into good & parfyte Englysshe as it shude be.[49]

As the king's printer, Notary's role as an early editor and adaptor of the work into English literary culture is indicative of the work's readership in noble or educated circles despite its later reputation as a low-brow publication without scientific merit.[50] The interlocking nature of the various translations and printings is evident in that, in 1518, Notary's edition was "based mainly on Copland's translation [and] supplemented by that contained in Pynson's edition."[51]

To clarify the complex, early print cross-overs in terms of textual material and the illustrative scheme of these texts,[52] Table 5.1 and Table 5.2 present a list of editions as well as a transmission chart to aid in the visualization of the textual relations.

The trace of Ptolemy

English readers in the early sixteenth century, as evidenced by the great popularity of texts such as the *Kalender* developed a conceptual geography within the context of astronomy, astrology, and prognostication—information that was self-consciously fashioned from the cultural authority of Ptolemy's *Geography, Almagest* (his treatise on astronomy and familiar to English readers via Chaucer's translation), and *Tetrabiblos* (his treatise on astrology). In Robert Wyer's re-titled edition from the 1530s, *The Compost of Ptholomeus,* [54] he relies on the set phrase "as Ptholomeus sayth" regarding the 12 signs of the zodiac and the subject of "phyzonomye." Although the *Kalender* has been described as containing "nothing from Ptolemy," other than the "falsification" of authorial attribution,[55] it does, in fact, have a general articulation of some of the astrological matters set forth in Ptolemy's *Tetrabiblos*, including, for example, planetary and zodiacal influences, the relationship of zonal geography and ethnography, and the reckoning of a healthy man's life expectancy. And while scholars have tended to discount the connection between the *Kalender, Compost*, and the scholarly texts of Ptolemy, the separation between astrology and astronomy was less absolute in the fifteenth and sixteenth centuries. Ptolemy's works themselves include astrology alongside mathematical astronomy and cartography; as F.E. Robbins observes, "no one thought it inconsistent with Ptolemy's reputation as a scientific astronomer that he should also have written upon astrology."[56]

Just as Ptolemy's *Tetrabiblos* engages with a calculation of life expectancy, the opening of the *Kalender* similarly establishes an estimated life expectancy based on astrological interpretation. In the *Tetrabiblos*, Ptolemy writes, "The consideration of the length of life takes the leading place among inquiries about events following birth …"[57] and from this initial premise, Ptolemy provides a complex and esoteric analysis of life expectancy based on the zodiac, fortune, and astronomy, with the number of degrees traversed on the zodiac wheel predictive of the years of life. The *Kalender*, too, is based on this Ptolemaic principle of prognostic astrology but simplifies Ptolemy's complex excursus on a "system of prorogations"[58] into a straightforward declaration of 72 years as the ideal lifespan.

> How be it þ^{t} lyuynge & dyenge be all at þ^{e} pleasure of almyghty god. Yet man may lyue by þ^{t} course of nature. lxxii. yere or more. this was his

Table 5.1 Editions of *The Kalender of Shepherds*[53] (arranged in chronological order)

Date	*Edition*
1493	*Le Co[m]post et kalendrier des bergiers.* Paris: Guy Marchant
1503	*The kalendayr of the shyppars.* Paris: Antoine Verard (STC 22407)
1506	*Here Begynneth the Kalendar of Shepherdes.* London: Richard Pynson (STC 22408)
c. 1508 [1516?]	*The Kalender of Shepeherdes.* London: Wynkyn de Worde (STC 22409)
c. 1510	[*Here Begynneth the Kalendar of Shepherdes*]. London: Richard Pynson (STC 22409.3)
1511	[*Here Begynneth the Kalender of Shepherdes*]. London: Wynkyn de Worde (STC 22409.5)
c. 1517	[*Here Begynneth the Kalendar of Shepherdes*]. London: [Richard Pynson] (STC 22409.7)
c. 1518	*Here Begynneth the Kalender of Shepardes.* [London: Julian Notary] (STC 22410)
1528	*The Kale[n]der of Shepeherdes.* London: Wynkyn de Worde (STC 22411)
1556	*Here Begynneth the Kalender of Shepardes. Newly Augmented and Corrected.* London: William Powell (STC 22412)
1559	[*Here Begynneth the Kalender of Shepardes. Newly Augmented and Corrected*]. London: William Powell, for John Wally (STC 22413)
c. 1570	*The Shepardes Kalender.* London: Thomas Este for John Wally (STC 22415)
c. 1580	[*The Shepardes Kalender.*] London: [John Charlewood for] John Wally (STC 22416)
c. 1585	*Heere Beginneth the Kalender of Sheepehards: Newly Augmented and Corrected.* London: [John Charlewood and George Robinson for] John Wally (STC 22416.5)
c. 1595	*Here Beg[inneth] the Kalender of She[pherds], The Shepheards Kalender.* [London: V. S[immes for T. Adams?] (STC 22418)
c. 1600	*The Shepheards Kalender.* [London: V. S[immes] for T. Adams] (STC 22419)
1604	*he Shepheards Kalender.* London: G. Elde for Thomas Adams (STC 22420)
1611	*The Shepheards Kalender.* London: for Thomas Adams (STC 22421)
1618	*The Shepheards Kalender.* London: Printed for Thomas Adams (STC 22422)
1631	*The Shepherds Kalender.* London: Printed for John Wright (STC 22423)

Table 5.2 Editions of *The Compost of Ptolemy*

Date	*Edition*
c. 1530	*The Compost of Ptholomeus*/ *Prynce of Astronomye*. London: Robert Wyer (STC 20480)[1]
c. 1540	*The Compost of Ptholomeus*/ *Prynce of Astronomye*. London: Robert Wyer (STC 20480a)
c. 1550	*The Compost of Ptholomeus*/ *Prynce of Astronomye*. London: Robert Wyer (STC 20481)
c. 1552	*The Compost of Ptholomeus*/ *Prynce of Astronomy*. London: Robert Wyer (STC 20481.3)
c. 1562	*The Compost of Ptholomeus Prynce of Astronomye*. London: T. Colwell (STC 20481.7)
c. 1600/1638?	*The Compost of Ptolomeus, Prynce of Astronomie*. London: M. Parson for Henry Gosson (STC 20482)[2]

[1] According to Bennett, Wyer produced at least three editions between 1530 and 1540; Bennett, 117.
[2] Only one copy is known to have survived, held by the British Library.

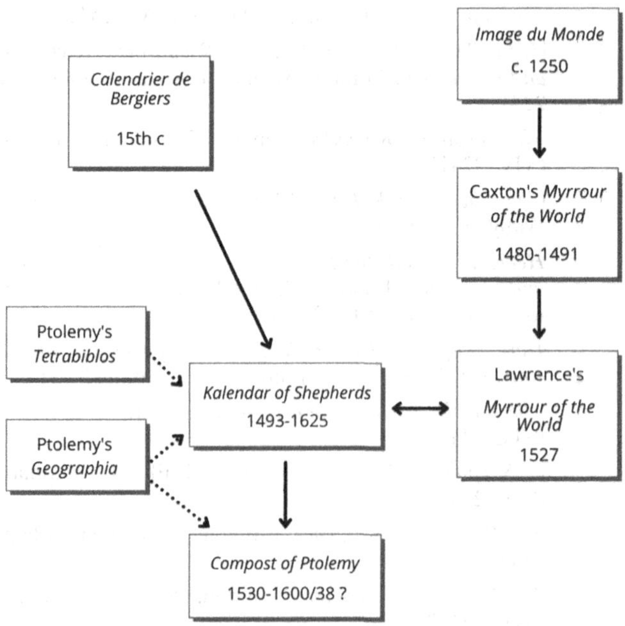

Figure 5.2 Relationship of the *Kalender*, *Myrrour*, and *Compost*
Source: Courtesy of Dana Hardy.

> reason. And he saith as moche tyme as a man hath to growe in beaute/ length/ bredeth/ and strength ... To this purpose of lyuynge and dyenge/ the sayd Sheparde sayth that the thynge that we moste fere is to dye soone/ thus he trauayled his vnderstandynge/ and made great dylygence to knowe and to do thynges possyble and requysyte for to lyue longe ...[59]

Thus, to say that the *Kalender* and the *Compost* have "nothing" to do with Ptolemy's writing is to erase the reception history and popularization of his works into vernacular form. Further, while the *Tertrabiblos* is primarily a treatise on astrology, it also contains "references to geography,"[60] albeit less directly than the *Geography* or the *Almagest*, in which Ptolemy discusses latitudinal-based climate zones and related ethnography.[61] Ptolemy's combination of astrology, astronomy, and geography speaks to the intermingling of these subjects in classical and, subsequently, medieval and early humanist geographical writing. As evidence of the regard accorded all three of these works, library records at Cambridge University show that some copies of Ptolemy's "tabulai" (editions of the *Geography* printed in Venice and Ulm) were owned by university scholars along with unspecified *mappa* or *mappamundi* and, more frequently, copies of Ptolemy's *Almagest* and *Tetrabiblos*.[62] The English clerics and scholars who owned copies of the *Almagest* and *Tetrabiblos* valued the study of astronomy for its role in establishing church feast days, but also "for its links with astrology and medicine."[63] That the *Geography* does not figure as prominently in university library records of the time may reflect the cost of such a richly-illustrated book of maps as well as the possibility that the moral and philosophical astronomy and astrology of the *Almagest* and *Tetrabiblos* spoke more powerfully to the English geographic imagination.

As Patrick Gautier-Dalché has insisted, astrology "should never be overlooked in a study of the reception of Ptolemy's geography."[64] The later disciplinary divide between poetics, astrology, and science, he argues, is not relevant to early humanist intellectual discourse. Indeed, Francesco Berlinghieri's (1482) edition of the *Geography*, printed in Florence, was adapted into verse.[65] Wyer's edition of the *Compost* brings the astrological portion of the text to a close with a 32-line aphoristic verse of similes comparing man to various animals, i.e. [a man is as] "Shrewed, as the Ape."[66] The aphoristic text concludes "And therefore he is called the lytell worlde."[67] This sense of man as a microcosm of the cosmos sums up the mapping of the heavens and the mapping of the body that forms the core of the *Kalender* and *Compost*, mirroring a long tradition of medieval geographic thought.[68] Geography did not merely represent a world "out there," but one that was intimately connected to the self through the practice of astrology, a practice framed by the interpretation of both textual and celestial signs—the co-habitus of "poetes and astronymers," a discourse not solely dependent on, but co-generative with, the emerging field of print cartography. As Keith Lilley notes, "all too often human geographers have emphasized discontinuities between the medieval and the modern,"[69] but for English readers and writers of this period, the "reading" of geography was not so much a break with the classical and medieval past as a confirmation and expansion, a process exemplified by the esteem with which the geography of Ptolemy was held and its integration into popular culture.

From shepherd to Ptolemy

A *longue durée* approach to the multi-text of the *Kalender* and *Compost* reveals the way in which a branch of the multi-text specifically reflects the introduction of Ptolemaic geography into English book culture. This introduction and integration are achieved primarily by way of the gradual substitution of Ptolemy in editions of the *Kalender* and *Compost* for what had previously been the authenticating figure of the philosopher or shepherd. As David Davis has shown in regard to woodcut images in the *Kalender* print history, any given image "retained much of its symbolic and iconic importance as well as its ability to ferry meaning and significance into new contexts."[70] This occurred both for visual images, as Davis discusses, as well as for narrative references. For example, in Pynson's (1506) edition of the *Kalender of Shepherds*, the source of the astrological wisdom is initially attributed to "auncyent Phylozophers:"

> To knowe vnder what planet a man or woman is borne it is nedefull to wete that there is vii. planettes on the skye … The auncyent Phylozophers saye that sol domyneth the sondaye/ the cause is they saye/ for the sonne amonge other planettes is moste worthy/ wherefore it taketh the worthyest daye/ that is sondaye.[71]

In the Prologue to Pynson's (1506) and Notary's (1518) editions, an unnamed, ancient shepherd is identified as the putative source of the astrological and health information of the text:

> As here before tyme there was a sheparde kepynge shepe in the feldes/ whiche was no clerke ne had no vnderstand*in*ge of þ[e] letterall sence/ nor of no manner of scripture nor wrytynge/ but of his naturall wytte & vnderst*a*n*d*i*n*ge sayd.[72]

The authentication for the information here is a pastoral figure of wisdom, one that is devoid of book learning and thus a "natural" wisdom source. Notary's Prologue also states that "this boke was made for them that be no Clerkes to brynge them to great vnderstandynge,"[73] thus offering access for lay readers to the very tradition of classical learning it obscures. There is also an intimation here of a connection between occult knowledge and the act of reading. The frontispieces in both Pynson's (1506) and Notary's (1518) editions depict a rustic shepherd, complete with bagpipe, shepherd's staff, and dogs under a night sky of moon and stars—an image which earlier appeared in Verard's (1503) edition and which proliferates through subsequent editions (Figure 5.3). Despite this "natural" source of authentication in text and

Figure 5.3 Here Begynneth the Kalender of Shepardes, frontispiece, Julian Notary, c. 1518 (STC 22410)
Source: RB 69474, The Huntington Library, California.

imagery, the immediately subsequent image is that of a scholar seated in his study, thus linking the tradition of natural philosophy as accessible to the lay person via the act of reading (Figure 5.4).

In Notary's (1518) text, an innovation addition to the text is made: Ptolemy is specifically cited in the rubrication (table of contents) in relation to the 12 signs of the zodiac: "To know the fortunes and destenyes of man borne vnder the.xii. sygnes after Ptholomeus prynce of astronomye."[74] What beforehand had been an anonymous master shepherd or philosopher now becomes specifically identified as a figure of classical learning: Claudius Ptolemy. Notary's chapter title regarding the 12 zodiac signs within the text itself reiterates the rubrication and adds the caption "Prince of Astronomy Ptholomeus"[75] above a cluster of three images: a shepherd engaged in astronomical measurement, a seated astronomer, and an image of a round Earth in the center of the

Figure 5.4 Here Begynneth the Kalender of Shepardes, frontispiece, Julian Notary, c. 1518 (STC 22410)
Source: RB 69474, The Huntington Library, California.

universe.[76] Notary's editorial innovation marks a visual and textual integration of the original holy day "computes" with the growing cultural interest in the astronomical and geographical work of Ptolemy.

Notary's edition was subsequently republished by Wynkyn de Worde in 1528.[77] In de Worde's edition, the chapter on the 12 zodiac signs (Ch. Liii) attributes the source of its authority to "auncynet Phylozophres" as had the earlier edition by Pynson, but de Worde's edition now includes a proximal image of Ptolemy and Dame Astronomy with an accompanying banderole identifying the figure below as "Ptholomeus." What had previously been, in Verard's first English edition, an image of a shepherd under the stars is now transformed into the specific figure of Ptolemy as both sage and scientist. Further, a textual reference to Ptolemy in the table of contents has entered into the paratextual apparatus within the book,[78] reinforcing the association with Ptolemy. The transference of the authentication of the text to Ptolemy intensifies in the 1530 and 1540 editions of London printer Robert Wyer.

The Compost of Ptolemy

Subsequent editions of the *Kalender* were published, in various forms, by a number of printers throughout the sixteenth century although the transference of authority to Ptolemy is neither linear or fully complete; editions continue with references to and images of an ancient shepherd alongside references to Ptolemy, depending on the edition. In the majority of *Kalender* editions, Ptolemy remains a minor but significant referent, associated primarily with the chapter on the 12 zodiac signs. Beginning in the 1530s, however, a strand of the multi-text breaks off: in Wyer's editions, the text is condensed, the Catholic feast day calendar is excised, and many of the images long associated with the text are eliminated. These changes suppress the earlier function of the text as a holy day calendar and prayer book and foreground, instead, the astrological and geographic qualities of the text. With these textual changes, Wyer retitles the work *The Compost of Ptholomeus/Prince of Astronomye, Translated out of Frenche in to Englysshe. for euery person, that wolde haue knowledge of the Compost* (Figure 5.5). Wyer's editions perform what Bernard Kline has called a "cartographic transaction,"[79] in which a cultural form, in this case the *Kalender*, is associated with cartographic information: as Shinn notes, "during the sixteenth century the English almanac was becoming more geographically specific."[80] As a result, the popular almanac genre becomes linked with the influential cartographic developments of the Ptolemaic revival—specifically the cultural esteem for Ptolemaic-based cartography and world maps. Via a new title, page headers with the work's title, textual references, and images of Ptolemy, Wyer closely links the text to one of the most revered classical figures of humanism.

In Wyer's derivative edition, the prologue basically reprints the text of the *Kalender* with a slight, but significant, alteration: "So as we understande Ptholomeus sayeth / that lyuynge and dyeng is all at the pleasure of almyghty god."[81] In its shifting of the authentication of the text, the introduction references Ptolemy liberally, with frequent phrases such as, "Ptholomeus sauth that the thynge that we desire moste in this worlde, is to lyue long," and "Wherefore this Ptholomeus thought the.Lvvii. yeres … is but a lytell and a small terme of lyfe," and "therefore sayth this Ptholomeus."[82] As Bennett notes, Wyer "gave his work an air of authority and novelty by replacing the word 'Shepharde' wherever it appeared by the word 'Ptholomeus' or 'Astrologian.'"[83] But more than a mere substitution, Wyer's re-shaping of the work expressly links the *Kalender* tradition with early modern cartography. If the textual association was not a sufficient declaration, the frontispiece and a full-page image of "Ptholomeus, prynce of Astronomye" make inordinately clear the connection between the astrological information of the *Kalender* and the

Herebegynneth
The Compost of Ptholomeus/
Prynce of Astronomye: Tran
slated oute of Frenche in to
Englysshe/ for them that
wolde haue knowlege
of the Compost.

Ptholomeus.

Compost. a.i.

Figure 5.5 The Compost of Ptholomeus/ Prynce of Astronomye, frontispiece, Robert Wyer, 1530 (STC 20480)
Source: RB 31739, The Huntington Library, California.

cultural cache of Ptolemy. The symbolic function previously vested in the images of scholar and shepherd are now conflated into the single figure of Claudius Ptolemy, "Prince of Astronomy." The frontispiece depicts Ptolemy with an astronomical quadrant, held to the sky, with Dame Astronomy behind him, pointing to the heavens. Like the previous images of the shepherd, Ptolemy does not hold a book, continuing the implication of a direct reading of the heavens, the purview of natural philosophy now with scientific implements.

In a subsequent image accompanying the introduction, Ptolemy is depicted holding a great book and staff—supplanting and integrating the role of philosopher and scholar. In a cross-textual twist, this woodblock image of Ptolemy in the (1540) edition of Wyer's *Compost* first appeared in Marchant's (1493) edition, *Le compost et Kalendrier des bergiers*, as an illustration of Prester John, a richly suggestive previous use, suggesting exoticism, travel, and spiritualized knowledge.[84] As Davis argues, such recycling exemplifies the process of image exchange, augmentation, and transmission of the *Kalender* tradition. Through this image, the attributes previously associated with Prester John are now repurposed and adhered to new interests in humanism and cartography. The textual and visual attributions to Ptolemy create an aura within the text that mystifies the diffuse authorship of the work, and that subsume a fascination with suppressed Catholic ritual into a pseudo-scientific discourse. In addition, the association with Ptolemy serves as symbol of humanist learning, now popularized as part of an emerging social identity of non-clerical literacy and vernacular-based learning. Through the appropriation and symbolic geographical capital of Ptolemy, the print history of the *Kalender* and *Compost* provides a hidden transcript[85] of populist cartographic transmission not defined by map production.

While earlier editions of the *Kalender* by de Worde and Pynson were produced for a well-to-do readership, Wyer's affordable octavo editions were aimed at a mass market readership. Wyer's re-purposing of the book to trade on the growing cache of "Ptholomeus" is cast by Bennett as a "theft" of the earlier material.[86] But this charge of thievery, aside from its anachronistic concept of authorial ownership in early modern print, does not acknowledge the public and collaboratively constructed nature of such a multi-text by "authors," publishers, and printers. Further, the almanac genre, like that of geography writing or map-making, constituted a flexible form such that printers continually integrated new material and expurgated the outdated, in much the same way that Caxton saw fit to update his edition of *The Myrrour of the World* (see Chapter 2). Indeed, with their scientific content, it was incumbent upon a printer to bring the work up to date, especially in an environment in which geographic findings were continually being brought to light as a result of new expeditions and new maps. As early as the Pynson edition of 1506, the claim is made that the text includes "many goodly addicyons newly putte therto"[87] and such claims are later incorporated into the titles of some editions, emphasizing this impulse: "Here beginneth the kalender of shepardes newly augmented and corrected."[88]

In Wyer's editions of *The Compost of Ptholomeus*, he not only strengthened the association of the verbal and visual text with Ptolemy, but also incorporated specifically geographical information. For example, in Wyer's (1540) edition, he concludes the astronomical section of the text with the phrase, "Thus endeth the Compost of Ptholomeus Prynce of Astronomye," but then continues the text with a new component: "hereafter foloweth the Rutter, of the dystaunces from one Porte or Countree to another. And fyrst of the compasse of Englande."[89] Wyer thus "augments" the traditional text with the addition of a "rutter," a navigational chart of the distances between various port cities. This addition intensifies the function of the text as specifically geographical and accelerates the process of a cartographic transaction. As Matthew Boyd Goldie notes, the inclusion of rutters in fifteenth-century manuscript anthologies suggests a late medieval reading audience who were neither "pilot or captain."[90] In a similar fashion, the rutters included in Wyer's edition suggest a loose geographical narrative genre for a general reading audience. Interestingly, while the (1528) publication of *The Rutter of the Sea* provided nautical distances in French leagues, Wyer's rutter is, instead, set forth in English "myles," thus continuing the process of naturalizing and "Englishing" the text. The text states that the compass of England is "rounde aboute, 16.iiii.D.iii.L. and iv. myles,"[91] then delineates the distances between Mediterranean port cities, following the practice of continental rutters. While David Waters ties the publication of English rutters with the nation's "oceanic expansion,"[92] the rutter in the *Compost* is less about actual navigational guidance than the signaling of an interest in geography, place names, and the construct of mental maps on the part of English readers. The addition of the rutter can be compared to the way in which Thomas Malory had earlier added quantifiable distances into his English "translation" of French romance sources to provide a geographical realism for his romance (see Chapter 1). Most significantly, the inclusion of the rutter in Wyer's *Compost* suggests that *even before* this specific geographic addition, the Kalender and the *Compost* were understood to be part of a broader geographical or spatial discourse, part of a continuum that initially mapped the sky and body, but that now also "mapped" the Earth. The coalescing factor is the link between astrology and geography, an association which locates these narratives within the popular reception of Ptolemy's *Almagest, Tetrabiblos*, and *Geography*.

Sometime between 1600 and 1638, M. Parson printed an edition entitled *The Compost of Ptolomeus* for Henry Gosson, of which a single copy is extant. Gosson had also printed, among other works, the geographically "restless" *Pericles, Prince of Tyre*, [93] partially attributed to Shakespeare. Strikingly, the Gosson edition of the *Compost* contains a newly introduced woodblock image of a world map (Figure 5.6), complete with a map legend listing continents and nations. The biblically inspired geography of "The Ten Christian Nations" that had been part of the *Kalender* tradition is now superseded by a secular spatial and geographical representation of a global Earth. Of particular interest, the text provides instructions to the reader on how to read a map:

> The world is duided into foure parts, which are these, Affrica, Europe, Asia, and America; Each of which parts, are againe diuided into seuerall Kingdomes or Countries, as you may perceiue in the following figure, or Mapp of the Earth.
>
> Looke where you see the figure of 2 that is Affrica, the figure of 3 is Europe, the figure of 4 is Asia, and the figure of 5 is America.
>
> Again, each of these aforesaid, are deuided into seuerall Kingdomes and Prouines, as you may perceiue by the Letters and figures there

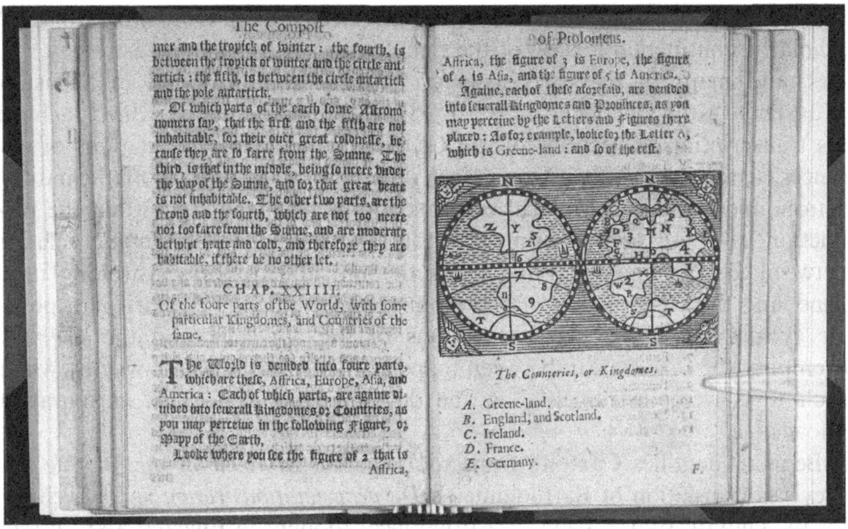

The Compost

mer and the tropick of winter: the fourth, is betweene the tropick of winter and the circle antartick: the fifth, is betweene the circle antartick and the pole antartick.

Of which parts of the earth some Astronomers say, that the first and the fifth are not inhabitable, for their ouer great coldnesse, because they are so farre from the Sunne. The third, is that in the middle, being so neere vnder the way of the Sunne, and for that great heate is not inhabitable. The other two parts, are the second and the fourth, which are not too neere nor too farre from the Sunne, and are moderate betwixt heate and cold, and therefore they are habitable, if there be no other let.

CHAP. XXIIII.

Of the foure parts of the World, with some particular Kingdomes, and Countries of the same.

The World is deuided into foure parts, which are these, Affrica, Europe, Asia, and America: Each of which parts, are againe diuided into seuerall Kingdomes or Countries, as you may perceiue in the following Figure, or Mapp of the Earth,

Looke where you see the figure of 2 that is Affrica,

of Ptolomeus.

Affrica, the figure of 3 is Europe, the figure of 4 is Asia, and the figure of 5 is America.

Againe, each of these aforesaid, are deuided into seuerall Kingdomes and Prouinces, as you may perceiue by the Letters and Figures there placed: As for example, looke for the Letter A, which is Greene-land: and so of the rest.

The Countries, or Kingdomes.

A. Greene-land.
B. England, and Scotland.
C. Ireland.
D. France.
E. Germany.

Figure 5.6 The Compost of Ptolomeus, M. Parson for Henry Gosson (1600/1638?) (STC, 2nd ed., 20482)
Source: © British Library Board/Bridgerman Images.

placed: As for example, looke for the Letter A, which is Green-land: and so of the rest.[94]

The map itself is a simple woodblock image far less sophisticated than earlier Continental map productions or near-contemporaneous English map production such as Wright's *Chart of the World.* But the significance of the addition of the map into the verbal text, like the addition of the rutter, is profound, signaling that the almanac tradition had *already* been considered spatialized discourse by its readers and printers. The didactic map-reading instructions recall the tone of John Rastell's geographical play (see Chapter 4) in which the "fygure" of the globe is displayed so as to educate the audience as to the sphericity of the world, proclaiming "The great worlde beholde."[95] In the *Kalender*, the traditional cyclical and time-based calendar of holy days that marked the earliest print editions was gradually supplanted by a visual and spatial map of the world. As Christopher Wild has noted, cartography resulted "in spatial representations of increasing distance from lived everyday experience," such that the legend requires a form of spatial translation for the reader. Despite the title of "*The Compost*,"[96] a measurement of time, the governing paradigm is no longer time, but space.

To be sure, the connection to specific Ptolemaic scientific principles in these popular texts is tenuous and popularized, but editions of the *Kalender* and *Compost* provide a record of the graduate textual and visual integration of the new geography into English popular culture in the era of Copernicus' (1543) *De revolutionibus oribum coelestium* [*Of the Revolutions of the Celestial*

Spheres], Cabot's world map of 1549, and Mercator's (1569) world map. A scholarly, continental analog of the rich context of astronomy, astrology, rutters, and maps of the early-sixteenth century can be found in Sebastian Münster's "Kollegienbuch" (1515–1518), a lecture notebook which includes his transcription of sections of the *Geography*, 44 maps, calculations of distances, a calendar with astronomical illustrations, as well as mathematical, astronomical and geographical excerpts. Gautier-Dalché argues that the production of these compilation texts form part of the "continuing relation between the Ptolemaic text and maps, theoretical astronomy, astrology, medicine, and the construction of astronomical instruments."[97] Albeit in a popular culture form, the English editions of the *Kalender* and the *Compost* exemplify the relationship between maps and astrology and the way by which such texts were instrumental in the development of populist cartographic knowledge.

Bennett identifies Caxton's translation of *The Myrrour of the Worlde* and Trevisa's translation of Bartholomew's *De proprietatibus rerum* as key vehicles for the cultural transmission of astronomical thought in England and, further, recognizes the continuation of that stream in the print trajectory of the *Kalender.*[98] In contrast to their shoddy reputation as moralizing, pseudo-science texts, when read chronologically and comprehensively, the multiple editions of the *Kalender* and the *Compost* provide a transcript of changing geographic thought in England within a populist-literary context. Their concern with popularized matter of "poets and astronomers" locates them as vital components of both English literary and spatial histories.

Class, learning, literacies

Although the *Kalender* and *Compost* are sometimes described as low-culture texts, Martha Driver has shown that editions of the *Le compost et Kalendrier des bergiers* "were owned in France and England by aristocratic and middle-class readers."[99] Pynson, official printer to Henry VIII, published editions of the *Kalender* in England as part of his printing œuvre, and Shinn's research reveals that the audience for these editions was one which "included nobility and common husbandmen."[100] From its initially elite beginning as a high-quality quarto, the *Kalender* formed part of the trajectory of the widening reach of print. Shinn, for example, situates English almanacs as part of "a publishing phenomenon closely associated with the rise of English print and the rise of the English reader."[101] Representing the early branch of this trajectory, Wynkyn de Worde's (1528) edition—a lovely, well-designed quarto edition presents the Catholic holy day calendar, complete with saints' days in red ink and the phases of the moon. As England moves more strongly toward Protestantism, the visual elements of the Catholic calendar are superseded by images of Ptolemy, Dame Astronomy, and astronomical instruments such as the quadrant, cross-staff, and armillary circle, a symbol of both cosmic harmony and scientific observation.[102]

Bennett notes that editions of the *Kalender* and *Compost* "with their many woodcuts and variety of information," served an audience similar to that of Caxton's *Myrrour* and "met a popular need for information and instruction."[103] In Copland's translation of the *Kalender of Shepherds*, printed by Wynkyn de Worde in 1508 and 1511,[104] evidence of an interest in new learning calls into question the oft-suggested premise of a geographically isolationist or intellectually conservative medieval perspective on the part of English readers.[105] In his rationale for the production of his text, Copland states that:

> remembrynge how the people desire to here and se newe thynges I shewed the sayd boke vnto my worhypful mayster Wynkyn de Worde at whose commaundement and Instygacyon I Roberte Coplande haue me applied directly to translate it out of frensshe agayne in to our maternall tonge after the capacyte of myne vnderstandynge accordynge to myne Auctoure.[106]

This trope of English interest in learning "newe thynges" is reinforced in subsequent editions. In Wyer's editions, for example, the frontispiece announces the transformation of new learning from foreign to domestic as well as the work's universal, non-elite appeal: "Translated out of Frenche in to Englyshe. for euery person, that wolde haue knowledge of the Compost."[107] In Gosson's edition of the *Compost* in the early 1600s, the appeal to a broad vernacular readership is again emphasized: the frontispiece proclaims that the text is "Very necessary and profitable for all such as desire the knowledge of the famous Art of *Astronomie*."[108] English editions of the *Kalender* and *Compost* formed a print genre of cosmological vernacular in which astrological and geographical space are linked to humankind *through the practice of reading*. Through the cultural work of these image-rich texts, the English reader is situated at the center of the universe.

Images and the practice of reading

Just as the verbal text is a fluid multi-text of the *Kalender* and *Compost* tradition, so too is the corpus of accompanying images. A strong inter-locking visual transmission runs through William Caxton's (1481) and (1490) editions of the *Myrrour*, reprinted by Andrew Lawrence,[109] and the myriad editions of the *Kalender* and *Compost*. Driver has demonstrated that the images in the *Kalender* "were particularly influential on English book illustration and were copied and recopied in a variety of contexts."[110] The "heavily illustrated" English editions, were derived "from woodcut models in the French editions."[111] In Lawrence's edition of the *Myrrour*, for example, he used Caxton's text but added a dense illustrative scheme of woodblock images, many of which were also associated with contemporary editions of the *Kalender*. Lawrence used stock images to populate his pages rather than Caxton's more careful practice of specifically illustrating the principles discussed in the text. Out of the early *Kalender* editions and Lawrence's *Myrrour*, a set of images—shepherd, scholar, and Ptolemy—demonstrate a progressive shift from the religious authority of natural philosophy to that of humanist astronomy.

The scholar, the shepherd, and Ptolemy

William Caxton authenticated his edition of the *Myrrour* with three images: the traditional medieval schoolmaster, the solitary scholar seated in his study, and the Eastern astronomer—possibly a representation of Ptolemy—all derived from images in his manuscript source. The early *Kalender* editions also utilized authenticating images of wisdom—both the learned scholar in his study but also the "natural" wise shepherd/astronomer under the night sky. The woodblock image of the seated scholar on the frontispiece of Lawrence's (1527) edition of the *Mirrour* had previously been used by Marchant for his (1493) edition of *Le Compost et Kalendrier de Bergiers* and was subsequently carried over into English editions of the *Kalender*, including Verard's (1503) edition of the *Kalender*, and de Worde's (1528) edition. In each of these editions, the image of the scholar augments and yet stands in tension with the authenticating figure of the verbal text—that of the shepherd. The Pynson (1506) edition of the *Kalender* includes the images of both scholar and shepherd,[112] despite the text's identification of the putative source of the *Kalender* as that of an uneducated shepherd, a kind of Caedmon-type persona,[113] "whiche was no clerke ne vunderstode no manere of scrypture nor wretynge but only by his natural wyt."[114] But via the image of the reading scholar, readers of the *Kalender* could participate in an aspirational identification with the process of scholarship, affirming that the reader, too, is engaged in the powerful symbolic action of contemplation and learning. The text and images suggest that even a lay reader could construct a "scholarly self," one who has the space or solitude requisite for reading and learning.[115] Reading itself is intimated as an occult process by which to gain the knowledge of the cosmos—via the transference of the philosopher, scholar, or ancient shepherd— that is now accessible to the everyday reader. As Gadi Algazi has observed, the scholar's study was a new architectural space in late medieval and early modern Europe such that "making room for engagement in private study" necessitated a concomitant negotiation of a "social arrangement" for such intellectual pursuits.[116] The study, Algazi argues, evolved from "a prestige object to a common element of middle-class housing arrangements" available to "humble teachers and country pastors" as well.[117] It is no surprise that the physical structure of a study in the homes of such middle-class readers developed, by and large, as a concurrent phenomenon with early print. The readers of the *Kalender* might not have had the actual, physical space of study, but, via the image of the scholar or Ptolemy in the text, they could participate in a symbolic and aspirational identification with the social role of humanist scholar in a space not directly mediated by an institution such as the monastery or the university.

The images of shepherd, scholar, and Ptolemy narrated an experience of new learning and the articulation of a scholarly self for a broad reading audience. Neither the woodblock image of the shepherd or scholar carries over once the text is renamed *The Compost of Ptholomeus*; instead, the symbolic function previously invested in the image of learning—that of the ancient shepherd or humanist scholar—is now conflated into the primary image of Claudius Ptolemy, "prynce of Astronomye" and associated with cartography. Thus, the accompanying woodcuts display a visual progression of the integration of Ptolemaic thought into English reading culture—through the medium of the book rather than the map.

The "crafty figure of the worlde"

The eventual inclusion of a world map in the Gosson edition of the *Compost* is pre-figured in the *Kalender* tradition by the "crafty figure of the worlde," an image that appears in many editions of the *Kalender* as well as in Lawrence's edition of *Myrrour of the World*—a pictorial representation of the Earth as represented by an agricultural scene set in the center of the universe.[118] In the 1506 edition by Pynson, the "crafty figure of the worlde" precedes the chapter concerned with "the.xii.synges goyne aboute/ and also of the mouynges of the heuens with the planettes."[119] This transitional astronomical/cartographic image merges various strands of global representation—the pictogram tradition of medieval maps, a geocentric universe, and the intimate connection between celestial and terrestrial cosmology, that is, between astronomy and cartography. The "crafty figure of the worlde" (Figure 5.7) makes clear why it is that editions of the *Kalender* and *Compost* prepared the ground, so to speak, for the introduction of a world map into later editions of the *Compost*.

Cartographic everyman and everywoman

Two of the most common images of the *Kalender* are that of a male and female in contemporary attire, often with blank banderoles above them (Figure 5.8). Driver points out how the factotum image of the male and female figures, composite woodcuts used and reused in multiple early print editions, "traverse the printed page over a sixty-year span—the leading man and leading lady, if you will, of early print illustration."[120] Driver further notes that their "persistent repetition implies a visual language with which sixteenth-century readers were familiar."[121] These male and female images appear so frequently throughout editions of the *Kalender* that Driver dubs them the "Everyman" and "Everywoman" of early modern English print.[122] These figures, I suggest, also indicate an implied reading community for the *Kalender* and *Compost* of both men and women. Ultimately, they stand in for the English map reader, a cosmological everyperson.

In addition to the images of the "Everywoman," the figure of "Dame Astronomy" appears alongside Ptolemy's image on the frontispiece of the

Figure 5.7 “Crafty figure of the world,” in *Here Begynneth the Kalender of Shepardes*, Julian Notary, c. 1518 (STC 22410)
Source: RB 69474, The Huntington Library, California.

Wyer editions, a visual muse, collaborator, and partner. She is denied any titular credit, but her pairing with Ptolemy visually suggests that the work is indeed for “to euery person who would haue knowledge of the Compost,” both male and female. While upper-class women in the Tudor period were often literate, basic female literacy as represented by the ability to write one’s name is estimated at only 8 percent (25 percent for males) by 1600.[123] However, the strong visual cues supplied by the woodcut images and the ballad-like verses would augment the usability of the *Kalender* for readers with basic or partial literacy skills. In addition, a growing share of the population—both men and women—could read “simple passages from ballads and elementary religious texts”[124] congruent with the many woodcut images and verse passages in the *Kalender*. An implied female audience (readers or listeners) is

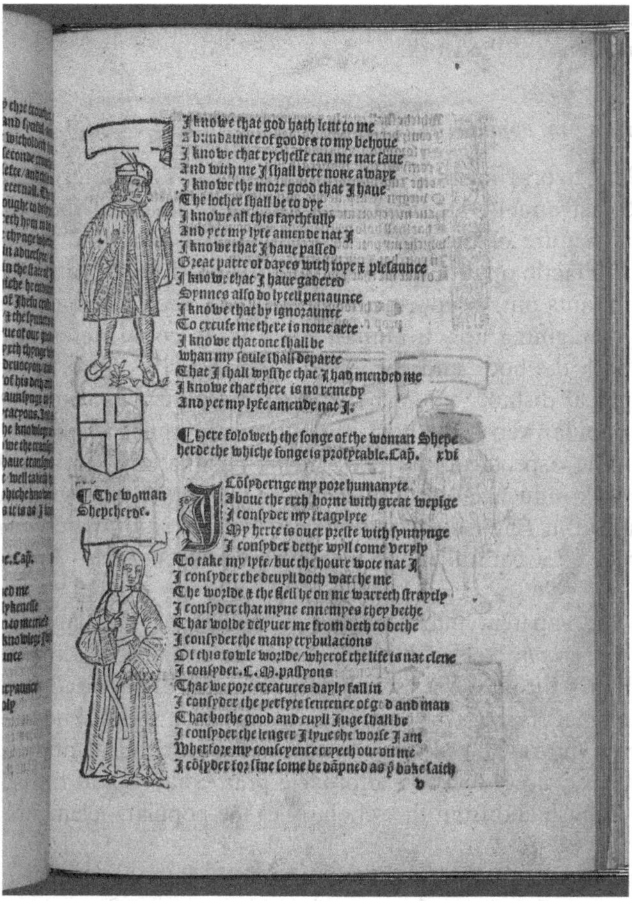

Figure 5.8 "Everyman and Everywoman," in *Here Begynneth the Kalender of Shepardes*, Julian Notary, c. 1518 (STC 22410)
Source: RB 69474, The Huntington Library, California.

suggested in the introduction of the poem "The Horner:" "Nowe euere man and woman ought to cease of theyr sinnes at the sownynge of a dredable horne."[125] In the multiple occasions where the reader is directly addressed, there are similarly inclusive references to both men and women as the recipients of the moral and intellectual matter of the text. In addition, along with "the ballade of a wyse man," some editions include the gender balancing verse, "The ballade of the woman Shepherde."[126]

In sum, the images that accompany editions of the *Kalender* and *Compost* are potent co-agents of English readership, giving evidence of an increasing interest in Ptolemaic geography and indicative of an inclusion of female as well as male readers. Into this rich textual and visual polystream, populist cartography is introduced in the form of a world map, one meant for the very

"Everyman" and "Everywoman" who illustrate the text and who were the implicit "readers" of the work.

Science writing or popular literature?

Despite their hybrid form as "miscellanies" and their intermingling of the poetic with astronomy, Driver argues for the primacy of the "didactic and salvational" nature of the work over that of the scientific.[127] Certainly, its moral and didactic qualities are deeply embedded in the text, but as Kellie Robertson points out, poetry of the period was "seen to be a highly suitable vehicle for imagining how a human subjectivity could potentially be influenced by both the body and the world it inhabits."[128] Thus, the itemization of geographical distance could be contained within a text inclusive of morality and popular verse: in short, the moralistic milieu does not negate the quasi-scientific aspects of the text, including anatomy and health advice. While the folk-influenced, moralistic genre has served to segregate the work from inclusion in English cartographic history, it is ultimately the scientific that emerges as a central focus in the derivative branch, *The Compost of Ptolemy.*

Despite the apparent public appetite for scientific learning, no edition of Ptolemy's *Geography* was published in England in the fifteenth or sixteenth centuries. Partly, the reason rests with the availability of imported Latinate books for educated readers, but also with the way in which the *Kalender*, followed by the *Compost*, provided general principles of astronomical and geographical thought in easily accessible and affordable print editions that translated—from French to English and from the scholarly to the popular—elements of the new geography.

The *Kalender* is equally considered outside the boundaries of "literature," a result of the moral and pseudo-scientific subject matter. Capp goes so far as to assert that "the appeal of the almanac did not lie in any literary value."[129] And yet, there can be little doubt that the *Kalender* functioned within the sphere of popular literature and thus this begs the question of the boundary of what constitutes a "literary" text. In urging the value of reading and knowledge acquisition for moral gain, the text includes moral as well as "light," some might say, doggerel, verse in the tradition of the ballad and broadside. For example,

> The songe of deth to all cresten people" (Though my pycture be nat to your pleasaunce/ And if ye thynke that it be dredable/ Take in worthe for surely in substaunce/ the sight of it may to you be profitable,)

But beyond "profitable" morality, the various poems also announce the text as an aesthetic experience of "pleasure" for its "Everyman" and "Everywoman" readers. For example, in "The Horner," the narrator calls out with a boisterous, gallows humor:

Ho, ho you blynde folke, derked in the clowde
Of ignoraunt fumes, thycke and mytsycall
Take heede of my horne, totynge all allowed
With boystrous sownes, and blastes boryall[130]

Spenser, it may be noted, includes a similarly boisterous "hey ho!" poem in his August "Aeglogue," suggesting the spirited sung ballad of folk gatherings. The potential fascination of the Horner poem to the early modern reader is heightened by the sole image of a Black person in the text, a complex racial exoticism. "The Horner" concludes with an alliterative stanza, drawing upon the traditional and by-now archaic English poetic form:

Flee faynt falshod, fyckell fowle and fell
Flee fatall flatterers, full of fayrenesse[131]

The matter of the poem is moral, but the rhymed and alliterative verse creates an aesthetic experience distinct from that of prose sermons or biblical reading, one that draws upon traditional English literary practices. Other poems in the almanac include what might be considered popular scientific matter; for example, "The gouernaunce of helth" ("Who wyll be hole & kepe himself fro sychenes/ And resyste stroke of pestilence"), and "the nature of the.vii. planettes" (To rede of shepardes besynes/ And special of the planettes seuen").[132]

With its witty and earthy agricultural association, the poem, "An assaute against a snayle,"[133] humorously locates the "power" of a woman against a snail, and notes that "The woman speaketh with an hardy courage."[134] In the accompanying image, the woman, distaff in hand, succeeds in vanquishing the garden intruder whereas the brute strength of the nearby men-at-arms has failed, a fitting symbol of the literary transition from the noble *chanson de geste* to that of an agricultural almanac.

In addition to the inclusion of verse in the *Kalender*, the reader is ever-present, placing reading itself as a meta-concern of the work and establishing the *Kalender* as a site of subjectivity for the common reader. In Notary's edition of the *Kalender*, the reader is addressed in verse by "the auctore" in the book's concluding verse:

And remember /Reders wherever þe go
That hony is sweet/ but connynge is sweeter
Caton the greate Clerke/ sometimes sayde so
Nowe gold is good/ and learnyng much better
Yet many ful good be/ that neuer knewe letter
And yet virtuous/ none can be lyuynge
But first of Priests and Clerkes/ they must haue learning.[135]

The allusion to the classical writer Cato reinforces the ethos that the work is drawn from both "poets and astronomers," a text that is scientific, moral and literary, to be read for both "profyte and pleasure." The use of popular verse reflects the propensity of fifteenth- and sixteenth-century writers and printers to view science and literature as from a common well and popular poetic forms as a suitable vehicle for the transmission of geographic thought.

The *Kalender* exhibits other features that may be considered "literary," including the proposition of a fictional third-person narrator of the work and the use of banderoles to signify dialogue. The *Kalender* retells biblical stories and narrativizes astrological knowledge via a central character (the shepherd or Ptolemy) who has "said" what is now ostensibly recorded wisdom. In addition, the blank banderoles above the woodblock images of the Everyman and Everywoman imply a dialogue within the text and the trope of learning via "conversation," a pedagogy in Socratic dialogue and utilized, for example, in the mid-century scientific texts *Cosmological Glasse* and *Castle of Knowledge*.[136] Through the "character" of the shepherd or of Ptolemy, a fictional dialogue with a master teacher is presented to non-clerical "Reders:"

> These prouerbs be good to marke
> The whiche foloweth in this boke
> Be thou neuer so greate a Clerke
> Disdayne not on them to loke[137]

Through the process of vernacularization, the text offers a cosmological equivalent to the cultural work being taken up by that of the English Bible. The argument for a broad readership is of central concern in the Preface to the *King James Bible* which emphasizes that "there should be one more exact Translation of the Holy Scriptures into the English tongue [in order to make] knowen unto the people," and that if God "sette his word before us," it is imperative for the "gentil Reader" to "reade it."[138] So, too, the *Kalender* and *Compost* "translated" Ptolemaic cosmology and geography into a narrative with recognizable literary forms—verse, character, dialogue—for English readers. This narrative, moreover, becomes *a story of reading*; that is, the text presents an account of a developing reading subjectivity, a hypothetical English autodidact gaining access, via the process of reading, to the knowledge of humanist scholars and of the cosmos, the properties of spatiality and soul.

That the *Kalender* intersected with literary culture is further attested in *Robert Laneham's Letter* which enumerated the purported books and ballads of "Captain Cox's List." Nineteenth-century literary editor Frederick J. Furnivall notes:

> In the list of the Coventry mason's library, Captain Cox's books, plays, ballads, etc., Laneham no doubt describes the contents of his own bookcase, the literature of which the middle-class man of his day was reared and fed, the books which men of Shakespeare's position, and many, if not most, of his audience would read.[139]

Indeed, Shakespeare seized upon the popularity of the almanac, as Shinn notes, such that it "was often portrayed on the stages of the Elizabethan playhouses as a prop for a foolish or gullible character who is too busy looking at the stars above his own head to see the gaping chasm looming beneath his feet."[140] While this may be true by the Elizabethan era, the *Kalender* was not initially the signifier for naïveté with which it came to be associated. Instead, it was part of a spiritualized practice of new literacy, opening to everyday readers the secrets of "heavenly" texts—be they scriptural or cosmological, positioning the reader at the center of the cosmos.

The *Kalender* tradition eventually inspired both scientific and literary progeny: the 1600 Gosson edition, replete with one of the earliest populist maps of the world printed in England, and Edmund Spenser's *Calender of the Shepheardes*, a literary re-imagination of the popular English almanac. The eventual separation of the poet and the cartographer that marks modernity was once conjoined in the textual history of the *Kalender.*

"the olde Astrologers and Philosophers:" Spenser's **Shepheardes Calender** *and the parting of literature and cartography*

Amidst the already complex and multi-faceted textual history of the *Kalender of Shepherds,* [141] near the end of its long print history, the multi-form work served as an inspiration for a new artistic expression. Fusing classical and English literary forms, Spenser drew upon the almanac tradition as inspiration for his innovative use of the structuring device of the calendrical form, the figure of the wise rustic as a narrator, and the tension between Catholicism and Protestantism.[142] In his bid to become "the new Poete" of England,[143] Spenser transformed the genre of the popular almanac into a classically-inspired poem that took to heart the great theme of the pastoral—that of love. As the shepherd Colin Cloute laments, "Tho couth I sing of love," (June, 41). With *The Sheppeardes Calender,* published in 1579, Spenser sought to "transform the almanac tradition"[144] and intuitively grasped the way in which this tradition *had itself* functioned as a part of popular English literary culture for over a century. Spenser utilizes some of the themes and motifs that formed an integral part of the agrarian *Kalender* tradition, among them the motif of "natural" or astrological wisdom, the attention to the development of the English language and the reader, the allegory of the year as the span of a man's life, the reliance on proverbial sayings, and the "mixed genre" compendium of verse and prose—a "gallimaufry or hodgepodge" as Spenser characterizes his pastoral eclogue.[145]

Lynn Staley Johnson argues that "Spenser gives evidence of seeking to create a new literature, a new aesthetic for a new age."[146] He does so, however, through the imaginative "re-discovery" of an archaic Englishness, in both language and themes, as spoken by the poem's narrators, including Colin, Cuddie, Hobinoll, and the other rustics. As Hobinoll declares:

Colin, to heare thy rymes and roundelayes,
Which thou were wont on wastfull hylls to singe.
(June, 49–50)

Though critics claim Spenser "saw his career in a Virgilian light,"[147] in addition to Virgil's *Eclogues*, Spenser's new aesthetic clearly takes inspiration from the folk-culture of the *Kalender of Shepherds* with its "rymes and roundelayes," suggesting that Spenser was attuned to the aesthetic possibilities of such a nativist and protean form. As Staley Johnson notes, Spenser's *Shepheardes Calender*, like its titular predecessor, presents a mixed genre of "religious, literary, social, and political" elements[148] that create a certain "discord" by including uncertainties and conflicting information,[149] a strategy, as we have seen, that harkens back to Caxton's editorial process for the *Myrrour* and *Polychronicon* (see Chapter 2). In Spenser's poem. Colin Cloute speaks to such uncertainties when he laments that "Art made a myrrhour to behold my plight" (January, 20), referring not only to the gloomy January landscape and his unrequited love but also to the poem's concern with the "plight" of the English language and Spenser's concern regarding the "infiltration" of foreign languages, the centrality of "fayth" amid the ongoing tension between Catholicism and Protestantism, the rural values of "the good shepherd" in tension with "proude and ambitous pastours," and, ultimately, the question of what constituted "old Britain" (June, Glosse) amid an emerging globalized English national identity. Here, the "myrrour of the world" promised in Caxton's geographical text has become fractured and subjective.

While Spenser's pastoral poem is most often read in the context of classical antecedents such as Virgil's *Eclogues*, it also provides a revealing insight into the reception of the *Kalender* as part of English literary culture. The classical eclogue tradition did not contain the division into monthly segments seen in Spenser's poem, a structural form clearly attributable to the Catholic tradition of Books of Hours and the *Kalender* tradition. As Shinn argues, not only does Spenser's *Shepheardes Calender* self-consciously distinguish itself from the classical eclogue and pastoral, but also "aligns itself to the contemporary debates surrounding calendrical reform."[150] In his effort to create a work of "authentic" archaism, Spenser refocuses his eclogue on the temporal cycle of months rather than on the spatial elements that had gradually become the domain of technical geographical writing. Tamsin Badcoe challenges Jeffrey N. Peter's assertion that "space appears to have lost its poetry" in the seventeenth century,[151] arguing that intense interest in the late sixteenth century in navigation and mapping, and in the "writing and re-writing of cosmos and nation," were central to shaping Spenser's "literary geographies" and his preoccupation with "moving forms" and "writing space."[152] However, Spenser's geographies focus on a different construction of knowledge and self distinct from the increasingly scientific measurement of space found in maps, gloss, and narratives of exploration and conquest. Badcoe relates an anecdote from Spenser's friend Gabriel Harvey in which Harvey notes that "Spenser himself is ashamed, though he is not completely ignorant of globe and astrolabe, of the difficulty he has with astronomical rules, tables, and instruments."[153] Unlike Chaucer before him, or the tradition of the poet/astronomer, Spenser's work is primarily in the realm of the poet/philosopher and, as Bernard Kline argues, "undertakes to question the epistemological significance of space."[154] Badcoe eloquently identifies Spenser's use of "wayward movements" in his major poems, his "worldmaking," and his questioning of "objective vision itself."[155] For Badcoe, Spenser is "a spatial maker."[156] But in contrast to works such as Cuningham's *Comographical Glasse*, Abraham Ortelius' (1570) world atlas, and even the purported homespun folk wisdom of the *Kalender*, Spenser's *Calender* appropriates the cartographic process of reckoning the self in the world to transform wayfinding into a journey of *poesis.* Spenser reverses the movement from time to space noted in the *Kalender* tradition and reconstitutes an "ecotonal," or transitional habitat[157] the poet occupies between country and city, amid the ancient wheel of a year's time. Shinn suggests that the almanac "was a format ripe for transformation"[158] and Spenser's *Calender* abjures the increasingly cartographical impulse of the *Kalender* tradition to reclaim its medieval emphasis on the marking of time and the integration of space, writing, and the imagination. Just as the *Kalender* should rightly be read as part of English cartographic writing, so its influence on Spenser's *Shepheardes Calender* suggests it must also be read as part of English literary history, the inspiration for Spenser to apply "an olde name to a new work."[159]

The composting of Ptolemy

At the close of the sixteenth century, the *métier* of the *Kalender* continued forward in two increasingly distinct strands: one in the form of Spenser's new poetic aesthetic of space and the other in the form of Gosson's cartographic edition of *The Compost of Ptolomeus.* From this juncture, literature and geography largely part company: the easy cohabitation of poets and astronomers that marked the *Kalender* tradition is gradually made unfamiliar amid the developing scientific discipline of geography. Cartography—at least in terms of the visual representation of world space—disengages from its once-literary, philosophical, and cosmological roots.

Although the term "compost" in its sixteenth-century use signified a computation of calendrical and astronomical information, the term "compost" in its modern sense is also an apropos descriptor of the multi-text that resulted in *The Compost of Ptolemy*: a slowly heated mass of old and new knowledge, forms, and images, continually redigested into contingent representations of the world. Intriguingly, the mixed-genre form employed in editions of the *Kalender* is familiar to postmodern readers via works such as W.G. Sebald's *The Rings of Saturn*, Juliana Bruno's *The Atlas of Emotion*, and Denis Wood's *Everything Sings: Maps for a Narrative Atlas*, works which explore the intermingling of the scientific and literary, the measured and the felt, the presence of the reader, and the collage effect of text and image.[160] Seeking an aesthetic of "poets and astronomers," Wood recalls the process of his experimentations in neighborhood-based cartography and his aesthetic breakthrough: "[t]hat's when I knew we could write poems in maps, and I began thinking seriously about a poetics of cartography."[161] In a similar vein, commenting on research findings in quantum physics, University of California

Berkeley mathematics professor Dr. Edward Frenkel writes that there "is really no escape from the mysterious—some might say, mystical—nature of the quantum world,"[162] a twenty-first-century perspective on science largely in accord with that of the *Kalender* and the *Compost*. Frenkel concludes his essay on the "paradoxes of quantum physics" with a quote from the sacred Vedas: "as is the human body so is the cosmic body,"[163] echoing the conclusion of many editions of the *Kalender* and *Compost:* "And therefore is Man called the Little World."[164] After the long disciplinary separation of cartography and literature during the Enlightenment, the *Kalender* and the *Compost* deserve a reappraisal: together they trace an important poetic geography which generated new literary forms and new maps for English readers in the early modern period.

Acknowledgments

Heartfelt thanks to the Huntington Library for the Gilbert and Ursual Farfel Fellowship which allowed me to work with their collection of early modern scientific books and to the National Endowment for the Humanities Summer Seminar in Antwerp and Oxford, under the direction of Dr. John King, which enabled me to study multiple copies of the *Kalender of Shepherds* and *The Compost of Ptolemy.*

Notes

1 For an explication of the term "multi-text," see Iain Macleod Higgins, *Writing East: The "Travels" of Sir John Mandeville* (Philadelphia, PA: University of Pennsylvania Press, 1997). Higgins uses this term to refer to the body of editions, translations, and redactions that make up the *Travels* of John Mandeville.

2 For the sake of consistency, the title the *Kalender* is used throughout this chapter; however, the spelling of the title varies between editions.

3 A.E.B. Coldiron, *Printers Without Borders: Translation and Textuality in the Renaissance* (Cambridge; Cambridge University Press, 2015), 10.

4 Ibid., 49.

5 Takami Matsuda, *The Kalender of Shepardes*, Keio University. Available at: www.flet.keio.ac.jp/~matsuda/ks/185b/html/about.html#KS. See also Takami Matsuda and Satoko Tokunaga, "A Composite Copy of the *Kalender of Shepherdes* in Keio University Library," in *Codices Keionenses on Western Manuscript and Early Printed Books in Keio University Library*, ed. Takami Matsuda (Tokyo: Keio University Press, 2005), 119–208.

6 Oskar H. Sommers, *The Kalender of Shepherdes: The Edition of Paris 1503 in Photographic Facsimile, A Faithful Reprint of R. Pynson's Edition of London 1506*, vol. 1: *Prolegomena* (London: Kegan Paul, Trench, Trubner & Co, Ltd, 1892), 55.

7 Jacques Derrida, "Signature, Event, Context," *Glyph* 1 (1977): 172–197.

8 *The Compost of Ptolomeus*, printed by M. Parsons for Henry Gosson (London, 1600/1638?), STC 20482, sig. D.8(r).

9 David Woodward, "Cartography and the Renaissance: Continuity and Change," in *The History of Cartography*, vol. 3: *Cartography in the European Renaissance*, Part 1, ed. David Woodward (Chicago: University of Chicago Press, 2007), 3–24; 7.

10 Ibid., 7.
11 Maureen Perkins, *Visions of the Future: Almanacs, Time, and Cultural Change 1775–1870* (Oxford: Clarendon Press, 1996), 3.
12 *The Compost of Ptholomeus*, Robert Wyer (London, 1540?), STC 20480a, sig. O. iii(v).
13 I use this term to refer to the period dating from the early fifteenth century to the mid-sixteenth century during which time Ptolemy's *Geography* was translated into Latin from the Greek, accelerating its influence via early print.
14 Perkins, 3. For a discussion of the various editions of the *Kalender of Shepherds*, see Martha Driver, "When Is a Miscellany Not Miscellaneous? Making Sense of the *Kalender of Shepherds*," *The Yearbook of English Studies* 33 (2003): 199–214.
15 For a complete listing of editions, see Patrick Gautier-Dalché, "The Reception of Ptolemy's *Geography* (End of the Fourteenth to Beginning of the Sixteenth Century)," in *The History of Cartography*, vol. 3: *Cartography in the European Renaissance*, Part 1, ed. David Woodward (Chicago: University of Chicago Press, 2007), 285–364.
16 David Waters, *English Navigational Books, Charts, and Globes Printed Down to 1600* (Coimbra: University of Coimbra, 1985).
17 Henry Stanley Bennett, *English Books and Readers, 1475–1557: Being a Study in the History of the Book Trade from Caxton to the Incorporation of the Stationers' Company* (Cambridge: Cambridge University Press, 1969). See especially Chapter VI, "The Variety of Books."
18 Waters, 250.
19 *Here Begynneth the Kalendar of Shepherdes*, Richard Pynson (London, 1506), STC 22408, sig. A.ii(r).
20 Robert K. Batchelor, *London: The Selden Map and the Making of a Global City, 1549–1689* (Chicago: University of Chicago Press, 2013).
21 Waters, 239. Waters also notes that the editions in the 1540s had an addition by Richard Proude called *A Rutter of the North Partes* which detailed the east coast of England, 241.
22 J.A. Williams, *The Cabot Voyages and Bristol Discovery under Henry VII* (Cambridge: Hakluyt Society, 1962).
23 For a discussion of the spatial imaginary of early English rutters and their inclusion in miscellany manuscripts, see Matthew Boyd Goldie, "An Early English Rutter: The Sea and Spatial Hermeneutics in the Fourteenth and Fifteenth Centuries," *Speculum* 90:3 (July 2015): 701–727. See also Matthew Boyd Goldie, *Scribes of Space: Place in Middle English Literature and Late Medieval Science* (Ithaca, NY: Cornell University Press, 2019).
24 Waters, 257.
25 Adam Max Cohen, "Englishing the Globe: Molyneux's Globes and Shakespeare's Theatrical Career," *The Sixteenth Century Journal* 37:4 (Winter 2006): 963–984; 968.
26 Ibid., 966.
27 Bernard Capp, *English Almanacs, 1500–1800: Astrology and the Popular Press* (Ithaca, NY: Cornell University Press, 1979), 23.
28 Martha Driver, *The Image in Print: Book Illustration in Late Medieval England and Its Sources* (London: British Library, 2004), 170.
29 Abigail Shinn, "'Extraordinary discourses of vnnecessarie matter': Spenser's *Shepheardes Calender* and the Almanac Tradition," in *Literature and Popular Culture in Early Modern England*, eds. Matthew Dimmock and Andrew Hadfield (Farnham: Ashgate, 2009), 137–149; 138.
30 Francis R. Johnson, *Astronomical Thought in Renaissance England: A Study of English Scientific Writing from 1500 to 1645* (Baltimore, MD: Johns Hopkins University Press, 1937, reprinted 1968), 3.

31 In addition to Shinn, see, for example, David Davis, "Images on the Move: The Virgin, the *Kalendar of Shepherds*, and the Transmission of Woodcuts in Tudor England," *Journal of the Early Book Society for the Study of Manuscripts and Printing History* 12 (2009): 99–132, Laura William Ambrose, "Travel in Time: Local Travel Writing and Seventeenth-Century Almanacs," *Journal of Medieval and Early Modern Studies* 43:2 (Spring 2013): 419–443, and William Kuskin, "'The Loadstarre of the English Language': Spenser's *Shepheardes Calender* and the Construction of Modernity," *Textual Cultures: Texts, Contexts, Interpretation* 2:2 (2007): 9–31.
32 Johnson, 3.
33 Pynson, STC 22408, sig. A(ii)r.
34 *The Kale[n]der of Shepeherdes*, Wynkyn de Worde (London, 1528), STC 22411, sig. O.viii(v).
35 Perkins, 1.
36 Driver, *The Image in Print*, 5.
37 Capp, 31.
38 Pynson, STC 22408, sig. O.vi(v).
39 Perkins, 4.
40 de Worde, STC 22411.
41 Pynson, STC 22408, sig. N.iv(v).
42 Gosson, STC 20482, sig. A.2(v).
43 Keith D. Lilley, *City and Cosmos: The Medieval World in Urban Form* (London: Reaktion Books, 2009).
44 Capp, 25.
45 Ibid., 20.
46 de Worde, STC 22411.
47 For a discussion of the earliest English editions, see Sommer.
48 Bennett, 116.
49 *Here Begynneth the Kalender of Shepardes*, Julian Notary (London, c. 1518), STC 22410, sig. A.ii(r).
50 Driver, "Miscellany."
51 Bennett, 116.
52 In the spirit of the many print versions of the *Kalender*, I must appeal to the reader and ask forbearance if this textual and visual history remains somewhat opaque. After ten years wandering the labyrinth textual history of these editions, the author welcomes "augmentation or correction."
53 For a discussion of the spelling variants within the multiple editions of the *Kalender of Shepherds*, see Hanna Rutkowska, "Typographical and Graphomorphic Features of Five Editions of the *Kalender of Shepherds* as Elements of the Early Printers' Community of Practice," in *Communities of Practice in the History of English*, eds. Joanna Kopaczyk and Andreas H. Jucker (Philadelphia, PA: John Benjamins, 2013), 123–150.
54 *The Compost of Ptholomeus*, Robert Wyer (London, 1530?), STC 20480.
55 Henry Burrows Lathrop, *Translations from the Classics into English from Caxton to Chapman, 1477–1620* (Madison, WI: University of Wisconsin Press, 1933), 330.
56 Ptolemy, *Tetrabiblos*. F.E. Robbins, ed and trans. The Loeb Classical Library (Cambridge, MA: Harvard University Press, 1940), ix.
57 Ibid., 271.
58 Ibid., 271.
59 Notary, STC 22410, sig. A.v(v)–A.vi(r).
60 See O.A.W. Dilke, with additional material supplied by the editors, "The Culmination of Greek Cartography in Ptolemy," in *The History of Cartography: Cartography in Prehistoric, Ancient and Medieval Europe and the Mediterranean*,

vol. 1, eds. J. Brian Harley and David Woodward (Chicago: University of Chicago Press, 1987), 177–200; 180.

61 See Ptolemy, *Tetrabiblos*, Book II, Chapter 3.

62 See *The Cambridge History of Libraries in Britain and Ireland*, vol. 1: *To 1640*, ed. Elisabeth Leedham-Green (Cambridge: Cambridge University Press, 2006).

63 Roger Lovatt, "Introduction," in *The University and College Libraries of Cambridge, Corpus of British Library Catalogues 10*, ed. Peter D. Clarke (London: The British Library, 2002), xxvii–xcii; lxxviii.

64 Gautier-Dalché, 352.

65 *Septe giornate della geographia* [The Seven Days of Geography] (Florence, 1482). For the context of Berlinghieri's verse edition, see Sean Roberts, *Printing a Mediterranean World: Florence, Constantinople, and the Renaissance of Geography* (Cambridge, MA: Harvard University Press, 2013).

66 Wyer, STC 20480a, sig. S.ii(r).

67 Ibid., STC 20480a, sig. S.ii(r).

68 See Lilley, *City and Cosmos*.

69 Keith Lilley, "Mapping Cosmopolis: Moral Topographies of the Medieval City," *Environment and Planning D: Society and Space* 22 (October 2004): 681–698.

70 Davis, 100.

71 Pynson, STC 22408, sig. N.ii(r).

72 Notary, STC 22410, sig. A.v(v).

73 Ibid., STC 22410, sig. A.vi(r).

74 Ibid., STC 22410, sig. A.iii(v).

75 Ibid., sig. N.i(r).

76 Notary, sig. N.i(r).

77 Bennett, 116. See de Worde, STC 22411.

78 For the influence of paratextual elements in a literary work, see Jerome J. McGann, *A Critique of Modern Textual Criticism* (Charlottesville, VA: University of Virginia Press, 1992) and Jerome J. McGann, *The Textual Condition* (Princeton, NJ: Princeton University Press, 1991). Specific to Caxton's editing practice and paratextual meaning, see William Kuskin, *Caxton's Trace: Studies in the History of English Printing* (South Bend, IN: Notre Dame University Press, 2006).

79 Bernhard Klein, *Maps and the Writing of Space in Early Modern England and Ireland* (New York: Palgrave Macmillan, 2001).

80 Shinn, 141.

81 Wyer, STC 20480, sig. b.i(r).

82 Ibid., STC 20480a, sig. b.i(r) – b.i(v)

83 Bennett, 117.

84 *Le Co[m]post et kalendrier des bergiers*, Guy Marchant (Paris, 1493), sig. A.1(r). For a discussion of the use of this image in various texts and the influence of the *Kalender of Shepherds* on English illustration, see Driver, *The Image in Print*. This image is cataloged as DeWorde no. 1081.

85 James C. Scott, *Domination and the Arts of Resistance: Hidden Transcripts* (New Haven, CT: Yale University Press, 2008).

86 Bennett, 117. For a similar view of Wyer's editorial work, see H.B. Lathrop, "Some Rogueries of Robert Wyer," *The Library*, Third Series, v (1914): 349–364.

87 Pynson, STC 22408, sig. A.ii(r).

88 *The Shepardes Kalender*, Thomas Este for John Wally (London, 1570?) STC 22415.

89 Wyer, STC 20480a, sig. S.ii(v).

90 Goldie, "An Early English Rutter," 716.

91 Wyer, STC 20480a, sig. S.ii(v).

92 Waters, 239. Tellingly, Waters does not include editions of the *Kalender* or *Compost* in his account of English navigational publication history.

93 "Introduction: Antony and Cleopatra," in *The Norton Shakespeare*, vol. 2: *Later Plays*, ed. Stephen Greenblatt (New York: W. W. Norton, 2008), 899.

94 Gosson, STC 20482, sig. D.8(r).

95 Richard Axton, ed. *Three Rastell Plays: Four Elements, Calisto and Melebea, Gentleness and Nobility* (Cambridge: D. S. Brewer, 1979), ln 162.

96 Christopher Wild, "Cartography and the Melancholic Self," in *Space and Self in Early Modern European Cultures*, eds. David Warren Sabean and Malina Stefanovska (Toronto: University of Toronto Press, 2012), 239–257.

97 Gautier-Dalché, 352.

98 Bennett, 116.

99 Driver, *The Image in Print*, 5.

100 Shinn, 138.

101 Ibid., 142.

102 Denis Cosgrove, *Apollo's Eye: A Cartographic Genealogy of the Earth in the Western Imagination* (Baltimore, MD: Johns Hopkins University Press, 2001), 11. For a catalog and images of medieval and renaissance scientific instruments, see Epact, available at: www.mhs.ox.ac.uk/epact/.

103 Bennett, 116.

104 De Worde's (1528) edition was a version of Julian Notary's (1518) edition, itself largely based on Copland's (1508) translation (printed by de Worde) but supplemented by Pynson's (1506) edition. Clearly, the publishing history is tightly interwoven. See Bennett, 117.

105 Kathy Lavezzo, *Angels on the Edge of the World* (Ithaca, NY: Cornell University Press, 2006).

106 [*Here Begynneth the Kalendar of Shepherdes*], trans. Robert Copland, Wynkyn de Worde (London, 1511), STC 22409.5.

107 Wyer, STC 20480a.

108 Gosson, STC 20482.

109 Andrewe Lawrence, *The Myrrour and dyscrypcyon of the Worlde* (London, 1527), STC 24764.

110 Driver, "Miscellany," 206.

111 Driver, *The Image in Print*, 5.

112 Pynson, STC 22408.

113 See Bede's *Ecclesiastical History of the English People* (Oxford: Oxford University Press, 1999) in which Caedmon, an uneducated shepherd, recites the Lord's Prayer.

114 Pynson, STC 22408, sig. A.iii(r).

115 Gadi Algazi, "At the Study: Notes on the Production of the Scholarly Self," in *Space and Self in Early Modern European Cultures*, eds. David Warren Sabean and Malina Stefanovska (Toronto: University of Toronto Press, 2012), 17–50; 38. See also Gadi Algazi, "Scholars in Households: Refiguring the Learned Habitus, 1480–1550," *Science in Context* 16:1–2 (2003), (special issue: *Scientific Personae*, edited by Lorraine Daston and Otto Sibum): 9–42.

116 Algazi, "At the Study," 18.

117 Ibid., 8.

118 See Pynson, STC 22408, sig. I.vi(r) and de Worde, STC 22411, sig. P.ii(v). See also Lawrence, STC 24764, sig. g.iii(v).

119 Notary, STC 22410, sig. A.iii(r).

120 Driver, *The Image in Print*, 75.

121 Ibid., 75.

122 Ibid., 9.

123 Robert Burcholz and Newton Key, *Early Modern England: A Narrative History*, 2nd edition (London: Wiley-Blackwell, 2008); 173.
124 Ibid., 173.
125 Notary, STC 22410, sig. M.vi(r).
126 See Pynson, STC 22408, and [*The Shepardes Kalender*], [John Charlewood for] John Wally (London: 1580), STC 22416.
127 Driver, "Miscellany," 211.
128 See Kellie Robertson, "Materiality and the Hylomorphic Imagination," in *Medieval Literature: Criticism and Debates*, eds. Holly A. Crocker and D. Vance Smith (London: Routledge, 2014), 367–375; 375.
129 Capp, 23.
130 Notary, STC 22410 sig. M.vi(v).
131 Notary, STC 22410, sig. M.vi(v).
132 All of the above verses are from ibid., STC 22410.
133 See Cap. xlvii in *The Shepheards Kalendar*, V. S[immes for T. Adams?] (London, 1600?), STC 22419.
134 Simmes, STC 22419.
135 Notary, STC 22410, sig. N.viii(r).
136 William Cunningham, *Cosmographical Glasse* (London: 1559), STC 20796; Robert Recorde, *The Castle of Knowledge*, printed by Reginald Wollfe (London: 1556), STC (2nd ed.) 6119.
137 Notary, 22410, sig. N.vi(r).
138 The *King James Bible*, published by Robert Barker (London: 1611).
139 F.J. Furnivall, ed. *Laneham's Letter*, The New Shakespeare Society, Series 6, No. 14 (London: Kegan, Paul, Trench, Trubner & Co., Limited, 1890), 3.
140 Shinn, 137.
141 Edmund Spenser, "General Argument," in *The Shepheardes Calender* in *The Yale Edition of the Shorter Poems of Edmund Spenser*, eds. William A. Oram, Einar Bjorvand, Ronald Bond, Thomas H. Cain, Alexander Dunlop, and Richard Schell (New Haven, CT: Yale University Press, 1989), line 51. All subsequent quotations from *The Shepheardes Calender* are from this text.
142 Robert Lane discusses Spenser's appropriation of the popular form of the calendar in *Shepheards Device: Edmund Spenser's Shepheardes Calender and the Institutions of Elizabethan Society* (Athens, GA: University of Georgia Press, 1993).
143 Spenser, dedication to the *Shepheardes Calender*.
144 Shinn, 144.
145 Spenser, "Epistle," in *The Shepheardes Calender*, 13.
146 Lynn Staley Johnson, *The Shepheardes Calender: An Introduction* (University Park, PA: Pennsylvania State University Press, 1990), 7.
147 Oram et al., 3.
148 Staley Johnson, 6.
149 Ibid., 1.
150 Shinn, 147.
151 Jeffrey N. Peters, *Mapping Discord: Allegorical Cartography in Early Modern French Writing* (Newark, NJ: University of Delaware Press, 2004), 18.
152 Tamsin Badcoe, *Edmund Spenser and the Romance of Space* (Manchester: Manchester University Press, 2019) 6; 2.
153 Ibid., 9.
154 Bernard Kline, *Maps and the Writing of Space in Early Modern England and Ireland* (Basingstoke: Palgrave Macmillan, 2001), 167.
155 Badcoe, 14.
156 Ibid., 15.
157 Badcoe uses this term as a metaphor for Spenser's interstitial spaces, 21.
158 Shinn, 149.

159 Spenser, "Epistle," 16.
160 W.G. Sebald, *Rings of Saturn* (New York: New Directions, reprint edition, 2016); Giuliana Bruno, *The Atlas of Emotion: Journeys in Art, Architecture, and Film* (New York: Verso, 2007); Denis Wood, *Everything Sings: Maps for a Narrative Atlas*, 2nd edition (Los Angeles: Siglo Press, 2013).
161 Wood, 19.
162 See Edward Frenkel, "The Reality of Quantum Weirdness," *The New York Times*, Sunday, February 2015, p. SR 9.
163 Ibid., p. SR 9.
164 Gosson, STC 20481, sig I.8(v).

Epilogue and analogue

Medieval geography and the Spatial Humanities

> It seems fitting to insert an epilogue after the foregoing.
>
> Claudius Ptolemy, *The Geography*[1]

Reader, a suggestion: go to YouTube, find "Ambient 2: The Plateaux of Mirror," by Harold Budd and Brian Eno[2] a work paired with cartographic metaphor and imagery. Allow this soundscape to create a sonic landscape as you read.

The act of inscription and the impulse of imagination connect the cosmographer and the writer.[3] Cosmography, the study of the universe and the Earth, explores both astronomy and geography, just as writing explores both the world and self. Each imagines a world and inscribes that world. This present study has surveyed the nexus of literature and cartography in England during the Ptolemaic revival when the dissemination of a book of maps or, rather, the textual and cultural "event"[4] of a network of books, images, globes, plays, and almanacs created the cultural context for the reception of Ptolemy's famed *Geography.* This network of book and geographic cultures proposed a measurable Earth and, beyond Ptolemy's work, an oceanic, navigable world. In addition to the better-known history of the learned and luxury transmission of Ptolemy's *Geography,* this study has considered the popular dissemination of spatial representations of a global world in a variety of material forms. Like the photograph "Earthrise," an image that powerfully captured the growing ecological, whole-Earth movement of the 1960s, the world map of Ptolemy's *Geography* stands as a visual icon for the clustered social processes that generated an emergent image of a global Earth, but the trajectory of those processes is more diffuse than any single image. In seeking the trace of Ptolemy's *Geography* in England, this study offers a geo-critical reading of literature and popular texts but also a social history of cartography and early print in England, one which reveals a geo-poetic aesthetic. No one can see the whole Earth; instead, it must be conceived in the imagination of writers and cartographers[5] and then disseminated in books and images.

Mirror of the World is activated within the interdisciplinary field of the Spatial Humanities, an emergent field that draws on the complex inquiry practices and cultural analysis of the humanities and the spatial reckoning of geography.[6] The intersection of the fields of geography and the humanities

DOI: 10.4324/9781003096245-7

and "the complexities of the entwinement of literature and cartography"[7] has received vital interest in the past two decades, generating overlapping sub-fields. "GeoHumanities," typically associated with digital data projects and Geospatial Information Systems (GIS),[8] might include data mining the toponyms, for example, in Samuel Taylor Coleridge's and Thomas Gray's journals that recount their walking tour of the Lake District in England.[9] Literary analyses using a geospatial lens have clustered under the term "geocriticism," as theorized by Robert T. Tally Jr;[10] the related term "geo-poetics" refers to an aesthetic of the biosphere and poetry rather than interpretive practice.[11] "Neogeography" references non-specialist use of digital mapping platforms, such as Google Earth or StoryMaps in the service of geo-spatial literary analysis,[12] while "literary mapping" projects such as the Map of Early Modern London identify the trajectory and connections of texts and landscapes, urban locales, or symbolic territory.[13] Geo-editions of literary texts or "enhanced e-books" such as Nicholas Bauch's *Enchanting the Desert*, which explores the "hidden geography" of a Grand Canyon photographic slide show, are creating a new geographic experience of reading.[14] All of these sub-fields, with overlapping fluidity, share a language of critical theory and cultural geography with an attentiveness to the subjectivity and symbolic representation of physical places and socially constructed spaces that give rise to maps, geographic texts, and literary spaces. They engender a critical dialogue "between dirt and discussion," between geographical representation and literary space, each a coded praxis of space as interpretive, culturally situated, and ideological.[15] The Ptolemaic revival was also distinguished through the use of practices and strategies that are currently defining research trajectories within the Spatial Humanities, a nexus of mathematically derived cartography with open-ended practices that are experimental, affective, multi-scalar, and attentive to the doubleness of place.[16] These literary and geographic practices were propelled, in the fifteenth century as in the late twentieth century, by the intersectional innovation of cartographic and book technologies.

GIS, a data-based system for analyzing spatial data and producing visualized, layered maps and images, produces complex modeling and reveals patterns to support problem solving and decision making. The limitations of GIS, as articulated by critical geographers, are its reliance on positivistic, bureaucratic, and abstract Cartesian coordinates. David Bodenhamer, John Corrigan, and Trevor M. Harris, whose collaborative work demarcated Spatial Humanities as encompassing both historical inquiry and GIS modeling, have since proposed a process of "deep mapping," a term initially coined by William Least Heat Moon to propose a different kind of mapping that allows for subjectivities and experience.[17] Current deep mapping projects utilize aspects of GIS or multimedia but also engage with the indeterminacy of the material record, multi-scalar spaces, spatially framed identities, and open-ended mappings.[18] Though this current study does not utilize computer aggregated data,[19] it operates within and has been theoretically informed by the dual frameworks of the fields of the Spatial Humanities and Book

History. Together, this dual framework is attentive to narrative and visual, symbolic and material spaces created by the intersection of literary and cartographic culture in the late fifteenth and early sixteenth centuries in Europe.

The current vitality of the broad field of the Spatial Humanities bridges the disciplinary divide between spatial technologies and the humanities, reactivating the collaboration between "poyetes and astronomiers" [20] that was central to geographic thought during the the Ptolemaic revival of the fifteenth and sixteenth centuries. From the vantage point of the twenty-first century, we can survey the considerable journey of geographic determinism from its origins on the far shores of the fifteenth century to find a surprising wealth of innovative practices in geographical thought long dismissed as non-scientific or quaintly irrelevant. That this contradictory, hybrid, spiritual, lyrical geography gave way to the mathematically ordered cartography of the early modern period was not a given. Rather, during the Ptolemaic revival, various paradigms flexibly conceptualized time and space before the ascendency of Renaissance linearity and Reformation certainty in literal Biblical truth.[21] As noted in the Introduction to this book, the context of this process of hybridity and assimilation has been poorly theorized, such that the reception of Ptolemy's *Geography* has been flattened to a near-single event in what was, instead, a complex geographical reception that included literature, writers, printers, and readers as well as mapmakers, scholars, and armchair travelers.

As Paul Smethurst has argued, the medieval and the postmodern both lack "an epistemological dominant, with its absolute faith in empirical method to deliver a consistent and unified 'real' world."[22] Compared to the scientific fixity of the Enlightenment, the fluid geographies of the fifteenth century contain the unique ethos and credibility of the Ptolemaic revival but also offer alternative literary and cartographic interventions for consideration in the postmodern geospatial moment, an "oceanic poetics" that is being rediscovered in current research into the Global Middle Ages and that potentially releases the medieval past from "nation-based silos."[23] The trajectory of this book through late fifteenth- and early sixteenth-century literature—Thomas Malory's Arthurian romance, William Caxton's geographical editions and image of globes, Wynkyn de Worde's cartographic landscapes, the Behaim globe as an edition of Mandeville's *Travels*, John Rastell's stage globe, and the "Everywoman reader" of the *Kalender of Shepherds*—proposes a question: which elements of geo-poetic thought from the Ptolemaic revival articulate principles of interest or utility to the current field of geospatial study? With their spirit of curiosity and learned enthusiasm, fifteenth-century cosmographers and mapmakers exemplify collaboration, revision, subjectivity, and augmentation. As an editor, Caxton appealed directly to his readers, "humbly requyryng alle them that shal fynde faulte, to correcte and amende."[24] The geographic writing, images, and maps of these texts augmented known information and included conflicting information, drawing the reader into a process of active interpretation rather than certainty. As contemporary investigations in the Spatial Humanities seek alternate models to solve complex problems and to loosen the dominant geographic framework that defined rationalist, Cartesian space, the era of the Ptolemaic revival is replete with creative examples to stir a new spatial imaginary.

In his second-century CE *Geography*, Ptolemy welcomed the generative conflict of old and the new information:

> We consider it necessary therefore for us to pay more attention to the newer records of our time, weighing, however, in our description these new records and those of former times and deciding what is credible and what is incredible.[25]

Relegated to the rear-view mirror of global navigation and modern mapping practices, pre-modern cartography and geographic thought, which assimilated old and new models of geographic thought, was often dismissed as naïve, chaotic, ignorant, even ridiculous. If we set aspersions of chaotic geographies aside, the Ptolemaic revival offers a capaciousness for "fluid geographies." Experimental, open-ended maps and documents reveal a convergence of alternative models, the continual augmentation of new information, and the possibility of heterotopic spaces. Portolan charts, for example, deployed variable cartographic scale across the surface of the map to create a two-dimensional representation of the intersection of ports and oceanic spaces. Flexible literary and cartographic spaces accommodated a sense of otherness, multiplicity, or simultaneity, familiar to postmodern discourses in which "the physical and the metaphysical, the everyday and the sacred, the historical and the eternal might overlap or impinge on each other."[26] The fifteenth-century hybrid geographic model speaks to the possibilities and tensions of a field such as the Spatial Humanities with its inquiries into polyvalent approaches and the interdisciplinary collaboration between geographic science and the humanities.

As we have seen, works such as Malory's *Le Morte Darthur*, de Worde's cartographic illustrations, Mandeville's *Travels*, and Behaim's globe offer affective geographies influenced by a cartographically inscribed world: powerfully emotional narratives that negotiate global spaces. In the twenty-first century, critical geographers call for "subaltern understandings and conceptions of place,"[27] for new representations of space that include subjectivity, experience, and emotion.[28] The urgency of the intersection of space and narrative is made manifest by the Black Lives Matter movement in the U.S. which has laid bare the geographies of structural racism as well as the risks of traversing privileged spaces as a black or brown body.[29] Structures of power and the enforcement of space are reliant, as they were in feudal medieval culture, on the perception of spatial and social fixity as "the real" or inevitable, either ordained by a divinely ordered cosmos, or "just the way things are." The presumption of cosmological or social fixity stands in tension with the pressures of migration,[30] statelessness, or global pandemics that reassert a fluid world.[31] Proposed border walls stand in contrast to Behaim's and Rastell's aspirations for and delight in the circulatory mobility of travel and worldly goods.

Cultural geographers have provocatively suggested in recent years that even the rationalist-scientific construct of scale is in need of reconsideration as a self-evident principle. Scale, the ratio of measured space to its representation on a map, began to aggregate as a mapping practice in the late fifteenth century but has come under critique as naturalizing a hierarchical system that interferes with the capacity for local rather than globalized problem solving,[32] and reinforces the hierarchical terminology of "First World" and "Third World" countries. In Malory's *Le Morte Darthur*, the slippage of mixed micro and macro scale allows the reader to explore the geo-political power of Camelot or Rome simultaneously with the otherworldliness of Avalon in a polyvalent scalar analysis of sovereignty of self and nation. Between the medieval and the postmodern eras, the intervening rationalization of space that underlaid the Ptolemaic grid and early modern maps gradually winnowed out such multi-scalar representations, heterotopic spaces, and polyvalent scale, perhaps not coincidentally with the waning of romance as a literary genre.

The fascination with the ancient place names of Ptolemy's "ancient tables" of the *Geography* stirred humanist scholarship into locating an ancient past amid the urbanizing present of the fifteenth century. Resonant with past histories and geographies, place names in Malory's and Caxton's texts created multi-layered and linguistically archeological geographies within the texts. Characterized by its preoccupation with ancient locations and toponyms, the Ptolemaic revival called attention to the doubleness of place and of the only partial erasure of Roman Europe, disrupting any singular correspondence of place and name. The current practice of land acknowledgment in the U.S. similarly resonates with the power of place naming, a harrowing of the present with the irruption and accountability of the past. Ultimately, such doubleness of place suggests there is no single mirror of the world; it is a hall of mirrors.

As in the fifteenth century, book and cartographic technological innovation are again dynamic co-generators of rapid and profound cultural change. The printing press and the attendant processes of woodcut illustration and copper engraving generated a new form of the book and created new images of a global Earth. Similarly, the linked technological innovation of books and cartography have generated a highly visual reading space and the use of "hypertext"—the digital links that have come to define the associative reading and research practices of the web—coincident with digital spatial technologies that are redefining the representation of global and local spaces. Released in 2005, Google Earth, Google Maps, and Google Street View are altering our imaginative, planned, and actual experiences of space in commercial, political, and cultural linkages.[33] In a corollary to the introduction of the Ptolemaic grid in the fifteenth century, GIS provides a system of modeling the Earth that is generating different ways of experiencing, problem solving, and interacting with the global Earth. Such representations of the Earth are not static documents but active instruments of cultural change, in all its complexity. If the trace of Ptolemy's *Geography* in England teaches us anything, it is that the paradigm shift to a new model of a global Earth was not the work of a single book or map, but rather a productive process in which "new thynges"[34] were gradually integrated into a radical reassessment of the known world through an ethos of experimentation, amalgamation, and structural openness.

As the fiction of "realistic" representation of space becomes increasingly fraught, fifteenth-century geographic thought—with its hybrid processes, discontinuous spaces, heterotopias, and polyvalent scale—offers fresh examples of alternate and confluent ways of understanding and representing the world that may delight, surprise, and inspire. While the utility of Gerardus Mercator's navigational projection of the world is undeniable, its distortions of near-equatorial landmasses are little known, even today, perpetuating a view of the Earth that minimizes the African landmass. The cartographic term "projection," the inherent distortions of any given model's ability to represent a three-dimensional object in two-dimensional space, is one that needs to come into greater cultural parlance just as the subjective work of editors has become foregrounded in the production of a literary edition. Consequently, the conception of English geographic thought in the early age of print as "belated" obscures a rich period of geographic thought; though ultimately left behind, the curious and passionate geographies of the Ptolemaic revival offer lively possibilities for new geographic reckonings.

The fabled inscription on the Pillars of Hercules, the mythic end posts of the straits of Gibraltar, cautioned sailors not to sail beyond the known world: *ne plus ultra*, no more beyond, they warned. Five centuries after vigorous, mathematical mapping, what are the limits to the mapping practices of the scientific era? "On the map," writes Campbell McGrath, "it is precise and rectilinear as a chessboard, though driving past you would hardly notice it, this boundary line or ragged margin … a tangled ditch grown up throughout with a fearsome assortment of wildflowers and bracken."[35] The astringent, anti-colonialist landscape of Jamaica Kincaid's *A Small Place* similarly contests the imposed graticule ("If you go to Antigua as a tourist, this is what you will see"),[36] and the ethereal soundscape of Brian Eno releases the listener from the "rectilinear map," a map, we have found, that can only take us so far. What could be more needed at this juncture of global migrations and climate change than to reimagine how we map the world? Perhaps the deep mappings of the Ptolemaic revival and of the Spatial Humanities, in conversation, offer less certainty but *plus ultra*—more beyond.

Notes

1 Claudius Ptolemy, *The Geography of Claudius Ptolemy*, trans. Edward Luther Stevenson (New York: New York Public Library, 1932, reprinted 2011), 162.
2 Harold Budd and Brian Eno, "Ambient 2: The Plateaux of Mirror," 1980. Available at: www.youtube.com/watch?v=mwJTwG5r5Ks. An evocative, extended sonic conversation.
3 Denis Cosgrove, *Geography and Vision: Seeing, Imaging, and Representing the World* (London: I.B. Taurus, 2008), 150.
4 Jerome McGann, *The Textual Condition* (Princeton, NJ: Princeton University Press, 1991).

5 Denis Cosgrove, *Apollo's Eye: A Cartographic Genealogy of the Earth in the Western Imagination* (Baltimore, MD: Johns Hopkins University Press, 2001).

6 Patricia Murrieta-Flores and Bruno Martins, eds. "The Geospatial Humanities: Past, Present, and Future," *International Journal of Geographical Information Science* 33:12 (2019): 2424–2429. Key funding and project support for work in the Spatial Humanities can be found in various universities internationally, including, in the U.S., the Scholars' Lab at the University of Virginia and, in the UK, the University of Lancaster.

7 Anders Engberg-Pedersen, "Introduction—Estranging the Map: On Literature and Cartography," in *Literature and Cartography: Theories, Histories, Genres*, ed. Anders Engberg-Pedersen (Cambridge, MA: The MIT Press, 2017), 1–18; 2.

8 See Charles Travis, "Digital Geohumanities," in *International Encyclopedia of Human Geography*, 2nd edition, vol. 3, ed. Aubrey Kobayashi (Amsterdam: Elsevier, 2020), 341–346.

9 *Mapping the Lakes: A Literary GIS*, a project of David Cooper, Ian Gregory, Sally Bushell, and Zoe Bolton, University of Lancaster. See also the journal *GeoHumanities*.

10 Robert T. Tally Jr, "Introduction: On Geocriticsm," in *Geocritical Explorations: Space, Place, and Mapping in Literary and Cultural Studies*, ed. Robert T. Tally Jr. (New York: Palgrave Macmillan, 2011), 1–9. See also Robert T. Tally Jr., ed. *The Routledge Handbook of Literature and Space* (Abingdon: Routledge, 2017).

11 Bernard Westphal, *Geocriticism: Real and Fictional Spaces*, trans. Robert T. Tally Jr. (New York: Palgrave Macmillan, 2000).

12 Peta Mitchell, "Literary Geography and the Digital: The Emergence of Neogeography," in *The Routledge Handbook of Literature and Space*, ed. Robert T. Tally Jr. (Abingdon: Routledge, 2017), 84–95.

13 Spatial Humanities projects that deploy spatial and literary analyses include Europe: A Literary History, 1348–1418; The Grub Street Project; and The Map of Early Modern London.

14 See Nicholas Bauch, *Enchanting the Desert: A Pattern Language for the Production of Space* (Stanford, CA: Stanford University Press, 2016). Bauch's interactive digital narrative "uses photographs to spatially narrate the cultural history of the Grand Canyon."

15 Kevin Bartoy, *Between Dirt and Discussion: Methods, Methodology, and Interpretation in Historical Archeology* (New York: Springer, 2006).

16 David J. Bodenhamer, John Corrigan, and Trevor M. Harris, eds. *The Spatial Humanities: GIS and the Future of Humanities Scholarship* (Indianapolis, IN: Indiana University Press, 2010).

17 William Least Heat Moon, *PrairyErth (a Deep Map)* (New York: Houghton, 1992).

18 David Bodenhamer, John Corrigan, and Trevor M. Harris, eds. *Deep Maps and Spatial Narratives* (Bloomington, IN: Indiana University Press, 2015).

19 Students in my "Maps and Literature" classes and I have developed a Google Earth project of Malory's Roman War itinerary, created StoryMaps (based on ArcGIS), and utilized Omeka to document the places of publication of books in our university's rare book library as a means of engaging in geo-spatial literary analysis.

20 This phrase is from *The Compost of Ptholomeus*, Robert Wyer (London, 1540?), STC 20480a, sig. O(iii)v. See Chapter 5.

21 Brian Cummings, *The Literary Culture of the Reformation: Grammar and Grace* (Oxford: Oxford University Press, 2007).

22 Paul Smethurst, "The Journey from Modern to Postmodern in the *Travels of Sir John Mandeville* and Marco Polo's *Divisament dou Monde*," in *Postmodern*

Medievalisms, Studies in Medievalism XIII, eds. Richard Utz and Jesse G. Swan (Cambridge: D.S. Brewer, 2004), 159–179; 173.

23 Jamie K. Taylor, "Toward Premodern Globalism: Ocean Exemplarity in Chaucer's Men of Law's Tale," *PMLA* 135:2 (March 2020): 254–272; 247. For a discussion of the way in which the West has been conceived as an oceanic civilization and of Freud's concept of "oceanic feeling," see Rob Wilson and Wimal Dissanayake, eds. *Global/Local: Cultural Production and the Transnational Imaginary* (Durham, NC: Duke University Press, 1996).

24 William Caxton, ed. *The Myrrour of the World*, ed. Oliver H. Prior, *Caxton's Mirrour of the World* (Berlin: A. Asher & Co. for the Early English Text Society, Extra Series CX, 1913), 7.

25 Ptolemy, *The Geography*, 29.

26 Smethurst, 175.

27 Murrieta-Flores and Martins, 2426. See also David Harvey, *Spaces of Hope* (Berkeley, CA: University of California Press, 2009).

28 See Giuliana Bruno, *The Atlas of Emotion: Journeys in Art, Architecture, and Film* (New York: Verso, 2007). See also my blog "Passionate Geography," available at: www.passionategeography.com.

29 Patrice Cullors, *When They Call You a Terrorist: A Black Lives Matters Memoir* (New York: St. Martin's Press, 2018).

30 The Climate Migrants StoryMap is a visual, cartographic, and empathetic story of the impact of climate on migration, using the ESRI GIS framework. Available at: https://storymaps.esri.com/stories/2017/climate-migrants/index.html.

31 Nishat Awan, "Introduction to Border Topologies," *GeoHumanities: Space, Place, and the Humanities* 2:2 (2016): 279–283. For a discussion of Mediterranean literature in the context of "the displacement of the nation as the default category of analysis," see Sharon Kinoshita, "Mediterranean Literature," in *A Companion to Mediterranean Literature*, eds. Peregrin Horden and Sharon Kinoshita (London: Wiley Blackwell, 2014) 314–329; 314.

32 Sallie A. Marston, John Paul Jones III, and Keith Woodward, "Human Geography without Scale," *Transactions of the Institute of British Geographers*, New Series, 30:4 (December 2005): 416–432. See also John Paul Jones III, Keith Woodward, and Sallie A. Marston, "Situating Flatness," *Transactions of the Institute of British Geographers*, New Series, 32:2 (April 2007): 264–276.

33 Cheryl Gilge, "Google Street View and Image as Experience," *GeoHumanities: Space, Place, and the Humanities* 2:2 (2016): 469–484.

34 [*Here Begynneth the Kalendar of Shepherdes*]. trans. Robert Copland, printed by Wynkyn de Worde (London, 1511) STC (2nd edition), 22409.5.

35 Campbell McGrath, *Prose Poem*, available at: https://poets.org/poem/prose-poem.

36 Jamaica Kincaid, *A Small Place* (New York: Farrar, Straus and Giroux, 2000), 3.

Bibliography

Primary sources (arranged in chronological order)

Caxton, William. *Chronicles of England.* Westminster: William Caxton, 1480 (STC 9991).

Higden, Ranulf. *The Discripcion of Britayne*, trans. John Trevisa, Westminster: William Caxton, 1480 (STC 13440a); reprinted as the *descrypcyon of Englonde.* Westminster: Wynkyn de Worde, 1498 (STC 13440b).

Caxton, William. *Myrrour and Dyscrypcyon of the Worlde.* Westminster: William Caxton, 1481 (STC 24762); 1490 (STC 24763).

Higden, Ranulf. *Polychronicon.* Translated by John Trevisa. Westminster: William Caxton, 1482 (STC 13438); Westminster: Wynkyn de Worde, 1495 (STC 13439).

Malory, Thomas. *Le Morte Darthur.* Westminster: William Caxton, 1485 (STC 801).

Le co[m]post et kalendrier des bergiers. Paris: Guy Marchant, 1493.

Anglicus, Bartholomeus. *De proprietatibus rerum.* Westminster: Wynkyn de Worde, 1495 (STC 1536).

Mandeville, John. *The Way to the Holy Lande* [*Travels*]. London: Richard Pynson, 1496 (STC 17246).

Dives & Pauper. Westminster: Wynkyn de Worde, 1496 (STC 19213).

Malory, Thomas. *Le Morte d'Arthur.* Westminster: Wynkyn de Worde, 1498 (STC 802); 1529 (STC 803).

Mandeville, John. *The Way to the Holy Lande* [*Travels*]. Westminster: Wynkyn de Worde, 1499 (STC 17247); 1515 (STC 14082).

Here Begynneth the Kalendar of Shepherdes. London: Richard Pynson, 1506 (STC 22408).

[*Here Begynneth the Kalendar of Shepherdes*]. Translated by Robert Copland. London: Wynkyn de Worde, 1511 (STC, 2nd ed. 22409.5).

Here Begynneth the Kalendar of Shepardes. London: Julian Notary, 1518 (STC 22410).

The Mirrour and Dyscrypcyon of the Worlde. London: Andrewe Lawrence, 1527 (STC 24764).

Glareanus, Henricus. *D. Henrici Glareani poetæ lavreati De geographia liber vnus.* Basle, 1527.

The Kale[n]der of Shepeherdes. London: Wynkyn de Worde, 1528 (STC 22411).

Le Morte d'Arthur. London: Wynkyn de Worde, 1529 (STC 803).

The Compost of Ptholomeus. London: Robert Wyer, 1530? (STC 20480).

Elyot, Thomas. *The Book Named the Gouernor.* 1531. Menston: Scholar Press, 1970.

The Compost of Ptholomeus. London: Robert Wyer, 1540? (STC 20480a).

Recorde, Robert. *The Castle of Knowledge.* London: Reginald Wollfe, 1556 (STC 20796).

Cuningham, William. *Cosmographical Glasse.* London, 1559 (STC (2nd ed.) 6119).

Mercator, Gerardus. *Nova et Aucta Orbis Terrae descriptio ad Usum Navigantium Emendata* [A New and Augmented Description of the Earth Corrected for the Use of Sailors]. 1569.

The Shepardes Kalender. London: Thomas Este, for John Wally, 1570? (STC 22415).

The Shepardes Kalender. London:[John Charlewood for] John Walley], 1580 (STC 22416).

Blunderville, Thomas. *A Briefe Description of Universal Mappes & Cardes and of Their Use.* London: Roger Ward, for Thomas Cadman, 1598.

The Shepheards Kalendar. London: V. Simmes, for Adams?. 1600? (STC 22419).

The Compost of Ptolomeus. London: M. Parsons, for Henry Gosson. 1600/1638? (STC 20482).

Secondary sources

Ackerman, Robert William. *An Index of the Arthurian Names in Middle English.* Vol. 10. Palo Alto, CA: Stanford University Press, 1952.

Ackroyd, Peter. *The Life of Thomas More.* New York: Nan A Talese, 1998.

Algazi, Gadi. "Scholars in Households: Refiguring the Learned Habitus, 1480–1550." *Science in Context* 16, no. 1–2 (2003), (special issue: *Scientific Personae*, edited by Lorraine Daston and Otto Sibum): 9–42.

Algazi, Gadi. "At the Study: Who Was With Early Modern Scholars When They Were Alone?" Lecture, UCLA, April 2008.

Algazi, Gadi. "At the Study: Notes on the Production of the Scholarly Self." In *Space and Self in Early Modern European Cultures*, edited by David Warren Sabean and Malina Stefanovska, 17–50. Toronto: University of Toronto Press, 2012.

Ambrose, Laura William. "Travel in Time: Local Travel Writing and Seventeenth-Century Almanacs." *Journal of Medieval and Early Modern Studies* 43, no. 2 (2013): 419–443.

Anglo, Sydney. *Spectacle, Pageantry, and Early Tudor Policy.* 2nd ed. Oxford: Clarendon Press, 1997.

Areford, David S. "Print Trouble: Notes on a Medium in Between." In *From Minor to Major: The Minor Arts in Medieval Art History*, edited by Colum Hourihane, 229–254. Philadelphia, PA: Penn State University Press, 2012.

Armstrong, Dorsey, and Kenneth Hodges. *Mapping Malory: Regional Identities and National Geographies.* New York: Palgrave Macmillan, 2014.

Awan, Nishat. "Introduction to Border Topologies." *GeoHumanities* 2, no. 2 (2016): 279–283.

Axton, Richard, ed. *Three Rastell Plays: Four Elements, Calisto and Melebea, Gentleness and Nobility.* Cambridge: D.S. Brewer, 1979.

Bachelard, Gaston. *The Poetics of Space.* 1964. Reprint, London: Penguin, 2014.

Bacon, Francis. *The New Organon.* (Book I, *Aphorism LXXXIV*). Translated by Lisa Jardine and Michael Silverthorne. Cambridge: Cambridge University Press, 2000.

Barber, Peter. "England I: Pageantry, Defense, and Government: Maps at Court to 1550." In *Monarchs, Ministers, and Maps: The Emergence of Cartography as a Tool of Government in Early Modern Europe*, edited by David Buisseret, 22–56. Chicago: University of Chicago Press, 1992.

Barber, Peter. "Mapmaking in England, ca. 1470–1650." In *The History of Cartography*, Vol. 3, *Cartography in the European Renaissance*, Part 1, *1589–1669*, edited by David Woodward. Chicago: University of Chicago Press, 2007.

Barron, Caroline M. "What Did Medieval London Merchants Read?" In *Medieval Merchants and Money: Essays in Honour of James L. Bolton*, edited by Martin Allen and Matthew Davies, 43–70. London: University of London Press, 2016.

Barron, W.R.J. and S.C. Weinburg, eds. *Laȝamon's Arthur: The Arthurian Section of La⊠amon's Brut (Lines 9229–14297)*. Austin, TX: University of Texas Press, 1989.

Barrows, Adam. *Time, Literature, and Cartography After the Spatial Turn: The Chronometric Imaginary.* New York: Palgrave, 2016.

Barthes, Roland. "The Death of the Author." In *Image Music Text*. Translated by Stephen Heath, 142–147. New York: Hill and Wang, 1977.

Bartoy, Kevin. *Between Dirt and Discussion: Methods, Methodology, and Interpretation in Historical Archeology.* New York: Springer, 2006.

Baskerville, C.R. "John Rastell's Dramatic Activities." *Modern Philology*, 13 (1915–1916): 557–560.

Batchelor, Robert K. *London: The Selden Map and the Making of a Global City, 1549–1689*. Chicago: University of Chicago Press, 2013.

Bauch, Nicholas. *Enchanting the Desert: A Pattern Language for the Production of Space*. Stanford, CA: Stanford University Press, 2016.

Bauman, Zygmunt. *Liquid Modernity.* Cambridge: Polity Press, 2000.

Bede. *Ecclesiastical History of the English People*. Oxford: Oxford University Press, 1999.

Bell, David N. "The Libraries of Religious Houses in the Late Middle Ages." In *The Cambridge History of Libraries in Britain and Ireland*, edited by Elisabeth Leedham-Green and Teresa Webber, 126–151. Cambridge: Cambridge University Press, 2006.

Bennett, Henry Stanley. *English Books and Readers.* Vol. 1, *1475–1557: Being a Study in the History of the Book Trade from Caxton to the Incorporation of the Stationers' Company.* Cambridge: Cambridge University Press, 1969.

Bennett, Josephine Waters. "The Woodcut Illustrations in English Editions of 'Mandeville's Travels'." *The Papers of the Bibliographic Society of America* 47, no. 1 (First Quarter, 1953): 59–69.

Bennett, Josephine. *The Rediscovery of John Mandeville.* New York: MLA, 1954.

Bennett, M.J. "Mandeville's Travels and the Anglo-French Moment." *Medium Aevum* 75, no. 2 (2006): 273–292.

Berggren, J. Lennart, and Alexander Jones. *Ptolemy's Geography: An Annotated Translation of the Theoretical Chapters.* Princeton, NJ: Princeton University Press, 2000.

Bevington, David. *Mankind to Marlow.* Cambridge, MA: Harvard University Press, 1962.

Bhabha, Homi. *The Location of Culture.* New York: Routledge, 1994.

Bishop, M. "Ambrosius Holbein's *Memento Mori* Map for Sir Thomas More's *Utopia.* The Meanings of a Masterpiece of Early Sixteenth Century Graphic Art." *British Dental Journal*, 199, no. 2 (2005): 107–112.

Blackburn, Elizabeth B. "The Legacy of 'Prester John' by Damião à Goes and John More." [A facsimile edition] *Moreana* 14 (1967): 37–98.

Blades, William. *The Life and Typography of William Caxton, England's First Printer.* Vol. 2. London, 1863, reprinted Cambridge: Cambridge University Press, 2014.

Blake, N.F . *Caxton: England's First Publisher.* London: Osprey Publishing, 1976.

Blake, N.F . *William Caxton and English Literary Culture.* London: The Hambledon Press, 1991.

Bodenhamer, David J ., John Corrigan, and Trevor M. Harris, eds. *The Spatial Humanities: GIS and the Future of Humanities Scholarship.* Indianapolis, IN: Indiana University Press, 2010.

Bodenhamer, David J ., John Corrigan, and Trevor M. Harris. *Deep Maps and Spatial Narratives.* Bloomington, IN: Indiana University Press, 2015.

Bolgar, R.R. *The Classical Heritage and its Beneficiaries.* Cambridge: Cambridge University Press, 1954.

Bolter, J. David and Richard A. Grusin. *Remediation: Understanding New Media.* Cambridge, MA: MIT Press1999.

Brooks, Gordon. Paper delivered at the International Arthurian Congress, Bristol, 2013. Private manuscript.

Brotton, Jerry. *Trading Territories: Mapping the Early Modern World.* London: Reaktion Books, 1997.

Bruno, Giuliana. *The Atlas of Emotion: Journeys in Art, Architecture, and Film.* New York: Verso, 2007.

Bryant, John. *The Fluid Text: A Theory of Editing for Book and Screen.* Ann Arbor, MI: University of Michigan Press, 2002.

Budd, Harold, and Brian Eno. "Ambient 1: Plateaux of Mirror." 1980. Available at: www.youtube.com/watch?v=mwJTwG5r5Ks

Burcholz, Robert and Newton Key. *Early Modern England: A Narrative History.* 2nd ed. London: Wiley-Blackwell, 2008.

Burnett, Charles. "Why Study Ptolemy's *Almagest*? The Evidence of MS Melbourne, State Library of Victoria, Sinclair." *The LaTrobe Journal* 81 (2008): 126–143.

Bury, J.B., ed. *The Cambridge Medieval History.* Cambridge: Cambridge University Press, 1924.

Campbell, Mary. "Introduction." In *A Knight's Legacy: Mandeville and Mandevillian Lore in Early Modern England,* edited by Ladan Niayesh. Manchester: Manchester University Press, 2011.

Campbell, Tony. "Red Names on the Portolan Charts (1311–1677)." *Map History/History of Cartography.* Available at: www.maphistory.info/RedNamesCommentary.html#conclusions.

Capp, Bernard. *English Almanacs, 1500–1800: Astrology and the Popular Press.* Ithaca, NY: Cornell University Press, 1979.

Carman, Christián C. "Rounding Numbers: Ptolemy's Calculation of the Earth-Sun Distance." *Archive for History of Exact Sciences* 63, no. 2 (2009): 205–242.

Casey, E.S. "Between Geography and Philosophy: What Does It Mean to Be in the Place-World?" *Annals of the Association of American Geographers* 91, no. 4 (2001): 683–693.

Castells, Manuel. *The Rise of the Network Society.* 2nd ed. Oxford: Blackwell, 2000.

Caxton, William. *The Dictes and Sayings of the Philosophers: A Facsimile Reproduction of the First Book Printed in England by William Caxton in 1477.* London: Elliot Stock, 1877.

Caxton, William. "Preface and Epilogue." In *Le Morte Darthur, by Sir Thomas Malory,* edited by Stephen Shepard, 814–819. New York: Norton, 2003.

Chambers, Iain. "Maritime Criticism and Theoretical Shipwrecks." *PMLA* 125, no. 3 (2010): 678–684.

Chartier, Roger. *Inscription and Erasure: Literature and Written Culture from the Eleventh to the Eighteenth Century.* Translated by Arthur Goldhammer. Philadelphia, PA: University of Pennsylvania Press, 2007.

Chase, Carol J. and Joan Tasker Grimbert. "Introduction." *Arthuriana* 19, no. 1 (2009): 3–6.

Chaucer, Geoffrey. "The Wife of Bath's Prologue." In *The Riverside Chaucer*, 3rd ed., edited by Larry Benson. Boston: Houghton Mifflin, 1987.

Cherewatuk, Karen. "Sir Thomas Malory's 'Grete Booke'." In *The Social and Literary Contexts of Malory's Morte Darthur*, edited by D. Thomas Hanks, 42–67. Cambridge: D.S. Brewer, 2000.

Childs, Edmund. *William Caxton: A Portrait in a Background.* London: Northwood Publications, 1976.

Chriqui, Sheri. "A 'Foreign' Queen in King Uther's Court: Fifteenth-Century Insular Xenophobia and Malory's Portrayal of Arthur's Mother." *South Atlantic Review* 81, no. 2 (2016): 72–92.

Clark, James G. "University Monks in Late Medieval England." In *Medieval Monastic Education*, edited by George Ferzoco and Carolyn Muessig, 56–71. London: Leicester University Press, 2000.

Clarke, Peter D., ed. *The University and College Libraries of Cambridge*, Corpus of British Medieval Library Catalogues, vol. 10. London: The British Library in association with the British Academy, 2002.

Cohen, Adam Max. "Englishing the Globe: Molyneux's Globes and Shakespeare's Theatrical Career." *The Sixteenth Century Journal* 37, no. 4 (2006): 963–984.

Cohen, Patricia. "Digital Keys for Unlocking the Humanities' Riches." *The New York Times*, November 16, 2010. Available at: www.nytimes.com/2010/11/17/arts/17digital.html.

Coldiron, Anne E.B. "Cultural Amphibians." *Yearbook of Comparative and General Literature* 51 (2003–2004):43–58.

Coldiron, Anne E.B. *English Printing, Verse Translation, and the Battle of the Sexes, 1476–1557*. Farnham: Ashgate, 2009.

Coldiron, Anne E. B. *Printers without Borders: Translation and Textuality in the Renaissance.* Cambridge: Cambridge University Press, 2015.

Collins, Marie. *Caxton: The Description of England, A Modern Rendering*. New York: Weidenfeld and Nicolson, 1988.

Conley, Tom. *The Self-Made Map: Cartographic Writing in Early Modern France.* Minneapolis, MN: University of Minneapolis Press, 1996.

Conley, Tom. "Early Modern Literature and Cartography: An Overview." In *The History of Cartography.* Vol. 3: *Cartography in the European Renaissance*, edited by David Woodward, 401–411. Chicago: University of Chicago Press, 2007.

Conley, Tom. *An Errant Eye: Poetry and Topography in Early Modern France.* Minneapolis, MN: University of Minnesota Press, 2010.

Cooper, Helen. "Introduction." In *Le Morte Darthur*, edited by Helen Cooper. Oxford: Oxford University Press, 1998.

Cooper, Helen. "Arthur in Transition: Malory's Morte Darthur." In *Romance and History: Imagining Time from the Medieval to the Early Modern Period*, edited by Jon Whitman, 120–133. Cambridge: Cambridge University Press, 2015.

Cooper, Helen. "Punctuating Malory's *Morte Darthur*." *Journal of the Early Book Society* 21 (2018): 171–184.

Cormack, Lesley B. "Ptolemy at Work: The Role of the *Geographia* in Geography and Mathematics Teaching in Early Modern England." In *Ptolemy's Geography in the Renaissance*, edited by Zur Shalev and Charles Burnett, 152–156. London and Turin: The Warburg Institute – Nino Aragno Editore, 2011.

Cosgrove, Denis. *Apollo's Eye: A Cartographic Genealogy of the Earth in the Western Imagination*. Baltimore, MD: Johns Hopkins University Press, 2001.

Cosgrove, Denis. *Geography and Vision: Seeing, Imagining and Representing the World*. London: I.B. Taurus, 2008.

Crofts, Thomas. "Thynges Foresayd Aledged." In *Reviewing Le Morte Darthur: Texts and Contexts, Characters and Themes*, edited by Kevin Sean Whetter and Raluca L. Radulescu, 49–64. Cambridge: D.S. Brewer, 2005.

Crofts, Thomas. *Malory's Contemporary Audience: The Social Reading of Romance in Late Medieval England*, Arthurian Studies LXVI. Cambridge: D.S. Brewer, 2006.

Crotch, W.J.B. *The Prologues and Epilogues of William Caxton*. Original Series, No. 176. London: Early English Text Society, 1928.

Cullors, Patrice. *When They Call You a Terrorist: A Black Lives Matters Memoir*. New York: St. Martin's Press, 2018.

Cummings, Brian. *The Literary Culture of the Reformation: Grammar and Grace*. Oxford: Oxford University Press, 2007.

Darnton, Robert. "What Is the History of Books?" In *The Book History Reader*, edited by David Finkelstein and Alastair McCleery, 65–83. London: Routledge, 2002.

Davidson, Roberta. "Prison and Knightly Identity in Sir Thomas Malory's *Morte Darthur*." *Arthuriana* 14, no. 2 (2004): 54–63.

Davis, David. "Images on the Move: The Virgin, the *Kalendar of Shepherds*, and the Transmission of Woodcuts in Tudor England." *Journal of the Early Book Society for the Study of Manuscripts and Printing History* 12 (2009): 99–132.

Dear, M.J., Jim Ketchum, Sarah Luria, and Doug Richardson, eds. *GeoHumanities: Art, History, and Text at the Edge of Place*. London: Routledge, 2011.

Dekker, Elly, and Kristen Lippincott. "The Scientific Instruments in Holbein's Ambassadors: A Re-Examination." *Journal of the Warburg and Courtauld Institutes* 62 (1999): 93–125.

Delano Smith, Catherine, and Roger J.P. Kain, eds. *English Maps: A History*. Toronto, University of Toronto Press, 2000.

Deleuze, Gilles, and Félix Guattari. *A Thousand Plateaus*. Translated by Brian Massumi. Minneapolis, MN: University of Minnesota Press, 1987.

Derrida, Jacques. "Signature, Event, Context." *Glyph* 1 (1977): 172–197.

Dictman, Mary. "The Tale of King Arthur and the Emperor Lucius." In *Malory's Originality*, edited by R.M. Lumiansky. Baltimore, MD: Johns Hopkins University Press, 1964.

Dilke, O.A.W. "The Culmination of Greek Cartography in Ptolemy." In *The History of Cartography*. Vol. 1, *Cartography in Prehistoric, Ancient and Medieval Europe and the Mediterranean*, edited by J. Brian Harley and David Woodward, 177–200. Chicago: University of Chicago Press, 1987.

Dillon, Janette. "John Rastell's Stage." *Medieval English Theater* 18 (1996): 17–18.

Doutrepont, George. *Inventaire de la "Librairie" de Philippe le Bon 1420*. Geneva: Slatkine Reprints, 1977.

Driver, Martha. "The Illustrated de Worde: An Overview." *Studies in Iconography* 17 (1996): 349–403.

Driver, Martha. "When Is a Miscellany Not Miscellaneous? Making Sense of the *Kalender of Shepherds*." *The Yearbook of English Studies* 33 (2003): 199–214.

Driver, Martha. *The Image in Print: Book Illustration in Late Medieval England and Its Sources*. London: The British Library, 2004.

Duff, E. Gordon. *William Caxton*. New York: Burt Franklin, 1970.

Echard, Siân. "Malory in Print." In *The New Companion to Malory*, edited by Megan G. Leitch and Cory James Rushton, 96–121. Cambridge: Boydell and Brewer, 2019.

Edney, Matthew. *Cartography: The Ideal and its History*. Chicago: University of Chicago Press, 2019.

Edson, Evelyn. *Mapping Time and Space: How Medieval Mapmakers Viewed Their World*. London: British Library, 1998.

Edwards, A.S.G . "*Geography and Illustration and Higden's Art into Life: Collected Papers from the Kresge Art Museum Medieval Symposium*, edited by Carol Garret Fisher and Kathleen Scott, 95–113. Grand Rapids, MI: Michigan State University, 1995.

Egan, Timothy. *Pilgrimage to Eternity*. New York: Viking, 2019.

Eisenstein, Elizabeth. *The Printing Press as an Age of Change: Communications and Cultural Transformations in Early-Modern Europe*. Cambridge: Cambridge University Press, 1979.

Ellis, Henry, ed. *Hall's Chronicles*, original title: *The Union of the Two Noble and Illustre Families of Lancastre & York*. London: 1809, reprinted New York: AMS Press, 1965.

Engberg-Pedersen, Anders, ed. *Literature and Cartography: Theories, Histories, Genres*. Cambridge, MA: MIT Press, 2017a.

Engberg-Pedersen, Anders. "Introduction—Estranging the Map: On Literature and Cartography." In *Literature and Cartography: Theories, Histories, Genres*, edited by Anders Engberg-Pedersen, 1–18. Cambridge, MA: MIT Press, 2017b.

Evans, John. "Extracts from the Private Account Book of Sir William More of Loseley, in Surrey, in the time of Queen Mary and Queen Elizabeth." *Archaeologia* 36 (1855): 284–310.

Falchetta, Pietro. *Fra Mauro's World Map*. Turnhout, Belgium: Brepols, 2006.

Fichte, Jorge. "Caxton's Concept of 'Historical Romance' within the Context of the Crusades: Conviction, Rhetoric, and Sales Strategy." In *Tradition and Transformation in Medieval Romance*, edited by Rosalind Field, 101–113. Woodbridge: D.S. Brewer, 1999.

Field, P.J.C. "Sir Thomas Malory, M. P." *Bulletin of the Institute of Historical Records* XXVII (1974): 24–35.

Field, P.J.C. "Caxton's Roman War." *Arthuriana* 5, no. 2 (1995): 31–73.

Field, P.J.C. *Malory; Texts and Sources*. Cambridge: D.S. Brewer, 1998.

Field, P.J.C. "Apparatus and Commentary." In *Le Morte Darthur by Sir Thomas Malory*, 2 vols., edited by P.J.C. Field. Cambridge: D.S. Brewer, 2013.

Field, Peter. *The Life and Times of Thomas Malory*. Cambridge: Cambridge University Press, 1993.

Ford, Margaret Lane. "Importation of Printed Books into England and Scotland." In *The Cambridge History of the Book in Britain*, vol. 3, *1400–1557*, edited by Lotte Hellinga and J.B. Trapp, 179–202. Cambridge: Cambridge University Press, 1999.

Foucault, Michel. "What Is an Author?" In *Modern Criticism and Theory*, edited by David Lodge. Harlow: Longman, 1969.

Foucault, Michel. "Of Other Spaces." Translated by Jay Miskowiec. *Diacritics* 16, no. 1 (1986): 22–27.

Frenkel, Edward. "The Reality of Quantum Weirdness." *The New York Times*, February 6, 2015, p. SR9.

Friedman, John B. "Cultural Conflicts in Medieval World Maps." In *Implicit Understandings: Observing, Reporting and Reflecting on the Encounters Between Europeans and Other Peoples in the Early Modern Era*, edited by Stuart B. Schwartz, 65–95. Cambridge: Cambridge University Press, 1994.

Furnivall, F.J. *Laneham's Letter*, edited by Frederick J. Furnivall for the New Shakespeare Society, Series 6, no. 14. London: Kegan, Paul, Trench, Trubner & Co.,1890.

Gaines, Barry. "Malory's Castles in Text and Illustration." In *The Medieval Castle: Romance and Reality*, edited by Kathryn Reyerson and Faye Powe, 215–218. Dubuque, IA: Kendal/Hunt Publishing, 1984.

Gautier-Dalché, Patrick. "The Reception of Ptolemy's Geography (End of the Fourteenth to Beginning of the Sixteenth Century)." In *The History of Cartography*, vol. 3, *Cartography in the European Renaissance*, Part 1, edited by David Woodward, 285–364. Chicago: University of Chicago Press, 2007.

Geoffrey of Monmouth. *The History of the Kings of Britain (Historia regum Britanniae)*, edited by Michael A. Faletra. Plymouth: Broadview editions, 2008.

Gilge, Cheryl. "Google Street View and Image as Experience." *GeoHumanities: Space, Place, and the Humanities* 2, no. 2 (2016): 469–484.

Gillespie, Alexandra. "'Folowynge the Trace of Mayster Caxton:' Some Histories of Fifteenth-century Printed Books." In *Caxton's Trace: Studies in the History of English Printing*, edited by William Kuskin, 167–195. Notre Dame, IN: University of Notre Dame Press, 2006.

Gillespie, Alexandra. "Caxton and the Invention of Printing." In *The Oxford Handbook of Tudor Literature, 1485–1603*, edited by Mike Pincombe and Kathy Shrank, 21–36. Oxford: Oxford University Press, 2009.

Gillespie, Stuart. *English Translation and Classical Reception: Towards a New Literary History.* Chichester: Wiley-Blackwell, 2011.

Gillies, John. *Shakespeare and the Geography of Difference.* Cambridge: Cambridge University Press, 1994.

Goldie, Matthew Boyd. "An Early English Rutter: The Sea and the Spatial Hermeneutics in the Fourteenth and Fifteenth Centuries." *Speculum* 90, no. 3 (2015): 701–727.

Goldie, Matthew Boyd. *Scribes of Space: Place in Middle English Literature and Late Medieval Science.* Ithaca, NY: Cornell University Press, 2019.

Goodman, Jennifer R. *Chivalry and Exploration (1298–1630).* Woodbridge: Boydell Press, 1998.

Goodman, Jennifer R. "Caxton's Continent." In *Caxton's Trace*, edited by William Kuskin, 101–123. Notre Dame, IN: University of Notre Dame Press, 2006.

Görz, Günther, and Norbert Holst. "The Digital Behaim Globe (1492)." In *Museum Interactive Multimedia 1997*, edited by David Bearman and Jennifer Trant, 155–172. Pittsburgh, PA: Archives & Museum Informatics, 1997.

Görz, Günther, and Martin Scholz. "Semantic Annotation for Medieval Cartography: The Example of the Behaim Globe of 1492." In *22nd General Conference of the International Council of Museums*, edited by H. Rengmann et al., 1–19. Shanghai: ICOM/CIDOC, 2010.

Gothic and Renaissance Art in Nuremberg, 1300–1550. Exhibition catalog. New York: Metropolitan Museum of Art, 1986.

Greenblatt, Stephen. *Marvelous Possessions: The Wonder of the New World.* Chicago: University of Chicago Press, 1991.

Greenblatt, Stephen. *The Swerve: How the World Became Modern*. New York: Norton, 2011.

Greenblatt, Stephen. *Hamlet in Purgatory*. Princeton, NJ: Princeton University Press, 2013.

Greetham, David. *Textual Scholarship: An Introduction*. New York: Garland, 1994.

Hallberg, Peter. "Thomas More's Cosmopolitan Civil Science: The 'New World' and 'Utopia' Reconsidered." *History of Political Thought* 33, no. 4 (2012): 578–606.

Hanks, Tom and Janet Jesmok, eds. *Malory and Christianity*. Kalamazoo, MI: Medieval Institute Publications, 2013.

Hanna, Ralph, III. "Mandeville." In *Middle English Prose: A Critical Guide to Major Authors and Genres*, edited by A.S.G. Edwards, 121–132. New Brunswick, NJ: Rutgers University Press, 1984.

Hardyng, John. *The Chronicles of John Hardyng*, edited by Henry Ellis. London: Rivington, 1812.

Harvey, David. *Spaces of Hope*. Berkeley, CA: University of California Press, 2009.

Harvey, P.D.A. *Mappa Mundi: The Hereford World Map*. London: British Library, 1996.

Hellinga, Lotte. *Caxton in Focus: The Beginnings of Printing in England*. London: British Library, 1982.

Hellinga, Lotte. *William Caxton and Early Printing in England*. London: British Library, 2010.

Henderson, John. "Knowing Someone Through Their Books: Pliny on Uncle Pliny (*Epistles* 3.5)." *Classical Philology* 97 (2002): 256–284.

Hiatt, Alfred. "Blank Spaces on the Earth." *The Yale Journal of Criticism* 15, no. 2 (2002): 223–250.

Hiatt, Alfred. *Terra Incognita: Mapping the Antipodes Before 1600*. Chicago: University of Chicago Press, 2007.

Higgins, Iain Macleod. *Writing East: The "Travels" of Sir John Mandeville*. Philadelphia, PA: University of Pennsylvania Press, 1997.

Higgins, Iain Macleod. *The Book of John Mandeville, with Related Texts*. Indianapolis, IN: Hackett, 2011.

Hodges, Kenneth. "Why Malory's Launcelot Is Not French: Region, Nation, and Political Identity." *PMLA* 2 (2010): 556–571.

Hodnett, Edward. *English Woodcuts, 1480–1535*. Oxford: Oxford University Press, 1973.

Hodnett, Edward. *Five Centuries of English Book Illustration*. Aldershot: Scolar Press, 1998.

Holinshed, Raphael. *Holinshed's Chronicles*, vol. 6, 1587. Holinshed Project. Available at: www.english.ox.ac.uk/holinshed/texts.php?text1=1587_6197 (accessed June 19, 2020).

Howe, Nicholas. *Writing the Map of Anglo-Saxon England: Essays in Cultural Geography*. New Haven, CT: Yale University Press, 2007.

Hughes, Robert. *Rome: A Cultural, Visual, and Personal History*. New York: Vintage, 2012.

Ingram, Clare. "Losing French: Translation, Nation, and Caxton's English Statutes." In *Caxton's Trace: Studies in the History of English Printing*, edited by William Kuskin, 275–298. Notre Dame, IN: University of Notre Dame Press, 2006.

Ingram, William. *The Business of Playing: The Beginnings of Adult Professional Theater in London*. Ithaca, NY: Cornell University Press, 1992.

Jardine, Lisa. *Worldly Goods: A New History of the Renaissance.* New York: Nan A. Talese, 1996.

Jardine, Lisa, and Jerry Brotton. *Global Interests: Renaissance Art Between East and West.* London: Reaktion Books, 2000.

Jensen, Kristian. "Universities and Colleges." In *The Cambridge History of Libraries in Britain and Ireland,* edited by Elisabeth Leedham-Green and Teresa Webber, 345–362. Cambridge: Cambridge University Press, 2006.

Johns, Adrian. *The Nature of the Book: Print, and Knowledge in the Making.* Chicago: University of Chicago Press, 2000.

Johnson, Francis R. *Astronomical Thought in Renaissance England: A Study of English Scientific Writing from 1500 to 1645.* Baltimore, MD: Johns Hopkins University Press, 1937. Reprint, 1968.

Johnson, W.R. *Lucretius in the Modern World.* London: Gerald Duckworth & Co., 2000.

Jonas, Andrew E.G. "Pro Scale: Further Reflections on the 'Scale Debate' in Human Geography." *Transactions of the Institute of British Geographers,* New Series, 31, no. 3 (2006): 99–406.

Jones, Alexander. "Ptolemy's Geography: A Reform that Failed." In *Ptolemy's Geography in the Renaissance.* Edited by Zur Shalev and Charles Burnett, 15–30. London and Turin:The Warburg Institute – Nino Aragno Editore, 2011.

Jones, John Paul, III, Keith Woodward, and Sallie A. Marston. "Situating Flatness," *Transactions of the Institute of British Geographers,* New Series, 32, no. 2 (2007): 264–276.

Kamps, Ivo. "Foreword." In *Maps and Memory in Early Modern England: A Sense of Place,* by Rhonda Lemke Sanford. New York: Palgrave, 2002.

Kelly, Robert L. "Malory's Argument Against War with France: The Political Geography of France and the Anglo-France Alliance in the *Le Morte Darthur.*" In *The Social and Literary Contexts of Malory's Morte Darthur,* edited by D. Thomas Hanks Jr. and Jessica G. Brogdon, 111–133. Cambridge: Boydell and Brewer, 2000.

Kennedy, E.J. "Lucretian Texture: Style, Metre and Rhetoric in the *De Rerum Natura.*" In *The Cambridge Companion to Lucretius,* edited by Philip Hardie and Stuart Gillespie, 76–91. Cambridge: Cambridge University Press, 2007.

Kennedy, Edward Donald. "Malory's Use of Hardyng's Chronicle." *Notes and Queries* 214 (1969): 167–170.

Kennedy, Edward Donald. "Malory's English Sources." In *Aspects of Malory,* edited by Toshiyuki Takamiya and Derek Brewer, 30–31. Cambridge: D.S. Brewer, 1981.

Kennedy, Edward Donald. "Caxton, Malory, and the 'Noble Tale of King Arthur and the Emperor Lucius'." In *The Malory Debate: Essays on the Texts of Le Morte Darthur,* edited by Bonnie Wheeler, Robert L. Kindrick, and Michael N. Salda, 217–232. Cambridge: D.S. Brewer, 2000.

Kennedy, Edward Donald. "Malory's Use of Hardyng's Chronicle: A Reconsideration." *West Virginia Philological Papers* 54 (2011): 8–15.

Kibler, William, ed. "Lancelot, Part V." In *Lancelot-Grail: The Old French Arthurian Vulgate and Post-Vulgate in Translation.* New York: Garland, 1995.

Kimble, George H.T. *Geography in the Middle Ages.* New York: Russell and Russell, 1938, reissued 1968.

Kincaid, Jamaica. *A Small Place.* New York: Farrar, Straus and Giroux, 2000.

Kinoshita, Sharon. "Mediterranean Literature." In *A Companion to Mediterranean Literature,* edited by Peregrin Horden and Sharon Kinoshita, 314–329. London: Wiley Blackwell, 2014.

Klein, Bernard. *Maps and the Writing of Space in Early Modern England and Ireland.* Houndsmills: Palgrave Macmillan, 2001.

Kline, Naomi Reed. *Maps of Medieval Thought: The Hereford Paradigm.* Woodbridge: Boydell Press, 2001.

Kohanski, Tamarah, ed., *The Book of John Mandeville: An Edition of the Pynson Text with Commentary on the Defective Version.* Tempe, AZ: Arizona Center for Medieval and Renaissance Studies, 2001.

Kohanski, Tamarah, and C. David Benson, eds. *The Book of John Mandeville.* Kalamazoo, MI: West Michigan Press for TEAMS, 2007.

Krishna, Valerie, ed. *The Alliterative Morte Arthure: A Critical Edition.* New York: Burt Franklin, 1976.

Kuskin, William. "Caxton's Worthy Series: The Production of Literary Culture." *ELH* 66, no. 3 (1999): 511–551.

Kuskin, William, ed. *Caxton's Trace: Studies in the History of English Printing.* Notre Dame, IN: University of Notre Dame Press, 2006.

Kuskin, William. "'The Loadstarre of the English Language:' Spenser's *Shepheardes Calender* and the Construction of Modernity." *Textual Cultures: Texts, Contexts, Interpretation* 2, no. 2 (2007): 9–33.

Kuskin, William. *Symbolic Caxton: Literary Culture and Print Capitalism.* Notre Dame, IN: University of Notre Dame Press, 2008.

Kuskin, William. *Recursive Origins: Writing at the Transition to Modernity.* Notre Dame, IN: University of Notre Dame Press, 2013.

Lakowski, Ronald Ian. "Geography and the More Circle: John Rastell, Thomas More and the 'New World'." *Renaissance Forum*, 4, no. 1 (1999): 83–103.

Lane, Robert. *Shepheards Device: Edmund Spenser's Shepheardes Calendar and the Institutions of Elizabethan Society.* Athens, GA: University of Georgia Press, 1993.

Larner, John. *Marco Polo and the Discovery of the World.* New Haven, CT: Yale University Press, 1999.

Larner, John. "Plucking Hairs from the Great Cham's Beard: Marco Polo, Jan de Langhe, and Sir John Mandeville." In *Marco Polo and the Encounter of East and West*, edited by Suzanne Conklin Akbari and Amilcare Iannucci, 133–155. Toronto: University of Toronto Press, 2008.

Larsen, Steven C., and Jay T. Johnson. "The Agency of Place: Toward a More-Than-Human Geographical Self." *GeoHumanities: Space, Place, and the Humanities* 2, no. 1 (2016): 149–166.

Lathrop, Harry Burrows. "Some Rogueries of Robert Wyer." *The Library*, Third Series, v (1914): 349–364.

Lathrop, Henry Burrows. *Translations from the Classics into English, from Caxton to Chapman, 1477–1620.* Madison, WI: University of Wisconsin Press, 1933, reprinted New York: Octagon Books, 1967.

Lavezzo, Kathy. *Angels on the Edge of the World: Geography, Literature, and English Community, 1000–1534.* Ithaca, NY: Cornell University Press, 2006.

Lawton, David. "Dullness and the Fifteenth Century." *English Literary History* 54, no. 4 (1987): 761–799.

Least Heat Moon, William. *PrairyErth (A Deep Map).* New York: Houghton, 1992.

Leedham-Green, Elisabeth. *Books in Cambridge Inventories: Book-Lists from Vice-Chancellor's Court Probate Inventories in the Tudor and Stuart Periods*, 2 vols. Cambridge: Cambridge University Press, 1987.

Leedham-Green, Elisabeth, ed. *The Cambridge History of Libraries in Britain and Ireland*, vol. 1: *To 1640*. Cambridge: Cambridge University Press, 2006.

LeFebvre, Henri. *The Production of Space*. New York: Wiley-Blackwell, 1992.

Lerer, Seth. "The Wiles of a Woodcut: Wynkyn de Worde and the Early Tutor Reader." *Huntington Quarterly* 59, no. 4 (1996): 381–403.

Lewis, C.S. "The English Prose Morte." In *Essays on Malory*, edited by J.A.W. Bennet, 7–28. Oxford: Clarendon Press, 1963.

Lewis, C.S. "The Genesis of a Medieval Book." In *Studies in Medieval and Renaissance Literature*. Cambridge: Cambridge University Press, 1998; originally published 1966.

Lilley, Keith. "Mapping Cosmopolis: Moral Topographies of the Medieval City." *Environment and Planning D: Society and Space* 22, no. 5 (2004): 681–698.

Lilley, Keith D. *City and Cosmos: The Medieval World in Urban Form*. London: Reaktion Books, 2009.

Lochrie, Karma. *Nowhere in the Middle Ages*. Philadelphia, PA: University of Pennsylvania Press, 2016.

Lovatt, Roger. "Introduction to the University and College Libraries of Cambridge," vol. 10, *Corpus of British Medieval Library Catalogues*. edited by Peter D. Clarke, xxvii–xcii. London: British Library, 2002.

Lozovsky, Natalia. *The Earth Is Our Book: Geographic Knowledge in the Latin West, 400–1000*. Ann Arbor, MI: University of Michigan Press, 2000.

Lozovsky, Natalia. "Roman Geography and Ethnography in the Carolingian Empire." *Speculum* 81, no. 2 (2006): 325–364.

Lucretius. *On the Nature of Things*, Book I. *The Internet Classic Archive*, MIT. Available at: http://classics.mit.edu/Carus/nature_things.1.i.html.

Lynch, Andrew. *Malory's Book of Arms*. Cambridge: D.S. Brewer, 1997.

Lynch, Andrew. "What Cheer? Emotions and Actions in the Arthurian World." In *Emotions in Medieval Arthurian Literature*, edited by Frank Brandsma, Carolyne Larrington, and Corinne Saunders. Cambridge: D.S. Brewer, 2015.

Lynch, Andrew. "Malory and Emotion." In *The New Companion to Malory*, edited by Megan G. Leitch and Cory James Rushton. Cambridge: D.S. Brewer, 2019.

Lyne, Raphael. *Ovid's Changing Worlds: English Metamorphoses, 1567–1632*. Oxford: Oxford University Press, 2001.

MacEachren, Alan M. "A Linear View of the World; Strip Maps as a Unique Form of Cartographic Representation." *The American Cartographer* 13, no. 1 (1986): 7–25.

MacGregor, Neil. *Shakespeare's Restless World*. London: Penguin Books, 2014.

Madden, Thomas F. *Venice: A New History*. New York: Viking Penguin, 2014.

Maddern, Phillipa C. "It Is Full Merry in Heaven: The Pleasurable Connotations of 'Merriment' in Late Medieval England." In *Pleasure in the Middle Ages*, edited by Naama Cohen-Hanegbi and Piroska Nagy, 21–38. Turnhout, Belgium: Brepols, 2018.

Malory, Thomas. *Morte Darthur*, edited by Sir Edward Strachey. London: Macmillan, 1919.

Malory, Thomas. *Le Morte Darthur*, edited by P.J.C. Field, 2 vols. Cambridge: D.S. Brewer, 2013.

Mann, Jill. "Taking the Adventure." In *Aspects of Malory*, edited by Toshiyuki Takamiya and Derek Brewer, 71–96. Cambridge: D.S. Brewer, 1981.

Marston, Sallie A., John Paul Jones III, and Keith Woodward. "Human Geography without Scale." *Transactions of the Institute of British Geographers*, New Series, 30, no. 4 (2005): 416–432.

Martin, Molly A. *Castles and Space in Malory's Le Morte Darthur*. Cambridge: D.S. Brewer, 2019.

Massing, Jean Michel. "Circa 1492 in Japan." In *Circa 1492: Art in the Age of Exploration*, edited by Jay A. Levenson, 228–229. New Haven, CT: Yale University Press, 1991.

Matheson, Lister M. "Printer and Scribe: Caxton, the *Polychronicon*, and the *Brut*." *Speculum* 60, no. 3 (1985): 593–614.

Matsuda, Takami, ed. *The Kalender of Shepardes*. Tokyo: Keio University Press. Available at: www.flet.keio.ac.jp/~matsuda/ks/185b/html/about.html#KS.

Matsuda, Takami and Satoko Tokunaga. "A Composite Copy of the Codices Keionenses in Keio University Library." In *Codices Keionenses: Essays on Western Manuscript and Early Printed Books in Keio University Library*, edited by Takami Matsuda, 119–208. Tokyo: Keio University Press, 2005.

Matthews, William. "Where Was Siesia-Sessoyne?" *Speculum* 49, no. 4 (1974): 680–686.

Matthews, William. "A Question of Texts." In *The Malory Debate: Essays on the Text of Le Morte Darthur*, edited by Bonnie Wheeler, Robert L. Kindrick, and Michael N. Salda, 65–107. Arthurian Studies 47. Cambridge: D.S. Brewer, 2000.

Mayhew, Robert J . "Geography, Print Culture and the Renaissance: 'The Road Less Travelled By'." *The History of European Ideas*, 27, no. 4 (2001): 349–369.

Mayhew, Robert J. "Material Hermeneutics, Textuality, and the History of Geography: Print Spaces in British Geography, c. 1500–1900." *Journal of Historical Geography* 33, no. 3 (2007): 466–488.

Mayhew, Robert J. "Cosmographers, Explorers, Cartographers, Chorographers: Defining, Inscribing and Practicing Early Modern Geography, c. 1450–1850." In *The Wiley-Blackwell Companion to Human Geography*, edited by John A. Agnew and James S. Duncan, 23–29. Chichester: Wiley-Blackwell, 2011.

McGann, Jerome. *The Textual Condition*. Princeton, NJ: Princeton University Press, 1991.

McGann, Jerome. *A Critique of Modern Textual Criticism*. Charlottesville, VA: University of Virginia Press, 1992.

McGrath, Campbell. *Prose Poem*. Available at: https://poets.org/poem/prose-poem.

McKenzie, D.F. *Bibliography and the Sociology of Texts*. Cambridge: Cambridge University Press, 1999.

McKitterick, D. *Print, Manuscript and the Search for Order, 1450–1830*. Cambridge: Cambridge University Press, 2003.

Medcalf, Stephen. "Classical Authors." In *The Oxford History of Literary Translations in English*, vol. 1, *To 1550*, edited by Roger Ellis, 364–389. Oxford: Oxford University Press, 2008.

Menna, F., et al. "High Resolution 3D Modeling of the Behaim Globe." *International Archives of the Photogrammetry, Remote Sensing and Spatial Information Sciences* XXXIX-B5, 2012.

Michelet, Fabienne L. "East and West in Malory's Roman War: The Implications of Arthur's Travels on the Continent." *Multilingua* 18, no. 2–3 (1999): 209–225.

Mitchell, Peta. "Literary Geography and the Digital: The Emergence of Neogeography." In *The Routledge Handbook of Literature and Space*, edited by Robert T. Tally, Jr., 84–95. London: Routledge, 2016.

Mittman, Asa. *Maps and Monsters in Medieval England*. New York: Routledge, 2006.

Moll, Richard J. *Before Malory: Reading Arthur in Later Medieval England*. Toronto: University of Toronto Press, 2003.

Moller, Violet. *The Map of Knowledge: A Thousand Year History of How Classical Ideas Were Lost and Found*. New York: Doubleday, 2019.

More, Thomas. *Utopia*. Edited by George M. Logan and Robert M. Adams. Cambridge: Cambridge University Press, revised edition, 2002.

Morgan, Victor. "The Literary Image of Globes and Maps in Early Modern England." In *English Map-Making, 1500–1650*, edited by Sarah Tyacke, 46–56. London: British Library, 1983.

Moseley, C.W.R.D. "Availability of Mandeville's Travels in England: 1356–1750." *The Library* 5, no. 2 (1975): 125–133.

Moseley, C.W.R.D. "Behaim's Globe and 'Mandeville's Travels'." *Imago Mundi* 33 (1981): 89–91.

Moseley, C.W.R.D. "Introduction." In *The Travels of Sir John Mandeville*. Translated by C.W.R.D. Moseley. London: Penguin, 1983.

Moseley, C.W.R.D. "*'Whet-Stone Leasings of Old: A Knight's Legacy: Mandeville and Mandevillian Lore in Early Modern England*, edited by Ladan Niayesh, 28–50. Manchester: Manchester University Press, 2011.

Mukai, Tsuyoshi. "De Worde's 1498 *Morte Darthur* and Caxton's Copy Text." *The Review of English Studies* 51, no. 201 (2000): 24–40.

Murrieta-Flores, Patricia, and Bruno Martins, eds. "The Geospatial Humanities: Past, Present, and Future." *International Journal of Geographical Information Science* 33, no. 12 (2019): 2424–2429.

Nall, Catherine. *Reading and War in Fifteenth-Century England: From Lydgate to Malory*. Cambridge: Boydell and Brewer, 2012.

Needham, Paul. "The Customs Rolls as Documents for the Printed-book Trade in England." In *The Cambridge History of the Book in Britain*, edited by Lotte Hellinga and J.B. Trapp, 148–163. Cambridge: Cambridge University Press, 1999.

Newstead, Helaine. "Romances, General." In *A Manual of the Writings in Middle English: 1050–1500*, fascicle 1. New Haven, CT: Connecticut Academy of Arts and Sciences, 1967.

Norris, Ralph. *Malory's Library: The Sources of the Morte Darthur*. Cambridge: D.S. Brewer, 2008.

Nugent, Elizabeth M. "Sources of Rastell's *The Nature of the Four Elements*." *PMLA* 57, no. 11 (1942): 74–88.

O'Gorman, Edmund. *Toward the Great Millennium Jubilee*. Leominster: Gracewing, 1998.

Ortenburg, Veronica. "Archbishop Sigeric's Journey to Rome in 990." *Anglo-Saxon England*, 19 (1990): 197–246.

Oxford English Dictionary, n.d. "Annihilate." Available at: www.oed.com (accessed April 22, 2018).

Pace, Richard. *De Fructu qui ex doctrina percipitur*, edited and translated by Frank Manley and Richard S. Sylvester. Renaissance Texts Series 2. New York: Frederick Ungar Publishing Co. for the Renaissance Society of America, 1967.

Parkes, George B. "The Geography of the Interlude of the Four Elements." *PQ* XVII (1938), 262.

Parr, Johnstone. "John Rastell's Geographical Knowledge of America." *Philological Quarterly* 27 (1948): 229–240.

Perkins, Maureen. *Visions of the Future: Almanacs, Time, and Cultural Change 1775–1870*. Oxford: Clarendon Press, 1996.

Peters, Jeffrey N. *Mapping Discord: Allegorical Cartography in Early Modern French Writing*. Newark, NJ: University of Delaware Press, 2004.

Petrina, Allesandra. "The *Aeneid* of the North: William Caxton's *Eneydos* and Gavin Douglas's *Eneados*." *The Mediaeval Journal* 7, no. 1 (2017): 145–162.

Pincombe, Mike and Kathy Shranks, eds. *Oxford Handbook to Tudor Literature, 1485–1603*. Oxford: Oxford University Press, 2009.

Pinet, Simone. "Diagrammatic Thought in Medieval Literature." In *Literature and Cartography: Theories, Histories, Genres*, edited by Anders Engberg-Pedersen, 173–197. Cambridge, MA: MIT Press, 2017.

Plomer, Henry R. *Wynkyn de Worde and His Contemporaries: From the Death of Caxton to 1535*. London: Grafton & Co., 1925.

Price, Derek J., ed. *The Equitorie of the Planetis, Edited from Peterhouse MS. 75 I*. Cambridge: Cambridge University Press, 1955.

Prior, Oliver H., ed. *Caxton's Mirrour of the World*. Extra Series CX. Berlin: A. Asher & Co. for the Early English Text Society, , 1913.

Ptolemy, Claudius. *The Geography of Claudius Ptolemy*. Translated by Edward Luther Stevenson. New York: New York Public Library, 1932. Reprinted by Cosimo Classics, 2011.

Ptolemy, Claudius. *Tetrabiblos*. Edited and translated by F.E. Robbins. The Loeb Classical Library. Cambridge, MA: Harvard University Press, 1940.

Ptolemy Project. University of California, Berkeley. Lead Researcher: Edward Lee. Available at: https://ptolemy.berkeley.edu/.

Raju, Alison. *Via Francigena: Pilgrim Trail Canterbury to Rome*. Milnthorpe, Cumbria: Cicerone Press, 2013.

Ravenstein, E.G. *Martin Behaim: His Life and His Globe*. London: George Philip & Sons, 1908.

Reid, Pauline. "Through a Looking-Glass: Invention and Imagination in the Visual Rhetoric of William Caxton's Mirrour and Description of the World and Stephen Hawes's Pastime of Pleasure." *Rhetorica* 34, no. 3 (2016): 268–300.

Richardson, Doug, Sarah Luria, Jim Ketchum, and Michael Dear, eds. "Introducing the GeoHumanities." In *GeoHumanities*, 3–4. New York: Routledge, 2011.

Richardson, Glenn. *The Field of Cloth of Gold*. New Haven, CT: Yale University Press, 2013.

Richmond, Colin. "Malory and Modernity: A Qualm about Paradigm Shifts." *Common Knowledge* 14, no. 1 (2008): 34–44.

Riddy, Felicity. "Contextualizing Le Morte Darthur." In *A Companion to Malory*, edited by Elizabeth Archibold and A.S.G. Edwards, 55–74. Cambridge: D.S. Brewer, 1996.

Roberts, R.J. "John Rastell's Inventory of 1538." *The Library*, 6th Series (1979): 34–42.

Roberts, Sean. *Printing a Mediterranean World: Florence, Constantinople, and the Renaissance of Geography*. Cambridge, MA: Harvard University Press, 2013.

Robertson, Kellie. "Materiality and the Hylomorphic Imagination." In *Medieval Literature: Criticism and Debate*, edited by Holly A. Crocker and D. Vance Smith, 367–375. New York : Routledge, 2014.

Roelof, Nicolai. "The Premedieval Origin of Portolan Charts: New Geodetic Evidence." *Isis* 106, no. 3 (2015): 517–543.

Roland, Meg. "From 'Saracens' to 'Infydeles:' The Recontextualization of the East in Caxton's Edition of *Le Morte Darthur*." In *Re-viewing Le Morte Darthur: Text and*

Contexts: Characters and Themes, edited by K.S. Whetter and Raluca Radulescu, 65–77. Cambridge: D.S. Brewer, 2005.

Roland, Meg. "Arthur and the Turks." *Arthuriana* 16, no. 4 (2006): 29–42.

Roland, Meg. "The Rudderless Boat: Fluid Time and Passionate Geography in (Hardyng's) Chronicle and (Malory's) Romance." *Arthuriana* 22, no. 4 (2012): 77–93.

Roland, Meg. "Malory's Sandwich: Alternate Arthurian Geographies and the Global Middle Ages." In *Re-thinking Medieval Margins and Marginality*, edited by Ann Zimo, Tiffany Sphrecher, Kathryn Reyerson, and Debra Blumenthal. New York: Routledge, 2019.

Roland, Meg. n.d. "The Utas of Seynt Hyllary: Time in the Roman War Campaign." Available at: www.passionategeography.com.

Ross, Elizabeth. *Picturing Experience in the Early Printed Book: Breyenbach's Peregrinatio from Venice to Jerusalem*. University Park, PA: Penn State University Press, 2014.

Rothbard, Murray Newton. *Economic Thought Before Adam Smith: An Austrian Perspective on the History of Economic Thought*. Auburn, AL: Ludwig von Mises Institute, 1995.

Rouse, Robert Allen and Cory James Rushton. "Arthurian Geography." In *The Cambridge Companion to the Arthurian Legend*, edited by Elizabeth Archibald and A.D. Putter, 218–234. Cambridge: Cambridge University Press, 2009.

Rundle, David. "The Scribe Thomas Candour and the Making of Poggio Bracciolini's English Reputation." *English Manuscript Studies 1100–1700* xii (2005), 1–25.

Rundle, David. "Humanism Across Europe: The Structure of Contact." In *Humanism in Fifteenth-Century Europe*, edited by David Rundle, 307–336. Oxford: Society for the Study of Medieval Languages and Literature, 2012.

Rushton, Cory James. "Malory and Form." In *The New Companion to Malory*, edited by Megan G. Leitch and Cory James Rushton, 125–143. Cambridge: Boydell and Brewer, 2019.

Russell, Jeffrey Burton. *Inventing the Flat Earth: Columbus and Modern Historians*. New York: Praeger, 1997.

Rutkowska, Hanna. "Typographical and Graphomorphic Features of Five Editions of the *Kalender of Shepherds* as Elements of the Early Printer' Community of Practice." In *Communities of Practice in the History of English*, edited by Joanna Kopaczyk and Andreas H. Jucker, 123–150. Amsterdam: John Benjamins, 2013.

Said, Edward. *Orientalism*. London: Penguin, 1991.

Sanford, Rhonda Lemke. *Maps and Memory in Early Modern England: A Sense of Place*. New York: Palgrave Macmillan, 2002.

Sanok, Catherine. *New Legends of England: Forms of Community in Late Medieval Saints Lives*. Philadelphia, PA: University of Pennsylvania Press, 2018.

Saunders, Corinne J. *The Forest of Medieval Romance: Avernus, Broceliande, Arden*. Cambridge: D.S. Brewer, 1993.

Scaffi, Alessandro. "After Ptolemy: The Mapping of Eden." In *Ptolemy's Geography in the Renaissance*, edited by Zur Shalev and Charles Burnett, 187–206. London and Turin: The Warburg Institute – Nino Aragno Editore. 2011.

Scott, James C. *Domination and the Arts of Resistance: Hidden Transcripts*. New Haven, CT: Yale University Press, 2008.

Scott, Kathleen L. "Past Ownership: Evidence of Book Ownership by English Merchants in the Later Middle Ages." In *Makers and Users of Medieval Books: Essays in Honour of A.S.G. Edwards*, edited by C. Meale and D. Pearsall, 150–177. Woodbridge: Boydell and Brewer, 2014.

Sebald, W.G. *Rings of Saturn*. New York: New Directions, reprint edition, 2016.

Sellink, Manfred. "A New Look on the World: The Invention of Landscape." In *The Age of Van Eyck: The Mediterranean World and Early Netherlandish Painting, 1430–1530*, edited by Till-Holger Borchert, 212–219. London: Thames and Hudson, 2002.

Seymour, Michael C. "The Early English Editions of Mandeville's Travels." *The Library* 5, no. 9 (1964): 202–207.

Seymour, Michael C., ed. *Mandeville's Travels*. Oxford: Clarendon Press, 1967.

Seymour, Michael C. *Sir John Mandeville*. Aldershot: Ashgate, 1993.

Seymour, Michael C. "Mandeville in England: The Early Years." In *A Knight's Legacy: Mandeville and Mandevillian Lore in Early Modern England*, edited by Laden Niayesh, 15–27. Manchester: Manchester University Press, 2011.

Shalev, Zur. "Main Themes in the Study of Ptolemy's Geography." In *Ptolemy's Geography in the Renaissance*, edited by Zur Shalev and Charles Burnett, 1–14. London and Turin: The Warburg Institute – Nino Aragno Editore, 2011.

Shalev, Zur and Charles Burnett, eds. *Ptolemy's Geography in the Renaissance*. London and Turin: The Warburg Institute – Nino Aragno Editore, 2011.

Shepherd, Stephen. *Le Morte Darthur*. New York: Norton, 2003.

Shinn, Abigail. "*'Extrardinary discourses of vnnecessarie matter': Spenser's Literature and Popular Culture in Early Modern England*, edited by Matthew Dimmock and Andrew Hadfield, 137–149. Farnham: Ashgate, 2009.

Shoeck, R.J. "The Geography of Erasmus." In *Northern Humanism in European Context: 1469–1625: From the 'Adwert Academy' to Ubbo Emmius*," edited by Fokke Akkerman, Arie Johan Vanderjagt, and A.H. Van Der Laan, 198–205. Leiden: Brill, 1999.

Short, John. *The World Through Maps: A History of Cartography*. Toronto: Firefly Books, 2003.

Siebold, Jim. "Behaim Globe." *Cartographic Images Network*. Available at: www.ca rtographic-images.net.

Silver, Larry. "Arts and Minds: Scholarship on Early Modern Art History (Northern Europe)." *Renaissance Quarterly* 59, no. 2 (2006): 351–373.

Skelton, R.A. "Introduction." In *Cosmographia: Bologna 1477*. Amsterdam: Theatrum Orbis Terrarum, 1963.

Skelton, R.A. "Introduction." In *Cosmographia: Rome 1478*. Amsterdam: Theatrum Orbis Terrarum, 1966.

Smethurst, Paul. "The Journey from Modern to Postmodern in the Travels of Sir John Mandeville and Marco Polo's Divisament dou Monde." In *Postmodern Medievalisms, Studies in Medievalism*, vol. XIII, 159–179. Cambridge: D.S. Brewer, 2004.

Smith, D.K. *The Cartographic Imagination in Early Modern England*. Aldershot: Ashgate, 2008.

Soja, Edward W. *Postmodern Geographies: The Reassertion of Space in Critical Social Theory*. New York: Verso, 1989.

Somerset, J.A.B., ed. *Four Tudor Interludes*. London: The Athlone Press, 1975.

Sommers, Oskar H. *The Kalender of Shepherdes: The Edition of Paris 1503 in Photographic Facsimile, A Faithful Reprint of R. Pynson's Edition of London 1506*. Vol. 1*Prolegomena*. London: Kegan Paul, Trench, Trubner & Co, 1892.

Sparke, Matthew. *In the Space of Theory*. Minneapolis, MN: University of Minnesota Press, 2005.

Sparke, Matthew. "Everywhere but Always Somewhere: Critical Geographies of the Global South." *The Global South* 1, no. 1 (2007): 117–126.

Sparke, Matthew. "Political Geographies of Globalization (3): Resistance." *Progress in Human Geography*, 32, no. 1 (2008): 1–18.

Spenser, Edmund. "The Shepheardes Calendar." In *The Yale Edition of the Shorter Poems of Edmund Spenser*, edited by William A. Oram, Einar Bjorvand, Ronald Bond, Thomas H. Cain, Alexander Dunlop, and Richard Schell. New Haven, CT: Yale University Press, 1989.

Springer, Simon. "Human Geography without Hierarchy." *Progress in Human Geography*, 38, no. 3 (2014): 402–419.

Spufford, Peter. *Power and Profit: The Merchant in Medieval Europe*. New York: Thames and Hudson, 2002.

Staley Johnson, Lynn. *The Shepheardes Calender: An Introduction*. University Park, PA: Pennsylvania State University Press, 1990.

Starkey, David, ed. *The Inventory of King Henry VIII: Society of Antiquaries MS 129 and British Library MS Harley 1419*. London: Harvey Miller, 1998.

Steiner, Emily. "Compendious Genres: Higden, Trevisa, and the Medieval Encyclopedia." *Exemplaria* 27, no. 1–2 (2015): 73–92.

Stevens, Henry. *Ptolemy's Geography: A Brief Account of All the Printed Editions Down to 1730*. London: Henry Stevens, Son, and Stiles, 1908; republished Theatrum Orbis Terrarum, 1972.

Stewart, Rory. *The Places in Between*. Orlando, FL: Harcourt, 2004.

Strachey, Edward, ed. "Introduction." In *Le Morte Darthur by Sir Thomas Malory*. London: Macmillan, 1919.

Summit, Jennifer. "Topography as Historiography: Petrarch, Chaucer, and the Making of Medieval Rome." *Journal of Medieval and Early Modern Studies* 30 (2000): 211–246.

Sutton, Anne F. "Malory in Newgate Prison: A New Document." *The Library*, 7th series, 1, no. 3 (2000): 243–262.

Sutton, Anne F. "The Acquisition and Disposal of Books for Worship and Pleasure by Mercers of London in the Later Middle Ages." In *Manuscripts and Printed Books in Europe 1350–1550*, edited by E. Cayley and S. Powell, 95–114. Liverpool: Liverpool University Press, 2013.

Sutton, Anne F. "Merchants." In *A Companion to the Early Printed Book in Britain, 1476–1558*, edited by Vincent Gillespie and Susan Powell, 127–133. Cambridge: D. S. Brewer, 2014.

Sykes, James. "The Terrestrial Globe in Raphael's 'The School of Athens'." *Globe Studies* 55/56 (2009): 53–72.

Takagi, Masako and Toshiyuki Takamiya. "Caxton Edits the Roman War Episode: *The Chronicle of England* and Caxton's Book V." In *The Malory Debate: Essays on the Texts of Le Morte Darthur*, edited by Bonnie Wheeler, Robert L. Kindrick, and Michael N. Salda, 127–167. Cambridge: D.S. Brewer, 2000.

Talbert, Richard A. *Rome's World: The Peutinger Map Reconsidered*. Cambridge: Cambridge University Press, 2014.

Tally, Robert T. Jr. "Introduction: On Geocriticism." In *Geocritical Explorations: Space, Place, and Mapping in Literary and Cultural Studies*, edited by Robert T. Tally, Jr., 1–9. New York: Palgrave Macmillan, 2011a.

Tally, Robert T. Jr., ed., *Geocritical Explorations: Space, Place, and Mapping in Literary and Cultural Studies*. New York: Palgrave Macmillan, 2011b.

Tally, Robert T. Jr., ed. *The Routledge Handbook of Literature and Space*. Abingdon: Routledge, 2017.

Taylor, E.G.R. *Tudor Geography: 1485–1583*. New York: Octagon Books, 1968.

Taylor, Jamie K. "Toward Premodern Globalism: Ocean Exemplarity in Chaucer's Men of Law's Tale." *PMLA* 135, no. 2 (March 2020): 254–272.

Taylor, Jane H.M. "'Hungry Shadows:' Pierre Sala and His Yvain." *Arthuriana* 19, no. 1 (2009): 7–19.

Tobienne, Francis Jr. *Mandeville's Travails: Merging Travel, Theory, and Commentary*. Newark, NJ: University of Delaware Press, 2016.

Tolkien, Christopher. "Introduction." In *The Fall of Arthur*, by J.R.R. Tolkien. Boston: Houghton Mifflin, 2013.

Tolkien, J.R.R. *The Fall of Arthur*, edited by Christopher Tolkien. Boston: Houghton Mifflin, 2013.

Tomasch, Sylvia. "Introduction: Medieval Geographical Desire." In *Text and Territory: Geographical Imagination in the European Middle Ages*, edited by Sylvia Tomasch and Sealy Gilles. Philadelphia, PA: University of Pennsylvania Press1998.

Tonry, Kathleen. "Reading History in Caxton's *Polychronicon*." *JEGP* 111, no. 2 (2012): 169–198.

Tonry, Kathleen. *Agency and Intention in English Print, 1476–1526*. Turnhout, Belgium: Brepols, 2016.

Trapp, J.B. "The Humanist Book." In *The Cambridge History of the Book in Britain*, edited by Lotte Hellinga and J.B. Trapp, 283–325. Cambridge: Cambridge University Press, 1999.

Travis, Charles. "Digital Geohumanities." In *International Encyclopedia of Human Geography*, 2nd ed. vol. 3, 341–346. Amsterdam: Elsevier, 2020.

Trevisan, Sara. "The Impact of the Netherlandish Landscape Tradition on Poetry and Painting in Early Modern England." *Renaissance Quarterly* 66, no. 3 (2013): 866–903.

Twomey, Michael W. "The Exemplary Environment of Bartholomaeus Anglicus." In *Reading the Natural World in the Middle Ages and Renaissance*, edited by Thomas Willard, 71–88. Turnhout, Belgium: Brepols, 2020.

Tyack, Sarah, ed. *English Map-making 1500–1650*. London: British Library, 1983.

Tydeman, William, ed. *Four Tudor Comedies*. London: Penguin, 1984.

Vaughan, Virginia Mason. "Preface: The Mental Maps of English Renaissance Drama." In *Playing the Globe: Genre and Geography in English Renaissance Drama*, edited by John Gillies and Virginia Mason Vaughan. Madison, NJ: Fairleigh Dickinson University Press, 1998.

Veltman, Kim H. "Ptolemy and the Origins of Linear Perspective." In *Atti del convegno internazionale di studi: la prospettiva rinascimentale, Milan 1977*, edited by Marisa Dalai-Emiliani, 403–407. Florence: Centro Di, 1980.

Vinaver, Eugène, ed. *The Works of Sir Thomas Malory*. 3 vols. revised by P.J.C. Field. Oxford: Clarendon Press, 1990.

Wade, James. "The Chapter Headings of the *Morte Darthur*: Caxton and de Worde." *Modern Philology* 111, no. 4 (2014): 665–667.

Wallace, David. "Imperium, Commerce, and National Crusade: The Romance of Malory's Morte." In *Medieval Literature: Criticism and Debate*, edited by Holly A. Crocker and D. Vance Smith, 321–332. London: Routledge, 2014.

Wallace, David, ed. *Europe: A Literary History, 1348–1418*. Vol. 1. Oxford: Oxford University Press, 2016.

Wallis, Helen M. "The First English Globe: A Recent Discovery." *The Geographical Journal* 117, no. 3 (1951): 275–290. Wallis, Helen M. "Further Light on the Molyneux Globes." *The Geographical Journal* 121 (1955): 304–311.
Waters, David. *English Navigational Books, Charts, and Globes Printed Down to 1600*. Coimbra: University of Coimbra Press, 1985.
Weiss, Benjamin. "The Geography in Print: 1475–1530." In *Ptolemy's Geography in the Renaissance*, edited by Zur Shalev and Charles Burnett, 91–120. London and Turin: The Warburg Institute – Nino Aragno Editore, 2011.
Weiss, Roberto. *Humanism in England During the Fifteenth Century*, edited and prepared for electronic publication by David Rundle and Anthony John Lappin. Oxford: Society for the Study of Medieval Languages and Literature, 2009.
Westfall, Suzanne R. *Patrons and Performance: Early Tudor Household Revels*. Oxford: Clarendon, 1990.
Westphal, Bernard. *Geocriticism: Real and Fictional Spaces*. Translated by Robert T. Tally Jr. New York: Palgrave Macmillan, 2007.
Westphal, Bertrand. "A Geocritical Approach to Geocriticism." *American Book Review* 37, no. 6 (2016): 4–5.
Wheeler, Bonnie. "Romance and Parataxis and Malory: The Case of Sir Gawain's Reputation." In *Arthurian Literature XII*, edited by James Carley and Felicity Riddy, 109–132. Cambridge: D.S. Brewer, 1993.
Whetter, Kevin S. *Understanding Genre and Medieval Romance*. Aldershot: Ashgate, 2008.
Whetter, Kevin S. *The Manuscript and Meaning of Le Morte Darthur: Rubrication, Commemoration, Memorialization*. Cambridge: D.S. Brewer, 2017.
Wild, Christopher. "Cartography and the Melancholic Self." In *Space and Self in Early Modern European Cultures*, edited by David Warren Sabean and Malina Stefanovska, 239–257. Toronto: University of Toronto Press, 2012.
Williams, J.A. *The Cabot Voyages and Bristol Discovery under Henry VII*. Cambridge: Hakluyt Society, 1962.
Wilson, Bronwen. *The World in Venice: Print, the City, and Early Modern Printed Books*. Toronto: University of Toronto Press, 2005.
Wilson, Catherine. *Epicureanism at the Origins of Modernity*. Oxford: Oxford University Press, 2008.
Wilson, R.H. "Malory and Caxton." In *A Manual of the Writing of Middle English, 1050–1500*, vol. 3, ed. A.E. Hartung, 757–807. New Haven, CT: Connecticut Academy of Arts and Sciences, 1972.
Wilson, Rob and Wimal Dissanayake, eds. *Global/Local: Cultural Production and the Transnational Imaginary*. Durham, NC: Duke University Press, 1996.
Withrington, John. "Caxton, Malory, and the Roman War in the *Morte Darthur*." *Studies in Philology* 89, no. 3 (1992): 350–366.
Withrington, John. "'He Telleth the Number of the Stars; He Calleth Them All By Their Names:' The Lesser Knights of Sir Thomas Malory's *Morte Darthur*." *Quondam et Futurus: A Journal of Arthurian Interpretations* 3, no. 4 (1993): 17–27.
Wogan-Browne, Jocelyn. "Reading the World: The Hereford mappa mundi." *Parergon* 9, no. 1 (1991): 117–135.
Wood, Denis. "Cartography Is Dead (Thank God!)." *Cartographic Perspectives* 45 (2003): 4–7.
Wood, Denis. *Everything Sings: Maps for a Narrative Atlas*, 2nd ed. Los Angeles: Siglo Press, 2013.

Woodward, David. "English Cartography 1650–1750, A Summary." In *The Compleate Plattmaker: Essays on Chart, Maps, and Globe Making in England in the Seventeenth and Eighteenth Centuries*, edited by Norman J.W. Thrower, 159–193. Los Angeles, CA: University of California Press, 1978.

Woodward, David. "Maps and the Rationalization of Geographic Space." In *Circa 1492: Art in the Age of Exploration*, edited by Jay A. Levenson, 83–88. New Haven, CT: Yale University Press, 1991.

Woodward, David. "Medieval Mappaemundi." In *The History of Cartography.* Vol 1, *Cartography in Prehistoric, Ancient, and Medieval Europe and the Mediterranean*, edited by David Woodward, 286–320. Chicago: University of Chicago Press, 2007a.

Woodward, David, ed. *The History of Cartography.* Vol. 3, *Cartography in the European Renaissance.* Chicago: University of Chicago Press, 2007b.

Woodward, David. "Cartography and the Renaissance: Continuity and Change." In *The History of Cartography.* Vol. 3, *Cartography in the European Renaissance*, edited by David Woodward, 3–98. Chicago: University of Chicago Press, 2007c.

Worms, Laurence. "Maps and Atlases." In *The Cambridge History of the Book in Britain*, vol. IV: *1557–1695*, edited by John Barnard and D.F. McKenzie, 228–245. Cambridge: Cambridge University Press, 2002.

Index

Locators in *italics* refer to figures and those in **bold** to tables.

9780367560560